Living within a Fair Share Ecological Footprint

According to many authorities the impact of humanity on the earth is already overshooting the earth's capacity to supply our needs. This is an unsustainable position. This book focuses not on the problem but on the solution, by showing what it is like to live within a fair share ecological footprint.

The authors describe numerical methods used to calculate this footprint, concentrating on low- or no-cost behaviour change rather than on potentially expensive technological innovation. They show what people need to do now in regions such as Europe, North America and Australasia where current lifestyles involve living beyond the available ecological means. The calculations focus on outcomes rather than on detailed analysis of the methods used. The main objective is to show that living with a reduced ecological footprint is both possible and not so very different from the way most people currently live in the West.

The book clearly demonstrates that change in behaviour now will avoid some very challenging problems in the future. The emphasis is on workable, practical and sustainable solutions based on quantified research, rather than on generalisations about the overall problems facing humanity.

Robert and **Brenda Vale** are Professorial Research Fellows in the School of Architecture, Victoria University of Wellington, New Zealand. They share common research interests in ecological footprinting and sustainable building design, and both are currently working on the new Foundation for Research, Science and Technology (FRST) project to deliver ecological footprinting and systems approaches to sustainable development of communities.

Living within a Fair Share Ecological Footprint

Edited by
Robert and Brenda Vale

This first edition published 2013
by Routledge
2 Park Square, Milton Park, Abingdon, Oxon, OX14 4RN

Simultaneously published in the USA and Canada
by Routledge
711 Third Avenue, New York, NY 10017

*Routledge is an imprint of the Taylor & Francis Group,
an informa business*

© 2013 Robert Vale and Brenda Vale for selection and editorial material; individual chapters, the contributors

The rights of Robert Vale and Brenda Vale to be identified as authors of this work has been asserted by them in accordance with sections 77 and 78 of the Copyright, Designs and Patents Act 1988.

All rights reserved. No part of this book may be reprinted or reproduced or utilised in any form or by any electronic, mechanical, or other means, now known or hereafter invented, including photocopying and recording, or in any information storage or retrieval system, without permission in writing from the publishers.

Trademark notice: Product or corporate names may be trademarks or registered trademarks, and are used only for identification and explanation without intent to infringe.

British Library Cataloguing in Publication Data
A catalogue record for this book is available from the British Library

Library of Congress Cataloging-in-Publication Data
Living within a fair share ecological footprint / edited by Robert Vale and Brenda Vale.
p. cm.
Includes bibliographical references and index.
1. Sustainable living. 2. Environmentalism. I. Vale, Brenda. II. Vale, Robert James Dennis.
GE196.L58 2013
640.28'6–dc23
2012033978

ISBN13: 978-0-415-50722-6 (hbk)
ISBN13: 978-0-415-50723-3 (pbk)
ISBN13: 978-0-203-12644-8 (ebk)

Typeset in Baskerville
by Saxon Graphics Ltd, Derby

Printed and bound in Great Britain by MPG Printgroup

Contents

Notes on contributors   ix
Preface   xiii

PART I
Introduction   1

1 **Ecological Footprints, Fair Earth-Shares and Urbanization**   3
   WILLIAM E. REES AND JENNIE MOORE

PART II
What Does Living within a Fair Earth Share Mean?   33

PART II.I
Personal Footprint   35

2 **The Ecological Footprint of Food**   37
   JAMES M. RICHARDSON

3 **Domestic Travel**   58
   ROBERT AND BRENDA VALE

4 **Consumer Goods**   73
   MAGGIE LAWTON

5 **The Dwelling**   84
   NALANIE MITHRARATNE

6 **Tourism**   96
   ABBAS MAHRAVAN

PART II.II
Collective Footprint 115

7 Infrastructure 117
   NING HUANG

8 Government 134
   JEREMY GABE AND REBECCA GENTRY

9 Services 147
   SOO RYU

PART III
Footprints in the Past 157

10 A Study of Wellington in the 1950s 159
    CARMENY FIELD (WITH BRENDA VALE)

PART IV
Footprints in the Present 183

11 A Study of China 185
    YUEFENG GUO

12 A Study of Suburban Thailand 201
    SIRIMAS HENGRASMEE

13 Kampung Naga, Indonesia 215
    GRACE PAMUNGKAS (WITH FABRICIO CHICCA AND BRENDA VALE)

14 A Study of Hanoi, Vietnam 224
    HAN THUC TRAN

15 A Study of Suburban New Zealand 240
    SUMITA GHOSH

16 The Hockerton Housing Project, England 262
    BRENDA AND ROBERT VALE

17 Education for Lower Footprints 275
    SANT CHANSOMSAK

18	**Footprints and Income**	287
	ELLA LAWTON	

19	**Sustainable Urban Form**	304
	FABRICIO CHICCA	

PART V
Conclusions 317

20	**'I Wouldn't Start from Here …'**	319
	BRENDA AND ROBERT VALE	

Index 322

Contributors

Sant Chansomsak, Lecturer, Naresuan University, Phitsanulok, Thailand. Following his PhD studies in New Zealand, Sant returned to teaching architecture at Naresuan University in Phitsanulok in Thailand. His research field is the sustainable design of buildings and in particular schools and educational buildings. He has also researched and written on community architecture and its role in creating sustainable communities.

Fabricio Chicca, PhD student, Victoria University of Wellington, New Zealand. Fabricio is a Brazilian architect and urban designer with over ten years' experience in all sectors of the real estate market in Brazil (industrial, retail, commercial buildings and residential). He is currently completing a PhD in architecture focused on urban sustainability at Victoria University of Wellington.

Carmeny Field, Tour Dispatcher, Allen Marine Inc., Juneau, Alaska, USA. Carmeny recently completed her Master's in Building Science at Victoria University of Wellington. Her Master's research estimated the ecological footprints of Wellingtonians in 1956 and 2006. In conjunction with this, research was conducted to understand the lifestyles and quality of life during the 1950s and today. This information formed the basis of comparisons between ecological footprints that are 50 years apart, and the relationship these have to lifestyle and quality of life.

Jeremy Gabe, PhD student, University of Auckland, New Zealand. Jeremy Gabe is a PhD student in the Department of Property at the University of Auckland Business School. His research interests are the motivations behind green property development and the growing use of non-financial indicators to signal environmental performance and improve market efficiency.

Rebecca Gentry, UC Santa Barbara, California, USA. Rebecca Gentry is currently a PhD Student at the Bren School of Environmental Science and Management at the University of California, Santa Barbara. Before commencing her studies she was a Program Associate with the California

Ocean Science Trust. She also worked for two years as a Policy Analyst for the Ministry of Fisheries in New Zealand.

Sumita Ghosh, Senior Lecturer, University of Technology, Sydney, Australia. As an architect and urban planner, Sumita's research focuses on sustainable urban form and morphologies, performance indicators, and local food production and suburban domestic gardens. Her previous research has developed an urban classification system, an environmental sustainability assessment method and descriptors for New Zealand urban forms. Her current research explores the extent to which existing morphological characteristics of different residential density developments influence their capacity to incorporate various sustainable practices and behaviour.

Yuefeng Guo, Researcher, Green World Solutions, Beijing, China. Yuefeng completed his PhD at the University of Auckland, New Zealand. He now works with a green design consultancy in Beijing. His main research interest is in how to promote sustainability of the built environment. He has published several academic papers on sustainability-related issues such as change in values and behaviour, happiness, ecological footprinting, intentional communities, urbanisation and lifestyles based on traditional cultures in rural areas of China.

Sirimas Hengrasmee, Lecturer, Naresuan University, Phitsanulok, Thailand. Sirimas is currently teaching architecture at Naresuan University, Thailand. She has written on sustainable architecture and design and her PhD research looked at the ecological footprint of living in suburban Thailand and how a more self-sufficient lifestyle could reduce environmental impact.

Ning Huang, Deputy Director, Green World Solutions, Beijing, China. Ning has a PhD from the University of Auckland. His thesis focused on ecological footprint calculation for the transport system of Auckland. He currently operates a consulting and design company in Beijing, Green World Solutions (Beijing) Co., Ltd, and its business includes green architecture design, sustainable planning and ecological footprint assessment. He has published several research papers and book chapters in English and in Chinese.

Ella Lawton, PhD student, Victoria University of Wellington, New Zealand. Ella has a dual role as both a PhD student and Project Manager for the Centre for Sustainable Practice at Otago Polytechnic. Ella works to liaise with Central Otago local government and communities in an effort to maximise project outcomes by incorporating end-users and beneficiaries in the research. This valuable community insight also supports the design and scenario development undertaken as part of Ella's PhD research.

Maggie Lawton, Director, Future by Design (formerly Braidwood Research and Consulting Ltd), Auckland, New Zealand. Until the recent amalgamation of Auckland councils, Maggie was Manager of Strategy and Policy at Manukau City Council. Originally a practising scientist and researcher, she was a member of the Executive Team at Landcare Research for 14 years until September 2006. She then ran her own business, Braidwood Research and Consulting Ltd, now renamed Future by Design. Maggie has a background in a wide range of rural and urban development issues, including climate change and policy related to sustainable development.

Abbas Mahravan, PhD candidate, Victoria University of Wellington, New Zealand. Prior to starting his PhD research, Abbas taught architecture at Razi University in Iran. Before this his work as an architect involved dealing with historic buildings in Iran and preparing tourism research plans. His current research is investigating the environmental, economic and cultural footprints of tourism.

Nalanie Mithraratne, Associate Professor, National University of Singapore. Nalanie is a trained architect and a researcher in the life-cycle assessment of construction systems, embodied energy assessment, and carbon and energy footprinting. Her research investigates whether proposed new ways of providing services, such as microgeneration of electricity using small-scale renewable energy systems or the use of domestic rainwater collection for water supply, are environmentally preferable to the current large-scale means of provision.

Jennie Moore, PhD student, University of British Columbia, Vancouver, Canada. Prior to commencing PhD studies, Jennie Moore was the Strategic Initiatives Division Manager for the Greater Vancouver Regional District, where she helped to launch the Sustainable Region Initiative. She has over a decade of experience in urban sustainability, including leading municipal efforts to address climate change, improve energy efficiency, advance the adoption of green building policies, and initiate eco-industrial networking projects to help industry and communities improve their environmental and economic performance. Her doctoral research uses the concept of the ecological footprint to examine options for creating sustainable cities.

Grace Pamungkas, PhD student, Victoria University of Wellington, New Zealand. Grace formerly worked in the area of historic buildings and vernacular culture in Indonesia. Her current research, based on a series of case studies in Indonesia, is looking at the symbiotic relationship between lifestyles and built environment in an attempt to examine the influence of traditional beliefs on sustainable behaviour.

William E. Rees, University of British Columbia, Vancouver, Canada. Bill is a professor and the former director of the School of Community and Regional Planning at UBC. He has taught at UBC for more than 40 years. He is the originator of the idea of the ecological footprint, on which this book is based. His primary interest is in public policy and planning with a focus on the ecological conditions required for sustainable socio-economic development.

James M. Richardson, PhD student, Victoria University of Wellington, New Zealand. James has a Master's degree in Landscape Architecture from the University of British Columbia. He has worked in permaculture and has written at length about the impact and land requirements of food growing in Vancouver. His current research is into how food can be supplied after oil, and is based on both theoretical studies and practical community involvement.

Soo Ryu, Architect, Jasmax, Auckland, New Zealand. Soo graduated from the University of Auckland majoring in Architecture with first class honours and continued her studies by undertaking a Master's in Architecture at Victoria University, gaining a distinction. Her research focused on carbon footprinting and carbon neutrality. She now works at one of New Zealand's largest architectural practices.

Han Thuc Tran, PhD student, Victoria University of Wellington, New Zealand. Han Thuc Tran is a PhD candidate in the School of Architecture, Victoria University of Wellington. Her research interests include sustainable patterns of living, ecological footprinting, sustainable housing and urban forms. She is currently working in Finland carrying out a study focusing on ecological footprint comparisons between Vietnam, New Zealand and Finland.

Robert and **Brenda Vale**, Victoria University of Wellington, New Zealand. Robert and Brenda are architects, academics and writers. They have worked in the field of sustainability since they were students in the late 1960s. They emigrated to New Zealand in 1996.

Preface

This is a book about the rule of law. However, it is not about the made-up law of economics. Many of those who govern the world seem to have an economist's view of it. In this view, growth is the first law of modern humanity, and our rulers (who have generally done very well out of it personally) have persuaded everyone to go along with belief in their law. Unfortunately, from the standpoint of physical reality, which is a growing population living on a finite planet, it is not possible to accommodate endless material growth. Growth contradicts the laws of thermodynamics. Modern humanity's first law is, in a very real sense, illegal. Politicians constantly change laws, and what was illegal becomes legal, or vice versa. Alcohol was legal, but became illegal in the United States for nearly 25 years in the twentieth century, while homosexuality, accepted in Ancient Greece, only became legal in England and Wales in the late 1960s. This ability to make fundamental changes in what is permitted or forbidden seems to have made our leaders believe that all laws are malleable. They appear to believe that they can repeal or overturn the laws of thermodynamics, which govern not only the finite system of the world but probably that of the universe. Believing that you can overturn the laws of the universe is what the Ancient Greeks called *hubris*, which always resulted in *nemesis*. Hubris has resulted in an entire global economy based on an annually increasing population, consuming annually increasing amounts of finite resources. Unfortunately society has yet to come to terms with the idea that there may be limits to this, but if we do not face up to these limits, humanity faces nemesis.

When limits are acknowledged, there is the matter of equity. In the end, all resources have to come from the earth. The planet is finite. In a finite system, if I want to have more, you will have to have less. Currently the wealthier societies take far more than their fair share of the earth's resources while the poorer societies take much less. But in a finite world, if the poor are to have more, the wealthy will have to have less. With their made-up laws economists like to pretend that we can all have more; if the rich get richer, the poor will also get richer – 'a rising tide lifts all the boats'. The reality in a world governed by the laws of thermodynamics is somewhat different.

xiv *Preface*

This book is an exploration of what modern societies might be like if people were to live within their means, with a fair share of the earth's resources. Unlike many texts promoting 'sustainability' it does not make unproven assumptions about the benefits of compact cities, or green buildings, or urban agriculture, or airship travel. Neither does it assume that miraculous new technologies will be the solution to all our problems. The authors show what life could be like in a truly sustainable future. They have backed their proposals with simple, clear calculations to reveal the reality of the situation, and to make clear that decisions need to be made right now about how we live.

The market-driven model, the only currently accepted way to organise societies, is clearly sending all the wrong messages. Just look at transport: cities are still planned around private cars rather than around public transport; in many places it is much cheaper to fly than to take the train; if you travel as a tourist you can import a much higher value of goods without paying duty than if you sent goods by mail. In all these examples, the market encourages the option which does much more damage.

The good news is that we can avoid nemesis not by implementing expensive new technologies but simply by changing our behaviour and our choices. The problem we face is that, in a market-driven world, this is also the bad news.

Part I
Introduction

1 Ecological Footprints, Fair Earth-Shares and Urbanization

William E. Rees and Jennie Moore

Introduction: Thinking about Urban Sustainability

Economic expansionists often argue that environmentalists exaggerate the human impact on earth. After all, humans are increasingly an urban species and cities occupy only 2 to 3 per cent of earth's land area; the entire human population could live in the US state of Texas in single-family dwellings with room to spare! Some economists even suggest that humanity's impact may actually be shrinking as people get richer and the economy 'dematerializes' or 'decouples' from nature.

One purpose of this chapter is to make the case that most such claims are the result of faulty accounting. The simple fact is that the geographic or political 'footprints' of human settlements bear scant relationship to the biophysical demands of their inhabitants on the ecosphere (Rees, 2010a, 2012). Contrary to common belief, urbanization is not further evidence that humanity is decoupling from nature. Urbanization merely separates people spatially from their supportive ecosystems without changing their functional dependence on those systems. In other words, urbanization effectively divides the human ecosystem into two distinct physical components – a widely dispersed productive 'hinterland' and a densely concentrated but wholly dependent consumptive core, the city itself. The increasingly global hinterland of a typical modern city – its true 'ecological footprint', or EF – is up to three orders of magnitude (a thousand times) larger than the city itself. From a human ecological perspective – all fantasies of Texas aside – we actually live in an ecologically full world (Daly, 2005).

Indeed, earth is full to overflowing. Recent estimates put humanity's aggregate ecological footprint at approximately 18 billion global hectares, or 2.7 global hectares (gha) per capita (a 'global hectare' represents a hectare of land or water ecosystems of world average productivity). Compare this to the earth's total stock of productive land and water ecosystems: 11.9 billion gha, or 1.7 gha per capita (WWF, 2010a). Humanity has already overshot global carrying capacity by roughly 50 per cent. This overshoot is made possible by using non-renewable resources; it means that our species is living and growing, in part, by depleting ecosystems and resource stocks, polluting air, water and soil, and undermining both local and global life-support functions. We are dissipating essential 'natural capital' and generally disordering the ecosphere. This situation demands that the world community thoroughly and urgently rethink its growth-based economic paradigm.

The primary cause of global degradation is resource over-consumption and excess waste production. Since, first, the wealthiest fifth of the world's people account for 76 per cent of private consumption and associated pollution (Shah, 2010) and, second, total human demand exceeds long-term bio-capacity by 50 per cent, the richest 20 per cent alone have effectively appropriated the entire bio-capacity of earth and contribute most to its degradation. This fortunate minority mostly comprises those living in industrial and post-industrial cities with average eco-footprints several times larger than their equitable shares of global carrying capacity.

Within this context, the major purposes of this chapter are to make the case, on both thermodynamic and ecological footprint (EF) grounds, that:

1   It is meaningless to plan for urban sustainability without ensuring the security and sustainability of the extra-urban ecosystems upon which cities are dependent.
2   Achieving equitable global sustainability in an urbanizing world will require significant reductions in energy and material throughput.

We argue that the world community has no choice but to reconcile the prevailing myth of unlimited economic growth with a rapidly deteriorating biophysical reality. Ecosystems can thrive in the absence of the economy, but no economy is possible without fully functional ecosystems. If we wish to extend the lifetime of global civilization on a finite planet, the world community must constrain its energy and material demands to comply with the limits set by the productive and assimilative capacities of our life-supporting ecosystems.

Clearly, sustainability is a serious business, more serious than most governments and international agencies have so far been willing to contemplate. The failure of the 2012 'Rio +20' conference is strong evidence of this (Monbiot, 2012). Starting with a world in overshoot, equitable sustainability implies that wealthy countries should already be implementing plans to reduce their per capita ecological footprints – by up to 80 per cent. Major reductions in throughput by the wealthy are necessary to create the ecological space needed for justifiable consumption growth in developing countries (Rees, 2008). This is the true biophysical meaning of the emergent concept of 'degrowth' that is now spreading virally around the world.

Cities, the Second Law and Urban Ecosystems

The modern city – its form, function and sheer scale in the landscape – has been made possible because of abundant cheap energy, particularly fossil fuels. There may be no better expression of human technical mastery of materials than the multiple transportation, communication, utility and other engineered systems that comprise the circulatory, digestive, nervous and waste disposal infrastructure of typical large cities.

Until recently, we have not had to think much about it, but the energy intensity and 'physicalness' of cities make them subject to the rigorous rule of the second law of thermodynamics, the entropy law. The second law is implicated in all real processes involving the use of energy and the transformation of materials.

Even if they have never heard of it, everyone is familiar with commonplace examples of the second law at work – every shiny new car eventually becomes a rusty old clunker; in plain English, the second law dictates that, without continuous maintenance, things naturally tend to erode, wear out and run down. And there are no known exceptions. Is anyone aware of a rusted-out old car that has reacquired its original showroom splendour all on its own?

Scientists describe the second law in slightly more technical terms: any spontaneous change in an isolated system (one not able to exchange energy or matter with its 'environment') increases the 'entropy' of that system. By this they mean that, over time, the system tends to lose form and function – 'randomness' increases as order erodes and energy dissipates. Eventually, an isolated system will reach 'thermodynamic equilibrium', a state of maximum entropy in which no further change is possible. Entropy is one of the more difficult concepts in physics. For our purposes, it is sufficient to define entropy as a measure of disorder or randomness. Thus, a completely dissipated system exhibits maximum entropy, while a highly ordered system is a low-entropy system. Concentrations of 'available' or useful energy or matter are sometimes called 'negentropy'.

Of course, the real world is full of complex systems, ranging from newborn infants through cities to the entire ecosphere, which are assuredly not decaying toward equilibrium. The ecosphere, for example, is a highly ordered system of extraordinary complexity, as represented by millions of distinct species, differentiated matter and accumulated biomass. Moreover, with the passage of geological time, its biodiversity, systemic complexity and energy/material flows have been increasing. In short, the ecosphere is seemingly evolving against the gradient of decay imposed by thermodynamic law. It has been rising ever further from equilibrium; the entropy of the system is decreasing (see Box 1.1). Prigogine (1997) asserts that this phenomenon may well be the measure of life: 'distance from equilibrium becomes an essential parameter in describing nature, much like temperature [is] in [standard] equilibrium thermodynamics'.

If all energy and material transformation is governed by the entropy law, how can we explain the ascent of the ecosphere? The paradox hinges on a single fact: all living systems, from the tiny organelles inside living cells to entire ecosystems, from cities to the ecosphere, are open systems that freely exchange energy and matter with their environments. In short, open systems are just as subject to degradation as isolated systems are but they are also able to import high-grade energy and raw materials for self-maintenance and growth, and to export the resultant low-grade wastes. In effect, open systems can shed their entropy into their 'environments' (see Box 1.1).

Box 1.1 The ecosphere and the second law

Imagine a homogenized world the surface layers of which contain exactly the same mass and mix of elements and stable compounds (for example water and carbon dioxide) as the real world, but where everything has been put through an entropic blender. In this imaginary world there are no concentrations of anything – no physical means of distinguishing any point in 'the system' from any other. Our randomized world is at thermodynamic equilibrium, a state of maximum local entropy.

Now imagine the enormous task of replicating the living world as we know it from this homogeneous mass of raw material: all the necessary 'stuff' is there, but consider the unfathomable quantity of external energy and the infinity of physical processes that would be required to assemble, molecule by molecule, the mind-boggling diversity and complexity (negentropy) of present-day earth from our simulated primordial soup and then maintain it there against the inexorable drag of the entropy law.

In fact, this massive expenditure of energy and effort has actually occurred, or else we wouldn't be here thinking about it! Where did the necessary energy come from? Initially, the process – the emergence of the first replicating molecules and single-celled organisms – was driven by residual chemical energy in the 'soup', but for the past 2.5 billion years the growth and evolution of the ecosphere has mostly been powered by solar energy. No sun, no ecosphere.

Green plants, using photosynthesis, assimilate and incorporate a small portion of this available high-grade solar energy (negentropy) as chemical energy in plant biomass. This concentration of energy and matter powers all other life-forms and living processes in the ecosphere. Much more of the energy falling on plants is used in evapotranspiration to cool them down. Thus, through photosynthesis, evapotranspiration and their own respiration, plants radiate vast quantities of low-grade, high-entropy waste heat into space. Consumer organisms (mostly animals, bacteria and fungi) produce themselves by consuming the surplus biomass (negentropy) generated by plants, but they also ultimately dissipate all this chemically bound solar energy back into space in degraded form. The ecosphere emerges in all its negentropic splendour, but the net entropy of the universe increases much more in the process.

Primary lessons? First, in a universe governed by the second law of thermodynamics, any system that achieves and maintains a highly differentiated, far-from-equilibrium dynamic steady state or which continues growing must have a constant, reliable source of external energy to enable it to resist the dissipative drag of the second law. Second, the increase in local negentropy represented by the growth and increase in complexity of any subsystem is purchased at the expense of a much larger increase in global entropy.

There are, of course, complications. Systems biologists recognize that living systems exist in overlapping, nested hierarchies in which each component subsystem (called a 'holon') is contained by the next level up and itself comprises a chain of linked subsystems at lower levels. (Think 'cell, tissue, organ, individual, population, community, ecosystem' or, if you prefer, contemplate a set of Russian 'matryoshka' nesting dolls.)

This organizational form is the basis for 'self-organizing holarchic open' (SOHO) systems theory (Kay and Regier, 2000). 'Holarchic' means a hierarchy of holons, and the key idea is that every holon in the hierarchy grows, develops and maintains itself using available energy and material (negentropy) extracted from its 'host' system one level up. It processes this energy and matter internally to produce and maintain its own structure and function, and exports its degraded energy and material wastes back into its environment. In short, living organisms maintain themselves in low-entropy far-from-equilibrium states at the expense of increasing global entropy, particularly the entropy of their immediate host system (Schneider and Kay, 1994, 1995). Because self-producing systems thrive by continuously importing, degrading and dissipating available energy and matter, they are called 'dissipative structures' (Prigogine, 1997).

Another expression of the second law is relevant to this discussion. No conversion of available energy and concentrated matter is 100 per cent efficient. Thus, the local negentropy 'created' when a living system produces itself is only a small fraction of the negentropy it must import and dissipate in the process. The simplest example in nature is the fact that the biomass of prey species (antelope, zebras and other herbivores) is generally greater than the biomass of predators (lions, for example). The available energy (negentropy) represented by the prey is always five to ten times greater than that contained by the predators. Thus, all so-called 'production' processes are actually mostly consumption processes that increase global entropy (Box 1.1).

Where do cities and sustainability come in? Like the ecosphere, individual people and entire cities – indeed, the entire human enterprise – are self-producing far-from-equilibrium dissipative structures. However, each is also a 'holon' contained within the SOHO hierarchy of the ecosphere. (Daly [1992] refers to the human enterprise as an open, growing, dependent subsystem of the materially closed, non-growing finite ecosphere.) Thus, the ecosphere evolves and maintains itself by 'feeding' on an extra-terrestrial source of energy, the sun, and by continuously recycling degraded matter. The sun is actually the 'next level up' in the hierarchy from the ecosphere. The ecosphere 'imports' high-grade light energy (negentropy) from the sun, uses some of it to produce itself, and exports low-grade infrared energy (entropic waste) back into space. Growing the ecosphere thus increases the entropy of the universe.

In a similar fashion cities grow and maintain themselves by 'feeding' on the rest of the ecosphere and ejecting their wastes back into it. And, again, the second law dictates that the order incorporated into cities (including urban humans with all their toys) is only a fraction of the order dissipated to produce them. In other words, as presently conceived and configured, cities can grow and increase their internal order (their local negentropy) only by 'disordering' the ecosphere, increasing global entropy.

While urban systems must inevitably consume and dissipate resources, the process need not be destructive of ecological integrity. Despite the constraints imposed by life in the SOHO hierarchy, there is nothing inherently unsustainable about cities. After all, humans are only one of thousands of so-called 'consumer' organisms, the majority of which thrive on earth without destroying the integrity of their ecosystems.

Empowered by virtually unlimited solar energy, ecosystems can self-produce indefinitely. In biological terms, constructive anabolism (such as photosynthesis) marginally exceeds destructive catabolism (respiration) in the non-humanized ecosphere, so biomass accumulates and natural 'sinks' (nutrient recycling) are capacious. Net primary production by producer species (mostly green plants) has always been more than adequate to sustain the world's entire complement of consumer organisms, including pre-industrial humans. The internal negentropy of the ecosphere has therefore generally increased over geological time, driven by the energy of the sun.

That said, the dynamics of perpetual human material growth contain the seeds of potential pathology. The human enterprise is thermodynamically positioned to consume and dissipate the ecosphere from the inside out. Problems must inevitably emerge with humanity's excessive scale and constantly rising demands. Renegade economist Nicholas Georgescu-Roegen (1971a, 1971b) was among the first to understand the implications of the second law for the human economy. Starting from the fact that all economic activity must draw low-entropy resources out of nature and dump useless high-entropy waste back in, he reasoned, first, that 'in a finite space there can be only a finite amount of low entropy and, second, that low entropy continuously and irrevocably dwindles away' (this is, entropy increases). Indeed, the accelerating pace of human-induced global ecological change suggests that humanity has already become dangerously parasitic on its planetary host.

Making matters worse, the spatial separation of city-dwellers from the distant ecosystems that fuel their consumer lifestyles blinds them to the inevitable eco-thermodynamic consequences of over-consumption and excessive pollution. Wealthy urbanites may experience a world of burgeoning cities and expanding economies, but it is also a world of depleting resources, degraded landscapes, declining biodiversity, dying oceans, greenhouse gas accumulation and climate change. Because of the sheer scale of human demand, wealthy cities are now as much the engines of global ecological decay as they are the engines of national economic growth. Far-from-equilibrium thermodynamics and SOHO theory thus provide a simple double-barrelled criterion for global sustainability: the human enterprise must not persistently consume more than nature produces nor generate more waste than nature can assimilate (with a generous allowance for the thousands of other consumer species with whom we share the planet).

Cities as ecosystems

As suggested above, modern humans have been forced seriously to consider their role as ecological agents only by a deepening 'environmental crisis'. An important by-product of this awakening is a growing interest in the notion of cities as ecosystems (e.g. Rees, 1997, 2003; Register, 2006; Newman and Jennings, 2008). There is even a relatively new journal called *Urban Ecosystems*.

The concept of 'urban ecosystem' remains ill-defined and ambiguous, as might be expected this early in the transition. To some analysts, humans are only incidental to the real story. Most natural scientists studying 'urban ecosystems' cast the city mainly as an unnatural habitat for other species. To them, the urban ecosystem consists of the assemblage of non-human organisms in the city and the purpose of their inquiries is to determine how these species have adapted to the structural and chemical characteristics of the 'built environment'. The majority of papers in *Urban Ecosystems* therefore focus on the impacts of urbanization on non-human plants and animals, or on remnant 'natural' ecosystems within the city (Rees, 2003).

This emphasis on other species' adaptations to an 'alien' urban environment is perfectly legitimate, but it also poses an absurdist paradox. After all, it is H. sapiens that created 'the city' and this most human of artefacts is clearly the most structurally remarkable habitat of any animal species on the planet. Moreover, the economic production and consumption (entropic throughput) required to satisfy the material demands of wealthy urbanites have become the major driver of global ecological change. Surely these realities should refocus attention on humans as the dominant ecological actor in all the world's cities (to say nothing of the world outside of those cities). Not to recognize H. sapiens as the keystone species of the urban ecosystem is a major cognitive lapse, and yet another consequence of Cartesian dualism, which fosters the perceptual separation of people and all things human from the natural world.

On the other hand, recognizing urbanites as ecological beings does not imply that the 'the city' per se comprises a human ecosystem. It is doubtless true that every city is a complex system (or, better, a complex of systems) and that cities represent an ecologically critical component of the human ecosystem. However, even if the entire population of H. sapiens lived in cities, this would not qualify 'the city' as a functional ecosystem from the human perspective.

To understand this paradox, consider Odum's (1971) notion that ecosystems should be conceived more in terms of structural properties and functional relationships than physical form. He defined a functionally complete ecosystem as a self-sustaining assemblage of living species existing in complementary relationships with each other and their physical environs. Thermodynamically speaking, ecosystems are also complex structures that continuously assimilate (and degrade) solar energy to (re)generate themselves from simple inorganic chemicals (water, carbon dioxide, nitrates, phosphates and a few trace minerals) that continuously (re)cycle through the system.

Clearly, by these definitions, no modern city qualifies as a complete human ecosystem. To achieve this standard, a city would have to include a sufficient complement of producer organisms (green plants), macro-consumers (animals, including humans), micro-consumers (bacteria and fungi) and non-living factors to support its human population indefinitely. As matters stand, some defining elements are missing from cities altogether or are insufficiently abundant to maintain the system's functional integrity. Consider, for example, that the largest and functionally most important components of urbanites' supportive ecosystems – the assemblage of organisms producing their food, fibre and oxygen, and most of the micro-consumers necessary to complete their nutrient cycles – are found mostly in rural environments scattered all over the planet.

Significantly, the distancing of urban consumers from the productive components of their de facto ecosystems (distant agricultural and forest lands) inhibits the replenishment of soil organic matter and the recycling of phosphorus, nitrogen and other nutrients contained in household and human wastes. These are typically washed out to sea, where they contribute to the emergence of marine 'dead zones'. Urbanization effectively transforms local, integrated, cyclical human ecological production systems into global, horizontally disintegrated, unidirectional throughput systems

(Rees, 1997). Ironically, the resultant continuous entropic 'leakage' of nutrients from farmland in shipments of food to cities threatens to undermine organic agriculture even as its products gain ground in the urban marketplace, as these lost nutrients are not available to form the nutrients for the next crop.

To summarize, in spatial, structural and functional terms, cities as presently conceived are, at best, incomplete ecosystems. They are nodes of intense energy and material consumption and waste generation entirely dependent for their survival on the productive and assimilative capacities of complementary producer ecosystems often located at great distance from the cities themselves (see Box 1.2). Urbanization thus creates a dramatic shift in city-dwellers' spatial and psychological relationships to the land, but there is no corresponding change in functional ecological relationships. Indeed, far from decoupling people from nature, urbanization generally increases their per capita load on cities' supportive ecosystems. Failure to understand these basic facts of urban human ecology increases cities' vulnerability to global ecological change and dooms humanity's quest for sustainability.

Box 1.2 The city as human feedlot

On a crude but illustratively useful level, 'the city' is the human analogue of a livestock feedlot (Rees, 2003). Industrial feedlots, common in North America, are large areas where livestock are brought together to be fattened on an intensive diet. Feedlots, like cities, are densely populated by a single macro-consumer species such as cattle (or pigs, or chickens). However, the fields that produce food for feedlot animals may be hundreds of kilometres distant from the feedlot operation. Also missing from feedlots are adequate populations of micro-consuming decomposers to deal with the waste from the animals. Thus, like cities, feedlots have separated the functionally inseparable, short-circuiting even the possibility of within-system decomposition and nutrient recycling. Feedlots thus generate vast quantities of manure containing vital nutrients that often are not redeposited on farm land for nutrient recycling, but are rather disposed of inappropriately, contaminating the atmosphere, soils and surface and subsurface waters at great distance and over large areas. Sounds rather like a typical city, doesn't it? Since feedlots are, in effect, subsystems of the human urban industrial system, and reflect the same paradigmatic framing, it is not surprising that they are eco-structurally similar to cities.

The Ecological Footprints (EFs) of Typical High-Income Cities

By early in the twenty-first century, more than half the human family was living in urbanized areas which, as noted at the outset, occupy only 2 to 3 per cent of the earth's total land area. This oft-repeated statistic is, however, functionally meaningless – urbanites ecologically occupy a vastly larger area of productive land and water which, for most high-income cities, is increasingly scattered all over the planet. Globalization and trade have enabled the spatial diffusion of the urban-centred human ecosystem but, in functional terms, this ecosystem includes both the built-up consumptive node we refer to as 'the city' and the widely dispersed productive hinterland.

We can approximate the true extent of the area of any city's demands on the earth using ecological footprint (EF) analysis. EF analysis uses data on material flows and ecosystems productivity to estimate, for any specified population, the area of productive land and water ecosystems required on a continuous basis to produce the resources that the population consumes and to assimilate its wastes. In second law terms, a population's eco-footprint approximates the area of natural solar collector required on a continuous basis to regenerate photosynthetically the biomass equivalent (negentropy) of the energy and material the population consumes and dissipates.

The magnitude of a population's eco-footprint depends on the number of people, their average material standards (resource consumption, which correlates with income) and average ecosystem productivity. (To facilitate international comparisons, analysts generally convert national eco-footprint estimates to their equivalent in hectares of global average productivity, gha.) Significantly, while human populations and incomes are generally increasing, the earth's area of productive ecosystems is constant or even declining. Hence, whether we are conscious of it or not, we are all competing for the earth's declining bio-capacity. The more people, the less we can each have. This sounds obvious but does not yet seem to have entered the thinking of most economists.

Just how large are the eco-footprints of modern urbanites? Consider the authors' home town, Vancouver, Canada. The City of Vancouver proper (not the metropolitan area) is home to approximately 578,041 people (at the 2006 census) and occupies an area of 11,467 hectares. We calculated the city's eco-footprint using data compiled by the City, the Metro-Vancouver region and the province for purposes of managing municipal and regional services. We also used national statistical data gathered for tracking population distribution, household types, food consumption and household consumer spending patterns. We employed the relatively conservative 'component' method of EF analysis which uses local data wherever possible to generate a finer-grained analysis of material consumption patterns than can be gained by the 'compound' method using national data.

We estimated Vancouver's eco-footprint at 2,352,627 gha, comprising the following contributing components:

- Food: 52 per cent (2.13 gha per person)
- Buildings: 16 per cent (0.66 gha per person)
- Transportation: 20 per cent (0.81 gha per person)
- Consumables: 12 per cent (0.47 gha per person)
- Water: less than 1 per cent.

For each component, the EF estimate includes the total area of various ecosystem types required by that component. For example, the food-related EF comprises growing land, land occupied by processing facilities and the carbon sink area needed to assimilate the carbon dioxide emissions associated with agriculture and the food processing industry.

This compilation yields an average EF of over 4 gha per person. Other studies based on the 'compound method' of compilation (e.g. Wilson and Anielski, 2005) produce a per capita EF of approximately 7 gha. Thus, depending on the starting assumptions and the method, the total ecological footprint of Vancouver is between 200 to 350 times larger than the city's physical footprint.

This result is fairly typical for high-income cities—despite methodological differences, urban eco-footprints are generally found to be two to three orders of magnitude larger than their respective official political or built-up areas. For example, Folke *et al.* (1997) estimated that the 29 largest cities of Europe's Baltic region depend on an area of forest, agricultural, marine and wetland ecosystems 565 to 1,130 times larger than the area of the cities themselves. The 7 million people of Hong Kong have a total eco-footprint of 332,150 to 478,300 square kilometres, which is 303 times the total land area of the Hong Kong Special Administrative Region (at 1,097 square kilometres), and 3,020 times the built-up area of the city (110 square kilometres; Warren-Rhodes and Koenig, 2001). In short, thanks to globalization and trade, high-income cities have been able to grow to enormous size by appropriating the biomass output and waste sink functions of productive ecosystem areas that are hundreds of times larger than the cities themselves and often located half a planet away.

High-density, compact cities such as Hong Kong have eco-footprint to built-up area ratios larger than those of more sprawling settlements like Vancouver. This points to the reality that what and how much we consume is much more a determinant of our ecological footprint than where we live. Certainly, higher density alone cannot solve the challenge of living within ecological carrying capacity.

The Biophysical Meaning of Sustainability and Society's Flaccid Response

At present, industrial and post-industrial countries live mainly by using (i.e. using up) vast stores of fossil fuels and resources that accumulated during previous geological eras. Fossil energy is particularly important because it not only serves us directly (for example by powering our transportation and heating our homes) but also enables us to extend our eco-footprints all over the planet to access (and over-exploit) reserves of other essential resources.

This situation is obviously not sustainable; we are depleting and dissipating renewable and replenishable 'natural capital' stocks (such as fish, soils and ground-water) faster than nature can produce them or, in the case of non-renewable resources, faster than human ingenuity can devise sustainable substitutes; our entropic detritus is filling nature's sinks to overflowing. To achieve sustainability humanity must learn to live on contemporary photosynthesis, on other forms of solar energy (hydroelectric, wind and photovoltaic energy) and on the continuous recycling of essential material elements through ecosystems. In second law terms, nature can no longer bear the entropic costs of further growing the human enterprise. We need to develop a steady-state economy in which human demands are compatible with sustainable rates of energy and material flow in nature, as Daly (1977) pointed out nearly 40 years ago.

But this is not how the world is responding. Most sustainability efforts today – such as 'green' buildings, hybrid vehicles and urban densification – are based on the assumed gains to be made from increased efficiency or on the belief that we can substitute human-engineered capital for depleted natural capital.

There is indeed significant scope for improved energy and material efficiency (von Weizsäcker *et al.*, 2009). However, historically, in the absence of policies to negate the so-called 'rebound effect', efficiency gains have actually stimulated consumption by lowering costs and enabling wage increases. This is not a new idea. In his 1865 book *The Coal Question*, William Jevons argued that efficient use of fuel tended to result in an increase in consumption of that fuel (Jevons, 1865: VII.3). Nor is substitution a complete solution. Consider just one example, biofuels: there is simply insufficient crop and forest land on earth to produce the biofuel equivalent of fossil energy. In any case, all available arable land will be needed for an anticipated doubling of food and fibre production. A just approach to sustainability demands improvement of the diets of the 3 billion people who are currently malnourished and accommodation of the needs of the additional 2 billion people expected by mid-century (Brown, 2008; see Box 1.3).

Box 1.3 Arable land, population and urban agriculture

Consider the following: there are only about 0.2 hectares of arable land per person on earth, according to World Bank data (Trading Economics, 2012), yet, if Vancouver is typical, as shown in Table 1.1 it takes approximately 1.6 hectares per person of crop and grazing land to produce the average high animal-protein North American diet (and this with contemporary high-input industrial agriculture). Obviously there is no chance whatsoever of achieving North American dietary standards for even the present world population, let alone an additional 2 billion people. Indeed, the increase in production needed to provide an adequate vegetarian diet for all will be difficult enough, in light of rising energy and other input costs, increasing water scarcity and the potentially negative effects of climate change, even if we ignore potential competition for cropland from biofuel production.

These numbers also have implications for proponents of urban agriculture. If the Vancouver diet is typically North American, 1.6 hectares per person implies that the growing-land component of the city's food footprint alone is over 900,000 real hectares, 86 times its political area. Even a globally equitable 0.2-hectare cropland footprint would cover 115,000 hectares, ten times the city's nominal area leaving no room for buildings. Increased interest in urban agriculture notwithstanding, it would obviously be physically impossible for urban populations to meet their food needs from within 'the city' as currently defined. The gap between possible local supplies and actual demand for arable land is just one measure of the adjustment needed to both our understanding of what constitutes the human urban ecosystem and our conception of local sustainability.

The 'Fair Earth-Share': Living Within Our Ecological Means

Eco-footprinting makes clear that cities can support large populations only by importing the capacities of ecosystems and rural hinterlands located far beyond their political boundaries. Sheer purchasing power enables wealthy urbanites to appropriate a disproportionate share of global bio-capacity. High-income city-dwellers have average eco-footprints ranging from 5 to 10 gha. However, with fewer than 12 billion hectares of biologically productive land and 7 billion people, there are only 1.7 hectares of biologically productive land per person on earth. This means that to support the entire human family at developed-world material standards would require three to six earth-like planets.

Since additional habitable planets remain out of reach for the foreseeable future, we must plan for sustainability by other means. To begin, we might ask what the world would look like if the global community agreed to work toward a more or less equitable distribution of global bio-capacity. This would require that the average person live on the sustainable productive and assimilative capacity of just 1.7 global average productive hectares. For the sake of this discussion, we will call this 1.7 gha the equitable or fair 'earth-share'. If everyone on earth adopted an equitable 1.7 gha lifestyle, the entire human family would be living within the means of nature. This is an ecological ideal that represents 'one-planet living'. We realize that absolute material equality is an impossibility – some inequity is probably a fact of life. Nevertheless, for long-term sustainability (one-planet living), the world will have to achieve greater equity and an average eco-footprint approaching the fair earth-share of 1.7 gha per person. And this fair earth-share is contracting with population growth and ecosystem degradation, so every year we can have less.

Setting ecological footprint reduction targets

We can now define the 'sustainability gap' for high-income countries as the difference between their populations' actual average per capita ecological footprint and the 1.7 gha fair earth-share (see Box 1.4). High-income lifestyles typically demand between 5 and 10 gha per person, so existing sustainability gaps range from 3.3 to 8.3 gha. Thus, the really inconvenient truth is that, to live fairly within earth's ecological means, Europeans should be implementing policies that will reduce their per capita ecological footprints by two-thirds. Average North Americans and Australians should be planning an 80 per cent reduction.

Box 1.4 Comparing societies at the fair earth-share with higher-income countries

More than half the world's population lives at or below the fair earth-share, mostly in Latin American, Asian and African countries. As shown in the table below, fair earth-share societies enjoy comparable longevity with the world average, but have somewhat larger households and lower per capita calorie intake, meat consumption, household energy use, vehicle ownership and carbon dioxide emissions. The differences between fair earth-share and high-income (3+ planet) countries are much greater still.

	Fair earth-share	World average consumption	3+ planet consumption
Per capita daily calorie supply	2,424	2,809	3,383
Per capita meat consumption, kg/yr	20	40	100
People per household	5	4	3
Household energy use			
GJ	8.4	12.6	33.5
kWh/yr	2,300	3,500	9,300
Motor vehicle ownership	0.0	0.1	0.5
Carbon dioxide emissions, t/yr	2	4	14
Life expectancy, yr	66	67	79

- One-planet societies include: Mali, Ecuador, Cuba, Guatemala, Uzbekistan, Vietnam, Philippines, Ethiopia, India and Haiti.
- Three-plus-planet societies include: USA, Canada, Australia, Kuwait, Sweden, Norway, Mongolia (because of the great demand for pasture there), Spain, Germany, Italy, UK, New Zealand, Israel, Japan and Russia.

Data for these estimates are taken from ethnographic field studies (FAO, 2012; Menzel, 1994) and statistical databases (World Bank, 2011; WRI, 2010; Worldmapper, 2010) and correlate to national ecological footprint assessments for 2007 (WWF, 2010b).

Deconstructing Vancouver's Eco-footprint

Let's return to our Vancouver example. With ecological footprints averaging at least 4 and up to 7 gha, this city's inhabitants should be planning a 58 to 75 per cent EF reduction in energy and material throughput to close the sustainability gap. (The lower figure reflects the component EF method, the higher figure the less locally sensitive but more comprehensive compound EF approach.)

We can examine this reduction across the major components of consumption to help envision how the city would fare under our 'one-planet living' model. Table 1.1 represents the consumption land-use matrix for Vancouver's ecological footprint. It shows the bio-capacity demand by consumption component, measured in global average hectares.

First, consider Vancouver's ecological footprint by land type (Figure 1.1). As is typical of most high-income societies, energy land accounts for half of Vancouver's ecological footprint. This is significantly more than for societies that are closer to the one-planet living target of 1.7 gha per person, where, on average, energy land accounts for 33 per cent of the footprint (WWF, 2010b). Energy derived from fossil fuels emits carbon dioxide when it is burned. Energy land mostly comprises the area of dedicated carbon-sink forest, at world average productivity, that would be required to sequester the carbon dioxide emissions of the study population (this is its true 'carbon footprint'). The accumulation of atmospheric carbon dioxide is empirical evidence that the global carbon sink is overflowing. High-income societies rely extensively on fossil fuels to support their lifestyles. For example, various liquid and gaseous fossil hydrocarbons power motor vehicles and aircraft; natural gas or electricity generated by burning fossil hydrocarbons heat, cool and supply electricity to our buildings; energy derived from fossil fuels heats and cools our water and cooks our meals; fossil fuel energy drives our factories and thousands of products, such as plastics, are made from petroleum and other hydrocarbons; natural gas is a major feedstock of

Table 1.1 Vancouver ecological footprint land-use matrix. Note that component figures have been rounded

EF consumption component	Cropland	Pasture land	Fishing area	Forest land	Energy land	Built area	Total
Food (gha/person)	1.47	0.13	0.12		0.41		2.13
Buildings (gha/person)				0.02	0.62	0.03	0.66
Consumables (gha/person)	0.02			0.15	0.31		0.48
Transport (gha/person)					0.80	0.01	0.81
Water (gha/person)					neg.	neg.	neg.
Services (gha/person)							n/a
Total (gha/person)	1.49	0.13	0.12	0.17	2.13	0.04	4.09
Total (gha)	863,889	75,145	69,365	96,309	1,233,967	23,116	2,361,792
Percentage of total	37%	3%	3%	4%	52%	1%	100%

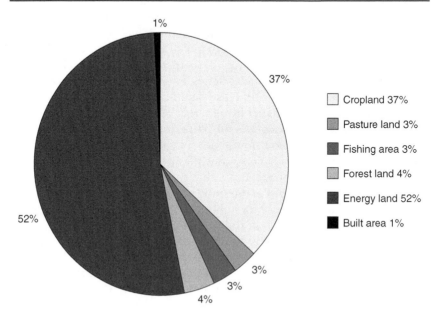

Figure 1.1 Proportional representation of Vancouver's ecological footprint by land type

fertilizers, essential for food production. Even uranium, although not derived from fossils, is a finite resource. Little wonder that finding ways to conserve energy is a critical aspect of becoming sustainable.

Cropland, at 37 per cent, is the other large component of Vancouver's ecological footprint. Cropland is the ultimate source of all our food and alcohol, and textile fibres such as cotton, hemp and flax. Much cropland, particularly in North America, is dedicated to producing animal feed for those feedlots (although 40 per cent of the corn crop in the US is now being diverted to 'feeding' the automobile fleet). Pasture land for grazing adds another 3 per cent of the Vancouver EF to the total land area required for food production.

Together, energy land, cropland and pasture account for over 90 per cent of Vancouver's ecological footprint. Also typical of most high-income cities, Vancouver's built area comprises only 1 per cent of its total ecological footprint.

Food production and processing comprise the largest consumption component of Vancouver's ecological footprint (Figure 1.2). Transport and buildings are next, followed by consumables. Water is omitted from further analysis because of its small impact (less than 1 per cent).

As noted, to achieve one-planet living, Vancouverites would need to reduce their average per capita ecological footprint by at least 58 per cent (i.e. from 4.09 gha to 1.7 gha). To understand what such a reduction might entail, we explore the ecological footprints of two eco-village communities, Findhorn in Scotland and Toarp in Sweden. These communities represent

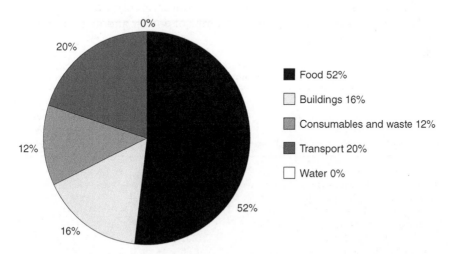

Figure 1.2 Proportional representation of Vancouver's ecological footprint by component

Source: The food footprint was prepared with the assistance of Dr Meidad Kissinger (2012) and relies predominantly on national data for Canada

some of the best performing eco-villages documented in the literature (Tinsley and George 2006; Haraldsson, Ranhagen and Sverdrup, 2001) and reveal that one-planet living is achievable in high-income countries.

Table 1.2 compares Vancouver's EF profile with the best performance attained in the two eco-villages. Findhorn data were used for the food, buildings and transport components (Tinsley and George, 2006) and Toarp data for consumables and services (Haraldsson, Ranhagen and Sverdrup, 2001). The resultant 'synthetic' eco-village has a combined EF less than the 1.7 gha 'one-planet' standard. We should exercise some caution in this comparison. While the Findhorn and Toarp studies, like our Vancouver analysis, used the component method, data were collected by community surveys rather than by estimating actual material flows. Also, the Findhorn and Toarp studies combine energy consumed by private sector businesses and infrastructure within a 'Services' category that also includes government services. In our study, we group the private sector businesses and infrastructure into the 'Buildings' category. Nevertheless, interpreting our Vancouver findings in light of actual eco-village performance is highly instructive.

The low food footprint for Findhorn residents is attributable to an almost exclusively vegetarian diet. In addition, most meals are prepared communally, further reducing demand for energy and supplies. Dairy products comprise approximately 65 per cent of Findhorn's food footprint, with fruit and vegetables accounting for the remainder (Tinsley and George, 2006). By comparison, Vancouver's food footprint, comprising more than half the total EF, is dominated by meat, fish and eggs followed by oils (including nuts and legumes) and dairy products (Figure 1.3).

The same data can be expressed in terms of energy and material inputs (Figure 1.4).

Table 1.2 Comparing Vancouver's ecological footprint to eco-village performance

Component	Vancouver	'Best performance' in Findhorn and Toarp eco-villages	Gap between Vancouver and eco-village 'best performance'
Food (gha/person)	2.13	0.42	1.71
Buildings (gha/person)	0.66	0.29	0.37
Consumables (gha/person)	0.48	0.19	0.29
Transport (gha/person)	0.81	0.37	0.44
Water (gha/person)	0.002	n/a	—
Services (gha/person)	n/a	0.09	—
Total (gha/person)	4.09	1.36	2.72

Sources: Findhorn data represented eco-village best performance for food, buildings and transport (Tinsley and George, 2006); Toarp for consumables and services (Haraldsson, Ranhagen and Sverdrup, 2001)

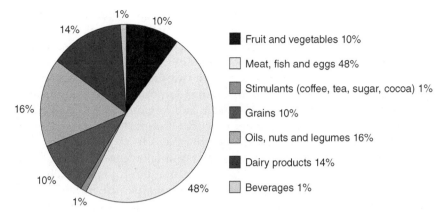

Figure 1.3 Vancouver's food footprint by food type

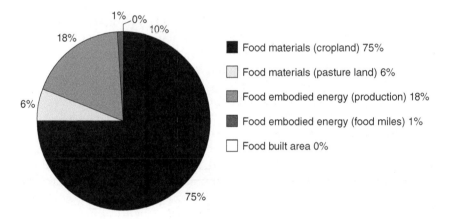

Figure 1.4 Vancouver's food footprint showing energy and material inputs

Figure 1.4 reveals that pasture land accounts for only 6 per cent of the food footprint. However, almost half the cropland (46 per cent, or 35 per cent of the entire food footprint) is dedicated to growing animal feed including corn, alfalfa and other types of hay. It is also worth noting that the 'food miles', around which people seem to have created a great deal of anxiety, are insignificant. It's not the miles, it's the food that matters.

Reducing Vancouver's Eco-footprint

What planning measures might be taken in Vancouver to bring the city's ecological footprint closer to the Findhorn-Toarp eco-village model or to one-planet living?

There is clearly plenty of scope in the average Vancouver diet for significant EF reductions while possibly even improving population health. However, City of Vancouver councillors and staff interviewed for this project saw changes to diet as a difficult policy question that is beyond the scope of municipal jurisdiction. They did identify raising awareness about the impacts of food choices and demonstrating corporate leadership through purchasing locally produced, in-season and organic food products as important changes to management practices. Councillors also recognized that policies to enable food production and sales from private homes, to encourage urban community gardens and farming on city-owned land, and to support farmers' markets could be important interventions. Finally, they recognized the need to engage with the community further, both at the level of government and of food producers, to pursue low-impact food production, processing, distribution and access.

City staff estimate that a 10 per cent reduction in high-impact foods, e.g. meat, could contribute a 3.4 per cent reduction in Vancouver's overall ecological footprint (City of Vancouver, 2011: 111). However, contributing significantly to closing the sustainability gap would require not only shifts in consumption patterns but also a radical change in the energy intensity of food production and processing. Options include:

i   incentives for locally produced organic meats and produce
ii shifting to more local production of most fruits and vegetables
iii adopting a predominantly vegetarian diet.

We estimate that these changes could reduce the food component of Vancouver's ecological footprint by approximately 70 per cent, to 0.66 gha.

The second largest consumption component of Vancouver's ecological footprint is transport (Figure 1.2; Table 1.2). Figure 1.5 reveals that operating energy (carbon dioxide emissions from combustion of fossil fuels) accounts for most of the transportation footprint. Privately owned and operated motor vehicles have more than double the operational demand of all other modes of travel, including air travel, combined.

The majority of materials associated with transportation (such as steel, concrete and plastics) are mined or manufactured predominantly from fossil fuels. Therefore, there is virtually no cropland or pasture land associated with the transportation footprint. Furthermore, these products are either highly recyclable (as in the case of steel) or have a very long service life (as in the case of concrete). Therefore, the negative impacts to bioproductive land from the original mining activity or the high levels of carbon dioxide emissions associated with making the cement used in concrete for roads are amortized over a long period of time. This results in a relatively small embodied energy value relative to operating energy.

Vancouver councillors and staff have recognized shifting personal transportation choices from private cars to walking, cycling and transit as a

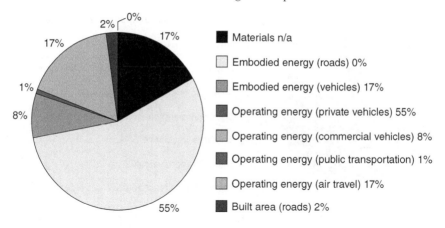

Figure 1.5 Vancouver's transportation footprint

priority policy objective. Supportive policies include improving cycling infrastructure and access to rental bicycles, increasing public transport capacity and tighter parking restrictions. These changes could reduce Vancouver's overall ecological footprint by 2 per cent (City of Vancouver, 2011: 112) but councillors see the city's lack of authority over public transport as a barrier to implementation.

In any event, a 2 per cent reduction is not significant. To narrow the sustainability gap significantly would entail:

i   a 50 per cent reduction in private vehicle use coupled with elimination of air travel, or
ii elimination of virtually all private vehicles.

We estimate that option (i) would reduce transportation's contribution to Vancouver's ecological footprint by 47 per cent to 0.38 gha; option (ii) would reduce it by 72 per cent to 0.23 gha.

Figure 1.6, which looks at buildings, reveals that, just as for transport, operating energy accounts for most of the building component's eco-footprint. Since Vancouver's electricity is predominantly from hydro-generation (which is low carbon intensity), the bulk of buildings' operating EF is attributable to carbon emissions from burning natural gas for water and space heating.

The city has identified zoning for district energy systems and building design guidelines that encourage passive heating and cooling as ways to reduce buildings' EF. One challenge is Vancouver's affinity to glass curtain wall construction that affords fantastic views but has inefficient thermal performance. City staff estimate that retrofitting buildings to improve thermal performance could reduce operating energy by 20 per cent across the existing building stock. This action, coupled with the introduction of

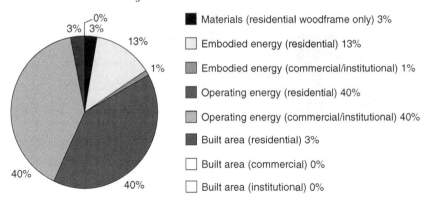

Figure 1.6 Vancouver's building footprint

district energy systems, could contribute to a 3 per cent reduction in Vancouver's overall ecological footprint (City of Vancouver, 2011: 111).

Once again, however, this is insufficient. We estimate that closing the sustainability gap will require a 62 per cent reduction in the operating energy of existing buildings. This would require

i comprehensive building retrofits coupled with district energy systems,
ii aggressive utility pricing incentives,
iii appliances that stimulate user conservation and
iv instant feedback mechanisms such as 'smart' meters

as policy options to induce conservation, although all this might be insufficient to achieve the necessary reduction without a change in the expectations of building performance. A 62 per cent reduction in operating energy would reduce the building component of Vancouver's EF by 50 per cent to 0.33 gha.

Embodied energy and materials account for 96 per cent of the EF of consumable products (Figure 1.7). Thus the EF is mostly the carbon sink and croplands dedicated to growing the bio-fibre in paper, textiles, rubber, etc. Of this amount, 68 per cent is attributed to non-recycled products and 28 per cent to recycled products. Carbon dioxide emissions from incineration and decomposition of wastes account for the remaining 4 per cent of carbon sink requirements. Less than half of 1 per cent is associated with solid and liquid waste management operations, including land dedicated to landfills.

Figure 1.8 shows the same data by product type. Paper comprises the majority share because of the mutually exclusive demand on forest lands for both wood-fibre and carbon sequestration. For recycled products only the energy used to remanufacture the product is counted here.

Ecological Footprints and Urbanization   25

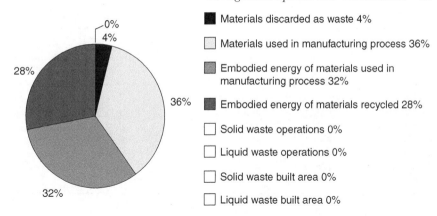

Figure 1.7 Vancouver's consumables footprint by energy/material category

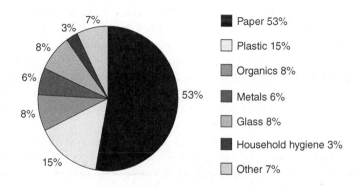

Figure 1.8 Vancouver consumables footprint by product type, including recyclables

Some city councillors and staff identified opportunities to reduce consumption through the promotion of smaller living spaces and by greater sharing of durable goods ('tool libraries' are already operating in parts of Vancouver). They also speculated on whether: (a) the city's role in issuing business licences could be explored as a means to encourage greener business practices and products; and (b) expanding the scope of waste diversion to increase current levels of composting and introducing recycling services to multi-unit residential developments would significantly reduce waste. More vendor take-back programmes that establish a value for waste products – called 'extended producer responsibility' or EPR – are one way to create market value for waste materials and encourage their diversion from the waste stream. EPR programmes for various products, from car batteries through paints to tyres, are already common in the region. However, EPR is beyond the scope of municipal jurisdiction and would require the city to work collaboratively with industry and senior levels of government.

Table 1.3 Closing the sustainability gap on Vancouver's ecological footprint

Component	Vancouver EF (gha/person)		
	Actual 2006	With proposed actions	Reduction from proposed actions
Food	2.13	0.66	1.47
Buildings	0.66	0.33	0.33
Consumables	0.48	0.30	0.18
Transport	0.81	0.38	0.43
Water	0.002	n/a	—
Services	n/a	n/a	—
Total	4.09	1.67 (59% below present)	2.41

City staff estimate that increasing recycling by 50 per cent (from the current 50 per cent diversion to a total resource recovery rate of 75 per cent) could contribute to a 0.5 per cent reduction in Vancouver's overall ecological footprint (City of Vancouver, 2011: 111). However, we argue that closing the sustainability gap requires reducing material throughput more generally. A strong emphasis on reusing and remanufacturing clothing, furniture, consumer durables, electronic equipment, etc., rather than recycling, could achieve a 30 per cent reduction in demand for consumable products. Coupled with a 60 per cent reduction in paper consumption this could reduce the consumables component of Vancouver's ecological footprint by 38 per cent to 0.3 gha.

To summarize, the total effect of actions proposed in Vancouver's Greenest City Action Plan might yield an 11.5 per cent reduction in Vancouver's ecological footprint (City of Vancouver, 2011: 111). This is far below the 33 per cent reduction in ecological footprint that was anticipated to be achieved through this initiative (Boyd, 2009), and a very long way off the idea of the 'Greenest City'.

There is, however, enormous potential for additional throughput reductions, as revealed in the comparison of Vancouver's EF with those of contemporary eco-villages. Table 1.3 summarizes the reductions in Vancouver's ecological footprint that might be achieved through a more concerted effort to achieve one-planet living using the additional actions noted above.

Discussion and Conclusions

This brief study demonstrates that becoming a 'one-planet city' would require fundamental changes in the average Vancouverite's lifestyle. The city could implement some important policy changes and management practices on its own, but it is clear that little can be achieved at the municipal level without the addition of active support at senior government level. Even so, much will depend on the cooperation of the private sector and on how to 'incentivize' low-EF lifestyle choices by citizens.

This is no trivial challenge. The fact is that no city or country has ever proposed to 'develop' by reducing its energy and material throughput if not its material standards. Nevertheless, some suggest that it should be possible to achieve a planned reduction in the eco-footprints of high-income countries of up to 75 per cent while actually improving average well-being and quality of life (von Weizsäcker *et al.*, 2009). Fully informed, and given proper economic and motivational incentives, people might willingly exchange further income growth for the greater economic security, ecological stability and social equity – coupled with more leisure time and family and community engagement – that are promised by true sustainability. This optimistic view is very much against the current power and interests of 'the market', so it is unlikely to gain much traction.

Some of the adjustments that would help bring Vancouver closer to the 'one-planet' target include:

i   Strong land-use, zoning and tax policies in support of producing most fruits and vegetables organically within the local area, the Lower Mainland of British Columbia. Citizens will also have to adopt a more vegetarian or even vegan diet.
ii Policies to induce radical reductions in, or elimination of, the use of personally owned and operated motor vehicles and to virtually eliminate fossil-powered air travel.
iii Renewed support for the city's eco-densification strategy accompanied by incentives to force a 60 per cent improvement in energy efficiency in buildings, specifically to reduce reliance on fossil fuels for both space and water heating.
iv Active promotion of a conserver society oriented towards reuse, coupled with a 60 per cent reduction in consumption of paper.

To elaborate in terms of the various consumption components considered in this analysis, an ecologically sustainable Vancouver might be described in the following terms:

- *Food* Vancouverites would have to adopt a radically restructured agri-food system and healthier diets. This implies relocalizing much of the city's food supply, a return to in-season consumption habits (for example, imports of many out-of-season fruits and vegetables might have to be abandoned), an almost total shift from animal to plant proteins and active policies to protect regional agricultural land for the production of local, organic foods. More citizens would participate in city-sponsored community gardening and fruit tree harvesting activities and in selling and buying produce through local food hubs and farmers' markets.

- *Transport* Vancouver would have to reorient residents' excessive reliance on private vehicles and air travel toward walking, cycling and public transport, in part by investing in the corresponding infrastructure. People cannot be expected to abandon their automobiles in the absence of satisfactory alternatives. This shift will require changes in tax and related incentive structures at all levels of government. For example, ending subsidies on fossil fuels and scheduling a steeply rising carbon tax (to the point where market prices tell the truth—we need to internalize ecological externalities) or implementing some form of cap-auction-trade for carbon emissions will be necessary both to induce essential economic structural changes and provide a revenue stream to help fund transportation alternatives. With appropriate positive and negative incentives, people will make greater use of city-sponsored bicycle infrastructure and regional rapid transit services. The augmentation of public transport capacity should be coupled with the phasing out of motor vehicle lanes and parking to accommodate improved walking and cycling facilities. Eliminating minimum requirements for parking in residential and commercial zoning would help promote public transport while lowering costs. (This actually simply reduces an unfair hidden subsidy for motor vehicle owners.)
- *Buildings* Vancouver must renew its commitment to ensure that developers work with sound urban design and land consolidation principles to greatly increase average population density while capitalizing on the aesthetic delights of the Vancouver environment and enhancing access to urban amenities (the economic viability of which will be increased in the process). The adoption of the city's eco-density charter in 2010 was an encouraging first step (City of Vancouver, 2008).

Buildings are discrete structures which dissipate energy and materials, but they are readily manipulable. Thus, the design features of buildings must support energy efficiency objectives, including increased thermal performance of the building envelope. Vancouver's affinity for glass towers has resulted in: (a) excessive heat-loss from downtown high-rises during winter months, cancelling gains from energy efficiency improvements in building mechanical systems; and (b) excessive passive solar gain on south exposures during summer months. The city will have to work simultaneously with government to maintain continuous pressure aimed at improving the thermal performance of buildings and the general efficiency of lighting, appliances and personal electronics. Building codes and related design guidelines must be upgraded effectively to prohibit inefficient envelope and cladding materials at the same time as encouraging the use of passive systems that allow natural daylight, ventilation and solar gain. District energy systems coupled with networked renewable energy technologies must increasingly supplement energy flows in the urban environment.

Simultaneously, there will need to be a widespread programme of improvement of existing buildings and an increasing acceptance of wider temperature swings.
- *Consumables* Sustainability requires that society abandon the central tenets of the consumer-based 'throwaway' culture, including planned obsolescence and even shopping-as-leisure-activity. With true-cost pricing and related tax incentives (Rees, 2010b) we believe people will demand higher-quality, longer-lasting products. There should be much greater reliance on reuse and sharing of clothing, tools, appliances, sporting equipment and personal electronics and a concerted effort to reduce paper consumption. The latter includes avoiding printed correspondence (such as bank statements and utility bills), newspapers and packaging, both at home and at work.
- *Water* Further analysis of the water component was omitted because of its small ecological footprint. However, water is essential for life, so water quantity and quality considerations could affect community sustainability in ways that complement one-planet living. One can readily imagine both the land-use and water management strategies required for one-planet living being implemented within an overarching regional planning strategy such as bioregionalism.

Finally, this research illustrates how far-from-equilibrium thermodynamics and eco-footprint analysis can change perceptions about the biophysical requirements for sustainability. Wishful thinking aside, the human enterprise is not decoupling from nature. Growing our urban culture (raising the human enterprise further from thermodynamic equilibrium) necessarily consumes and dissipates ever larger volumes of low-entropy energy and materials. As a result, the ecological footprints of our increasingly global urban culture now exceed the bio-capacity of earth by up to 50 per cent. Humanity is now living by dissipating and degrading the very ecosystems that sustain us.

Since the second law of thermodynamics (unlike all human-made laws) cannot be revoked, the status quo is not a viable option. Trading off the ecosphere for additional growth is foolish and ultimately fatal. The global community must face the fact that sustainability demands an absolute reduction in energy and material throughput of 75 to 80 per cent in high-income countries.

Economic and material efficiency, the remedy most favoured by politicians and economists, or at least of those few who are prepared to admit that there is a problem, clearly does not meet this challenge. On the contrary, ecological footprint analysis (EFA) reveals a widening gap at the macro level between such illusory perceptions of progress and actual performance. Fortunately, such analysis also identifies potentially the most effective points for policy leverage and can be used at the micro scale to monitor the success or failure of specific policy and planning initiatives.

While the results and implications of EFA are far from universally recognized and accepted, supporting data and analyses from disparate disciplines are making it increasingly difficult to deny our basic conclusions. EFA is therefore gaining ground in sustainability assessments around the world. It is particularly gratifying that Vancouver has taken up the EF challenge as part of its Greenest City Initiative. City staff are using our findings both to (a) explore options for further reductions in the city's ecological footprint, and (b) engage the community in a creative dialogue to build support for the extraordinary policy initiatives that will be necessary for Vancouver to reach its goal of one-planet living by 2050. We are increasingly hopeful that other large cities will soon follow Vancouver's lead.

References

Boyd, D. (2009) *Vancouver 2020: A Bright Green Future*. Vancouver: City of Vancouver. http://vancouver.ca/greenestcity/PDF/Vancouver2020-ABrightGreenFuture.pdf (accessed June 3, 2011)

Brown, L. (2008) *Plan B 3.0: Mobilizing to Save Civilization*. New York: W.W. Norton and Company.

City of Vancouver (2008) *Ecodensity: Vancouver Ecodensity Charter*. Adopted June 10 2008. Vancouver: City of Vancouver. http://vancouver.ca/commsvcs/ecocity/pdf/ecodensity-charter-low.pdf (accessed January 27, 2012)

City of Vancouver (2011) *Greenest City 2020 Action Plan*. Administrative Report, July 5. Vancouver: City of Vancouver. http://vancouver.ca/ctyclerk/cclerk/20110712/documents/rr1.pdf (accessed December 10, 2012)

Daly, H.E. (1977) *Steady State Economics: The Economics of Biophysical Equilibrium and Moral Growth*. San Francisco: W.H. Freeman.

Daly, H.E. (1992) 'Steady-state economics: concepts, questions, policies', *Gaia* 6: 333–338.

Daly, H.E. (2005) 'Economics in a full world', *Scientific American* September: 105–107. http://steadystate.org/wp-content/uploads/Daly_SciAmerican_FullWorldEconomics(1).pdf (accessed January 20, 2012)

FAO (Food and Agriculture Organization) (2012) 'Nutrition country profiles'. Rome: Food and Agriculture Organization. http://www.fao.org/ag/agn/nutrition/profiles_by_country_en.stm (accessed January 30, 2012)

Folke, C., Jansson A., Larsson J. and Costanza R. (1997) 'Ecosystem appropriation by cities', *Ambio* 26(3): 167–172.

Georgescu-Roegen, N. (1971a) *The Entropy Law and the Economic Process*. Cambridge, MA: Harvard University Press.

Georgescu-Roegen, N. (1971b) 'Afterword', in J. Rifkin and T. Howard, *Entropy: A New World View*. New York: The Viking Press.

Haraldsson, H., Ranhagen, U. and Sverdrup, H. (2001) 'Is eco-living more sustainable than conventional living? Comparing sustainability performances between two townships in southern Sweden', *Journal of Environmental Planning and Management* 4(5): 663–679.

Jevons, W.S. (1865) 'Of the economy of fuel', chapter 7 in *The Coal Question: An Inquiry Concerning the Progress of the Nation, and the Probable Exhaustion of Our Coal-Mines*. London: Macmillan and Co.

Kay, J. and Regier, H. (2000) 'Uncertainty, complexity, and ecological integrity', in P. Crabbé, A. Holland, L., Ryszkowski, L. and Westra, L. (eds), *Implementing Ecological Integrity: Restoring Regional and Global Environment and Human Health*, NATO Science Series IV: Earth and Environmental Sciences 1. Dordrecht, the Netherlands: Kluwer Academic Publishers, pp. 121–156.

Kissinger, M. (2012) 'International trade related food miles: the case of Canada', *Food Policy* 37(2): 171–178.

Menzel, P. (1994) *Material World: A Global Family Portrait*. San Francisco: Sierra Club Books.

Monbiot G. (2012) 'The end of an era', *The Guardian* 25 June. http://www.monbiot.com/2012/06/25/end-of-an-era/ (accessed June 27, 2012).

Newman, P. and Jennings, I. (2008) *Cities as Sustainable Ecosystems: Principles and Practices*. Washington, DC: Island Press.

Odum, H.T. (1971) *Environment, Power and Society*. New York: Wiley.

Prigogine, I. (1997) *The End of Certainty: Time, Chaos and the New Laws of Nature*. New York: The Free Press.

Rees, W.E. (1997) 'Urban ecosystems: the human dimension', *Urban Ecosystems* 1: 63–75.

Rees, W.E. (2003) 'Understanding urban ecosystems: an ecological economics perspective', in A.R. Berkowitz, C.H. Nilon and K.S. Hollweg (eds), *Understanding Urban Ecosystems*. New York: Springer-Verlag.

Rees W.E. (2008) 'Human nature, eco-footprints and environmental injustice', *Local Environments* 13(8): 685–701.

Rees, W.E. (2010a) 'Getting serious about urban sustainability: eco-footprints and the vulnerability of twenty-first-century cities', in T. Bunting, P. Filion and R. Walker (eds), *Canadian Cities in Transition: New Directions in the Twenty-first Century*. Toronto: Oxford University Press, pp. 70–86.

Rees, W.E. (2010b) 'True cost economics', in C. Lazlo, K. Christensen, D. Fogel, G. Wagner and P. Whitehouse (eds), *Berkshire Encyclopedia of Sustainability, Volume 2: The Business of Sustainability*. Great Barrington, MA: Berkshire Publishing Group.

Rees, W.E. (2012) 'Cities as dissipative structures: global change and the vulnerability of urban civilization', in M. Weinstein and R. Turner (eds), *Sustainability Science: The Emerging Paradigm and the Urban Environment*. New York: Springer.

Register, R. (2006) *Ecocities: Rebuilding Cities in Balance with Nature*. Gabriola Island, BC, Canada: New Society Publishers.

Schneider, E.D. and Kay, J.J. (1994) 'Complexity and thermodynamics: toward a new ecology', *Futures* 26: 626–647.

Schneider, E.D. and Kay, J.J. (1995) 'Order from disorder: the thermodynamics of complexity in biology', in M.P. Murphy and L.A. O'Neill (eds), *What is Life? The Next Fifty Years*. Cambridge, UK: Cambridge University Press.

Shah, A. (2010) 'Poverty facts and stats', Global Issues. http://www.globalissues.org/article/26/poverty-facts-and-stats (updated September 20, 2010; accessed January 19, 2012)

Tinsley, S. and George H. (2006) *Ecological Footprint of the Findhorn Foundation and Community*, Forres, UK: Sustainable Development Research Centre. http://www.ecovillagefindhorn.com/docs/FF%20Footprint.pdf (accessed July 14, 2011)

Trading Economics (2012) 'Arable land (hectares per person) in world', Trading Economics. http://www.tradingeconomics.com/world/arable-land-hectares-per-person-wb-data.html (accessed July 4, 2012) von Weizsäcker, E., Hargroves, K., Smith, M., Desha, C. and Stasinopoulos, P. (2009) *Factor 5: Transforming the Global Economy through 80% Improvements in Resource Productivity*. London: Earthscan.

Warren-Rhodes, K. and Koenig, A. (2001) 'Ecosystem appropriation by Hong Kong and its implications for sustainable development', *Ecological Economics* 39: 347–359.

Wilson, J. and Anielski, M. (2005) *Ecological Footprints of Canadian Municipalities and Regions*. Ottawa: Federation of Canadian Municipalities.

World Bank (2011) 'World development indicators'. Washington, DC: World Bank. http://data.worldbank.org/indicator (accessed January 18, 2012)

Worldmapper (2010) 'Reference maps'. Sheffield, UK: Worldmapper. http://www.worldmapper.org/ (accessed January 20, 2012)

WRI (2010) 'Earth trends: environmental information', databases, World Resources Institute. http://earthtrends.wri.org/ (accessed January 18, 2012)

WWF (2010a) *Living Planet Report 2010*. Gland, Switzerland: World Wide Fund for Nature.

WWF (2010b) 'Footprint interactive graph'. Gland, Switzerland: World Wide Fund for Nature. http://wwf.panda.org/about_our_earth/all_publications/living_planet_report/living_planet_report_graphics/footprint_interactive/ (accessed December 26, 2011)

Part II

What Does Living within a Fair Earth Share Mean?

Part II.I
Personal Footprint

2 The Ecological Footprint of Food

James M. Richardson

Eating Food

People consume an average of 2,550 kcal per day to rebuild cells, think, activate muscles and carry out all the other functions associated with life. Men consume roughly 2,900 kcal per day and women approximately 2,200 kcal (National Research Council, 1989). As a comparison, this is roughly the same amount of energy required to power two 60-watt light bulbs – a fact that will have bearing later in this chapter. In order to meet food needs, governmental and non-governmental organizations (NGOs) alike have developed food policy statements that resemble this one, written by the Food and Agriculture Organization of the United Nations (FAO). They define food security as a state in which 'all people, at all times, have physical, social and economic access to sufficient, safe and nutritious food which meets their dietary needs and food preferences for an active and healthy life' (FAO, 2011c).

The ecological footprint concept is well suited to measuring the biophysical component of food security and facilitating conversations about the more subjective cultural quality of food. At upwards of one quarter of an individual's ecological footprint, and 8 per cent of New Zealand's primary energy use, food is significant (Cardiff Council, 2005; Ewing, Moore, *et al.*, 2010; Ewing, Reed, *et al.*, 2010; Ministry of Economic Development, 2011b; Vale and Vale, 2009). As you will read below, the way we grow, process, move, eat and compost food does make a difference.

What Kiwis Eat

In New Zealand, roughly 830 kilograms of food are supplied per capita per year for human consumption (calculated from FAOSTAT, 2011a). While not all of this is consumed, it is probably the best indicator available to measure food needs. It is calculated by adding the mass of imports (I) to food produced (P) within New Zealand and subtracting exports (E). Food extracted from stockpiles (X) is added and food used for seed (S) or fed to livestock (L) is subtracted and accounted for in the yield section. When divided by the population of New Zealand (p) we arrive at the average food supplied per capita per year (F):

$$F = \frac{P + I - E + X - S - L}{p}$$

At 833 kilograms per person per year, domestic food supply in New Zealand is low when compared with countries like Canada, which supplies roughly 1,026 kilograms per person per year, or Australia, at 934 kilograms per person per year. The United States tops the list at 1,038 kilograms per person per year; Fijians, by comparison, have only 621 kilograms per person per year (FAOSTAT, 2011b). FAOSTAT trade data compares remarkably well with local LINZ data (Parnell *et al.*, 2010), which found that adult Kiwis consumed 925 kilograms per year in total and 664 kilograms per year when non-alcoholic drinks (tea, coffee, soft drinks, etc.) were excluded. The disparity probably originates from different ways of measuring the mass of food with respect to added water, which clearly makes up a large portion of our non-nutritional diet.

Equally interesting is the difference between calories available to people in developed countries like New Zealand compared with those in developing regions (Table 2.1).

While supply exceeds minimum dietary needs according to the 1,800 kcal per person per day threshold, there is certainly a disparity within regions, giving rise to the acute famines seen on the news from time to time.

Table 2.1 Daily calories available in local food supplies for selected countries and regions, 2006

Country or region	Daily food supply (kcal/person/day)	Annual food supply (MJ/person/year)
Minimum daily requirement (FAO, 2011b)	1,800	2,751
Canada	3,533	5,402
Cuba	3,286	5,060
Fiji	3,015	4,635
New Zealand	3,146	4,806
United States of America	3,767	5,756
European Union	3,464	5,280
Least developed countries	2,112	3,268
Asia	2,627	4,041
World	2,778	4,245

Source: FAOSTAT (2011b)

Food Waste

Agricultural waste, measured 'up to the house door', is relatively low, at 3.5 per cent, when compared to waste that likely occurs in the kitchen or on the dining room table. For example, Statistics Canada estimates that 36 per cent of food purchased by Canadians is not consumed (Statistics Canada, 2002). While wastage is a problem, it is difficult to identify what quantity of wastage is avoidable. Though finishing one's peas and carrots seems reasonable, what about the broccoli stems, onion skins or chicken bones which contribute to the overall mass of food purchased, but perhaps contribute less to its nutritional value? For the purposes of this study, food supply plus pre-consumer waste seem an appropriate metric to calculate an individual's foodprint.

As a net exporter of food, New Zealand produces very little food for domestic use in comparison with its exports (Figure 2.1). This fact has a bearing when determining what food and farmland Kiwis are actually responsible for.

The calculations used to produce Figure 2.1 assume Kiwis are 'responsible' for the agricultural land required to supply the domestic market as well as

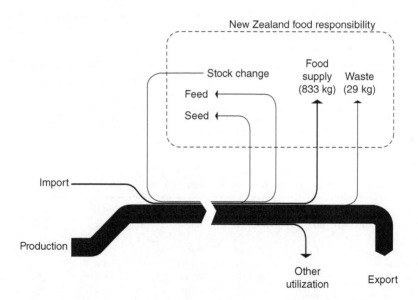

Figure 2.1 New Zealand food system balance. Notice the food dedicated to domestic consumption and waste (833 kg plus 29 kg per person per year), which pales in comparison with either production (5,284 kg per person per year) or exports (3,877 kg per person per year). 'Other utilization' includes commodities used for non-food purposes or consumption by tourists. Foods processed are included in food supply

Source: FAO (2011a)

the land required for feed and seed for domestic consumption, which amounted to roughly 20 per cent of food produced and imported. This attribution factor was applied to domestic food freight and food production energy inputs, though it is slightly less than factors used by Andrew and Forgie (2008), who attributed a responsibility factor of 35 per cent.

Types of Land

A range of land types are needed to produce food: cropland for annual and perennial crops, pasture land to graze animals and forestry land to offset the carbon emissions associated with agriculture. Similar to other consumption categories, a 'foodprint' (Ef) is the mass of food (m) an individual or community purchases divided by its local yield (y).

$$Ef = \frac{m}{y}$$

Cropland

Crop yield is the mass of food produced, less the seed required to grow it, divided by the area required to grow it. In this regard, those foods that people consume a lot of or those that have particularly low yields contribute most to a community's foodprint. For instance, sesame seed production requires over 20 square metres of land to produce 1 kilogram of seed, whereas wheat requires only 1.25 square metres. Table 2.2 describes the foodprint of crops based on the average Kiwi diet and local yield.

Table 2.2 Ecological footprint of crops. Weighted yield refers to the yield after seed requirements and processing losses are accounted for

Food group	Food supply (kg/person)	Weighted yield (kg/ha)	Direct footprint (ha/person)	Direct footprint (m²/person)	Percentage of crop footprint
Cereals	91.3	7,135	0.01	128.0	16%
Starchy roots	60.8	35,594	0.00	17.1	2%
Sugar and sweeteners	56.4	5,786	0.01	97.5	12%
Pulses	6.0	6,832	0.00	8.8	1%
Treenuts	3.5	1,113	0.00	31.4	4%
Oilcrops	5.5	2,107	0.00	26.1	3%
Vegetable oils	9.8	370	0.03	265.1	33%
Vegetables	147.1	38,904	0.00	37.8	5%
Fruits	105.7	12,266	0.01	86.2	11%
Stimulants	4.6	783	0.01	58.8	7%
Spices	0.6	803	0.00	7.5	1%
Alcohol	89.0	24,959	0.00	35.7	4%
Total	580.3	—	0.10	799.8	100%
Average	—	11,388	—	—	—

Animal land

To calculate the ecological footprint (EF) of animal products, one must first consider the amount of land required per animal, a function of the land required for pasture, housing and any additional land required for feed. Livestock unit coefficients (*LSU*) help compare the amount of pasture land required per animal for grazing, or its inverse, stock density. These vary from region to region based on climatic variations, and national estimates are generated by multiplying the local stock density (d) by the percentage of farms with this stock density (x) for each livestock type. Internationally all densities are normalized against sheep, which have an *LSU* of 1.0 (RuralFind, 2011).

$$LSU_{weighted} = d_1 x_1 + d_2 x_2 + d_3 x_3 \dots$$

Ted Christopher (2010) of Landcare Research expanded on this. He found that 63 per cent of dairy farms have an *LSU* of 8, 18 per cent have an *LSU* of 6.5 and the remaining 19 per cent had an *LSU* of 3.5. Thus, the weighted *LSU* in New Zealand is 6.9 for dairy and 4.75 for beef cattle, while the weighted *LSU* for sheep is 0.91. These figures compare well with international standards (AsureQuality New Zealand Ltd, 2012; RuralFind, 2011) though they differ for irrigated and dryland pasture conditions. In New Zealand, one livestock unit (*LSU*) requires 520 kilograms dry weight of fodder per year (RuralFind, 2011) and pasture grasses grow at roughly ten tonnes of dry matter per hectare per year (Baars, 1981), thus 1 hectare of pasture could feed roughly 21 sheep, 4 cattle or almost 3 dairy cows. Animal husbandry codes of welfare developed by the Ministry of Agriculture and Forestry were used to determine best practice densities of poultry and pigs, typically barn-raised at 172,727 chickens per hectare (0.06 square metres per chicken) and 21,315 pigs per hectare (0.46 square metres per pig; Buijs *et al.*, 2009; Honeyfield-Ross *et al.*, 2009, Ministry of Agriculture and Forestry, 2003; National Animal Welfare Advisory Committee, 2010; Ravindran, undated).

To complicate matters further, some livestock (broiler chickens, pigs, etc.) can be grown to their full mass (m) in less than a year, increasing their annual yield (y_{annual}) according to the number of cycles (c) run per year. Animal cycles were calculated by dividing the number of animals processed per year by the number of stock for each animal category according to FAO statistics (FAOSTAT, 2010). For instance, there were almost eight times more broiler chickens processed in 2006 than were counted (stock) at the time of the FAO survey. Thus the annual yield must increase by a factor of seven for broiler chickens. Conversely, the ratio of processed animals to stock for beef was roughly 0.7, meaning it takes on average 1.4 years (the inverse of 0.7) before beef cattle reach an age at which they can be processed, and the annual yield should decrease by this factor. In addition, a processing coefficient (p) between 0 and 1 was used to discern final products from raw materials. For example, the carcass weight of the average beef cow in New Zealand is 168 kilograms (FAOSTAT, 2010), but assuming a processing rate of 70 per cent the dressed weight is only 118 kilograms. Milk fat (MF) concentration was used to calculate the amount of raw milk (and the associated pasture land) required for products such as cream or butter. Assuming a concentration of 4 per cent MF, 1 kilogram of whole milk is able to produce 160 grams of cream at 25 per cent MF or 47 grams of butter at 85 per cent MF.

$$y_{annual} = m \times c \times p$$

Additional feedland was calculated by identifying the ecological footprint of the feed consumed in New Zealand and dividing it by the number of stock in each animal class (New Zealand Feed Manufacturers Association, 2012). Not surprisingly, the majority of it was used for poultry and pigs, since cattle and sheep are largely grass fed. Like cropland, a New Zealand attribution factor was applied to feedland, but since most poultry and pig products are bound for domestic use, New Zealand's feedprint is largely its own. After the math, feedland amounted to roughly 200,000 hectares of additional cropland, or 0.05 hectares per person. Table 2.3 compares the land-use requirements for the core animal-based food groups.

The average energy yield of animal products is at best comparable to fruit and is generally less than one quarter of the yield of cereals.

Table 2.3 Ecological footprint of animal products. Annual yield and ecological footprints account for multiple rotations per year or rotations that take longer than one year. Methane emissions are largely attributed to dairy and beef production

Food group/ component	Food supply (kg/ person)	Energy yield (kcal/ha)	Weighted yield (kg/ha)	Footprint (ha/ person)	Direct footprint (m^2/ person)	Percentage of animal foodprint
Meat	109.8	815,751	527	0.209	2,085.1	78%
Animal fats	14.1	5,081,552	657	0.021	214.8	8%
Eggs	10.4	2,166,448	1,543	0.007	67.4	3%
Milk	92.2	4,897,697	9,444	0.010	97.6	4%
Methane emissions	—	—	—	0.022	221.4	8%
Total	226.5	—	—	0.246	2,686.3	100%
Average	—	3,240,362	3,043	—	—	—

Sources: Adapted from Buijs *et al.*, 2009; Chilonda and Otte, 2006; National Animal Welfare Advisory Committee, 2004; RuralFind, 2011

Fishing grounds

Though oceans, lakes and rivers are volumetric in their shape (rather than flat, on a single plane), most of the primary production that supports the aquatic food web occurs within 80 metres of the surface, in the *euphotic* zone, where light is able to penetrate and support photosynthesis. With this in mind, the surface area of aquatic environments can be treated in the same way as land area, a methodology consistent with international protocol (Ewing, Reed *et al.*, 2010). Here, yield is taken as the average or allowable catch divided by the catch area. Given the dynamic nature of aquatic environments, food production is a non-exclusive service, making it different from cropping or forestry where most of the primary production is appropriated for human use.

Oceanic and freshwater fishing systems are an important component of the New Zealand diet, contributing species such as snapper, red cod, blue moki and paua from open-water and near-shore systems and trout, salmon, whitebait and freshwater eel from New Zealand's lakes and rivers. Of New Zealand's 440 million hectares (a) of exclusive marine fishing grounds, roughly 686 million kilograms of fish (m) are allowed to be caught (Ministry of Fisheries, 2012) and an average of 629 million kilograms are actually caught. This discerns an allowable yield of 1.56 kilograms per hectare and an actual five-year average yield of 1.49 kilograms per hectare, with New Zealand freshwater systems contributing a similar yield of roughly 1.70 kilograms per hectare. This appears extremely low when compared with the gross yield of wheat, at 7,952 kilograms per hectare, or even mutton, at 279 kilograms per hectare, and serves to illustrate the difference between a cultivated agriculture and 'wild food'. Consuming just over 20 kilograms of fish per capita per year, the average Kiwi demands nearly 18 hectares of fishing grounds – 17 hectares in the ocean and another hectare of inland waters. Fortunately for New Zealand, there is a lot of sea out there – nearly 100 hectares per person within New Zealand's 200-mile exclusive economic zone, over 60 times the area of arable cropping land (class 1–4) and 30 times the area available for grazing or forestry (class 5–7; Landcare Research NZ Ltd, 1973). This does not mean the oceans are not vulnerable to exploitation in New Zealand or abroad. One study on the seafood footprint of the Baltic region showed that citizens living within the drainage basin of the Baltic Sea would appropriate 300 per cent of its bio-capacity if the seafood needs of local inhabitants were to depend exclusively on its waters (Folke *et al.*, 1998; Jansson *et al.*, 1999). Indeed, if the world's population were to consume seafood in the same way that those living in large cities do, global consumption would well exceed the ocean's bio-capacity (Folke *et al.*, 1998). One might argue that farmed fish would have a smaller footprint given the efficiency of production methods and ease of harvest. However, it appears the embodied footprint of the food fed to farmed fish (fish pellets, etc.) make the footprint of fish such as farmed salmon roughly the same whether contained in cages or out at sea (Folke, 1988; Folke *et al.*, 1998).

To help compare the impact of organisms that live high up the trophic ladder versus bottom feeders or producers, members of the Global Footprint Network developed a methodology that accounts for the embodied primary production of various aquatic products (Ewing, Reed *et al.*, 2010; Gulland, 1971; Pauly and Christensen, 1995). They reveal the effective footprint of fish products accounting for by-catch and the trophic level of the food group, where top carnivores have much greater effective footprints than primary consumers or producers. Kilogram for kilogram, herring appropriate a footprint four times that of shrimp based on their place on the food chain (Pauly and Christensen, 1995). While this work enables comparisons between marine products, with other food groups (crops) and with other countries, it is based on global 'sustainable' yields collected and collated by Gulland in 1971 before the collapse of many global fisheries and probably misrepresents global bio-capacity. Needless to say, it is helpful to consider that for every kilogram of tuna consumed many kilograms of primary producers and primary consumers are required, not to mention the by-catch. In this regard, the lower on the food chain the food, the lesser its effective ecological footprint.

Energy Use in Food Production

Agricultural production for the New Zealand diet costs 4.7 petajoules (assuming a domestic responsibility factor of 20 per cent), producing 0.08 tonnes of CO_2 equivalent (CO_2e) per person and requiring 0.006 hectares of forest and renewable energy land for sequestration or to produce electricity. Another 0.22 tonnes of CO_2e are produced from methane and nitrous oxide emissions, requiring 0.02 hectares of forest land per capita (calculated from Saunders and Barber, 2007). While this accounts for moving farm machinery it probably misses the carbon or energy land required to build equipment, manufacture fertilizers, build roads, produce fodder, and so on. Several studies have explored the energy and carbon intensity of farm elements (Saunders and Barber, 2007, 2008; Wells, 2001), which Saunders, Barber and Taylor (2006) used to evaluate the relative energy inputs and carbon emissions of dairy, meat production (lamb), onion production and apple farming in New Zealand and the UK. They developed impressive comparisons between the direct (in fuel consumption), indirect (in producing fertilizer and pesticides) and capital (embodied in vehicles, buildings, fences, stock water supply, etc.) energy expenditures and carbon emissions for each farm type. If applied to the New Zealand footprint, with apples acting as a proxy for fruit, lamb for meat, onions for horticultural products and dairy representing dairy, taking into account the larger weighting of horticultural products and grains in the Kiwi diet, the ratio of direct, indirect and capital-related carbon emissions was 1 : 1.3 : 0.4.

Table 2.4 Ecological footprint of the energy expenditures for food production in New Zealand

Type of energy use	Energy use (MJ/person)	Domestic carbon emissions (tCO$_2$e/person)	Domestic energy land (ha/person)	Percentage of total
Direct	1,121	0.08	0.006	34%
Indirect	1,306	0.10	0.008	45%
Capital	248	0.03	0.002	14%
Fishing	229	0.02	0.001	7%
Total	2,904	0.23	0.017	100%

Sources: Calculated from Saunders et al. (2006) and Saunders and Barber (2007)

That is, for every tonne of CO$_2$e emitted in operating a farm, another 1.7 tonnes are emitted in manufacturing materials or building it. Given carbon intensity from direct energy use was 0.08 tonnes CO$_2$e per capita, indirect and capital-related carbon emissions summed to 0.13 tonnes CO$_2$e per person or 0.01 hectares (100 square metres) of additional forest land per person. A similar algorithm was applied to renewable energy land (the land required to supply electricity), which demands an additional 0.00013 hectares (1.3 square metres) of renewables land per person for indirect or capital-related energy expenditure.

In addition to the direct and indirect footprint from agriculture, the direct energy used for fishing bound for domestic markets amounted to 0.95 petajoules (230 megajoules per person) in 2006, requiring 0.00077 hectares (7.7 square metres) of forest land (for carbon sequestration) and 0.00001 hectares (0.1 square metres) of renewable energy production land (for electricity) per person to power it. Table 2.4 summarizes these findings. Note that, while fishing consumes 17 per cent of direct food energy inputs, it supplies only 3 per cent of the food by weight or 1.5 per cent of the food energy, at 71 megajoules per person. Put another way, for every unit of food energy harvested from fish, 5.7 units of direct energy are invested in fishing. This energy imbalance is consistent with findings from other studies (Biswas and Biswas, 1976; Leach, 1975; Stanhill, 1974) and gives us cause to question the viability of the fishing industry – and other low-output, high-input foods – in an energy-scarce future.

Energy Use in Food Processing

Processing energy inputs into food can be roughly categorized as industrial processing, or the work done to refine milk, bake bread and boil jams before the consumer gets them, retail processing, including the refrigeration and cooking that occur at retail outlets or restaurants, and post-purchase processing, which includes storage, refrigeration and heating associated with meal preparation, whether at home or in a restaurant.

National energy inputs in food processing were 16.12 petajoules in 2006, which, if we assume 20 per cent is dedicated to the New Zealand diet, amounts to 0.0056 hectares (56 square metres) of domestic energy land for electricity production or carbon offsets, twice that required for food production.

Statistics New Zealand conducts energy-use surveys for primary, manufacturing and service sectors. In 2010, energy use by accommodation and food services totalled 1,698 megajoules per person (Bascand, 2011; FAO, 2011c). This amounted to 0.0004 hectares (4 square metres) of renewable electricity land and 0.0096 hectares (96 square metres) of forest land per person for carbon sequestration, but should be taken with a grain of salt given the inclusion of accommodation (hotels, etc.) in the survey. Built-up land areas required for supermarkets and restaurants were not taken into account, nor were the materials needed to build them.

A study of household energy use in 2010 showed that 22 per cent of the 7,872 kilowatt hours of annual household electricity use is for cooking and refrigeration (Isaacs et al., 2010). This expenditure would require 0.0098 hectares (98 square metres) of carbon land and 0.0009 hectares (9 square metres) of renewable electricity land per person to offset demands. Combined processing inputs at 0.02 hectares per capita compare well with Field's (2011) 0.016 hectares per capita or an adapted version of Andrew and Forgie's (2008) estimate of 0.03 hectares per capita. An astute reader will note that processing inputs following harvest and at home are nearly double those of production and fishing inputs combined, giving us a good place to focus attention to reduce our ecological footprint.

Table 2.5 Ecological footprint of the energy expenditures from food processing in New Zealand

Energy destination	Energy use (MJ/person)	Carbon emissions (tCO_2e/person)	Domestic energy land (ha/person)	Percentage of total
Industrial processing	767	0.07	0.0056	22%
Pre-purchasing processing	1,787	0.13	0.0100	39%
Post-purchasing processing	2,150	0.12	0.0103	40%
Total	4,704	0.32	0.026	100%

Moving Food

Transport inputs are difficult to approximate given the convoluted nature of the global food system and difficulties in attributing the direct footprint of highways or rail lines to food (versus housing, commuting, etc.). However, the freight mass distance of food can be attributed to the sum of international freight, interregional freight and the domestic transport costs associated with picking up groceries. It is expressed in terms of tonne-kilometres shipped, reflecting the fact that both the quantity and distance are important when calculating freight costs. Furthermore, the mode-specific cost of transport is even more important. Direct and embodied carbon emissions combined are 0.321, 0.053, 0.043 and 0.321 kilograms CO_2e per tonne-kilometre for transport by road, rail, coastal shipping and air freight, respectively (Facanha and Horvath, 2007; Natural Resources Canada, 2010). Canadian statistics have been used here for coastal shipping and air freight and are increased by one third to account for embodied emissions.

In this regard, shipping 1 tonne of milk 10 kilometres by road is actually more environmentally costly than shipping it 60 kilometres by rail. It is also interesting to note that air travel 'costs' the same as shipping by road, albeit planes typically fly much further than trucks, earning them a reputation for high-impact transport. Multiplying the mass distance (md) of each food group by the mode-specific cost of transport (C) divided by the sequestration potential of forests (13.2 tonnes per hectare for Pinus radiata; Hollinger et al., 1993) yields the ecological footprint of food freight ($EF_{freight}$).

$$EF_{freight} = \sum \frac{md \times C}{13.2}$$

With 90 per cent of the mass of all foodstuffs coming to New Zealand from 14 countries (including 56 per cent from Australia), we can take a good stab at the cost of international freight which comes from either air (1 per cent) or sea (99 per cent) by mass (Bascand, 2006; see Figure 2.2).

Figure 2.2 International food freight to New Zealand. The line weight is proportional to the energy expended in moving food from the country of origin to New Zealand

Table 2.6 Moving food to and within New Zealand

Mode	Energy use (MJ/person/ year)	Mass-distance (1000 tkm or 1000 Pkm)	Total emissions (tCO₂e)	Forest land (ha/ person)	Percentage of total
International shipping	1,259	8,780,332	401,649	0.007	20%
Interregional freight	923	896,000	267,818	0.005	13%
Intra-shopping store freight	1,586	1,380,000	442,980	0.008	22%
Domestic shopping	3,548	4,505,644	937,174	0.017	46%
Total	7,316	15,561,975	2,049,621	0.037	100%

Sources: Overseas cargo statistics (Bascand, 2006), the New Zealand Household Travel Survey (Ministry of Transport, 2011a), New Zealand vehicle fleet statistics (Ministry of Transport, 2011b), a national freight demands study (Richard Paling Consulting, 2008) and FAO statistics (FAO, 2011c) were used to discern the relative mass-distance costs of moving food to and within New Zealand. It is assumed that 50 per cent of 'shopping and personal business trips' were used to buy food

The remaining 10 per cent of food came from over 100 countries and was ignored in this study. National freight was easier to model accurately, with the majority of shipping by road (Richard Paling Consulting, 2008). Table 2.6 shows that transporting food for supermarkets and food retailing made up a similar proportion to international shipping, at roughly 20 per cent of the total. However, domestic shopping was the largest contributor to carbon emissions, demonstrating that reducing shopping trips or walking to the local market can make the largest impact on our ecological footprint as far as travel is concerned. Table 2.6 compares the relative cost of moving food to and within New Zealand.

Foodshed

With all these considerations in mind, the amount of space required for growing food amounts to roughly 0.382 hectares (3,820 square metres) per person, subtracting from the total the fishing grounds and the retail and accommodation footprint, which do not fit the EF paradigm (Table 2.7). Expressed in global hectares (gha), Kiwis require 0.71 gha each for food, accounting for better yields from New Zealand grazing and forestry land than the global average (Ewing, Reed et al., 2010).

Figure 2.3 shows New Zealand's food footprint set out like a farmer's field. Not surprisingly, animal products make up the lion's share of land-based appropriation, at 64 per cent of total land use, and the fuel-related transportation inputs at 10 per cent of the total amount to much less than one might guess, given the hype around food miles. To make sense of this figure requires placing food beside the other consumption categories listed

Table 2.7 Ecological footprint of food in New Zealand. Forest land is required for sequestering carbon emitted during production and processing. Accommodation and food retailing could not be decoupled in this study and should be observed with caution. Fishing grounds have multiple uses, thus probably misrepresent exclusive appropriation for human use that the EF methodology purports to accomplish. Animal products include methane production

	Cropland	Grazing land	Built-up land	Forestry land	Renewable energy land	Total
Crops (ha/person)	0.0800	—	—	—	—	0.080
Animal products, including methane (ha/person)	0.0343	0.1861	0.0001	0.022	—	0.243
Direct energy use, agriculture (ha/person)	—	—	—	0.006	0.000	0.006
Food processing, excluding retailer (ha/person)	—	—	—	0.015	0.001	0.016
Transportation (ha/person)	—	—	—	0.037	—	0.037
Total (ha/person)	0.1143	0.1861	0.0001	0.082	0.001	0.382
Global yield factor	0.7	2.5	1.0	2.0	1.0	—
Total (gha)	0.0800	0.4651	0.0001	0.1630	0.0011	0.709

Source: Global yield factors were derived from Ewing, Reed *et al.* (2010)

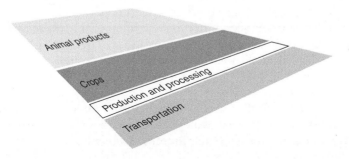

Figure 2.3 New Zealand's food footprint expressed in the form of a field

in this book, and to make use of it one should put the foodprint concept into the context of real land. The foodshed concept can help in the transition of the ecological footprint analysis from an abstract measurement to a land-use planning tool. It is an idea that was first developed by Hedden in 1929, in his text entitled *How Great Cities Are Fed*, and further refined by Getz in 1991, and is designed to help visualize a community's demand for farmland. Like a watershed, a foodshed is the land required to meet an individual's or community's food needs, whether it exists in one's own back yard or further afield. The area of a community's foodshed (S) is determined by multiplying the foodprint (Ef) of an average citizen by the number of people in that community (p):

$$S = Ef \times p$$

Peters et al. (2009) applied the foodshed concept to New York State, demonstrating that the population of the state eats more than can be provided by the state itself. Richardson (2010) applied a similar methodology to map the foodshed of Vancouver in the province of British Columbia, Canada, revealing that the demands of the city would hypothetically require most of the agricultural land of the whole province to meet its needs. In this way, the foodshed concept helps to reveal the relationship between demand and capacity. New Zealand boasts some of the best grass production in the world, at 2.5 times the global average (Ewing, Moore et al., 2010). In fact, GNS Science estimates the metabolizable energy supplied by New Zealand's pasture (the consumable energy in grass) sums to 800 petajoules per annum (Baisden et al., 2010), exceeding the total 764 petajoules of primary energy supplied to the whole of New Zealand in 2006, and far more than the 537 petajoules actually consumed (Ministry of Economic Development, 2011a). To say that New Zealand is powered by the sun isn't far from the truth. This land-use efficiency paired with the hilly nature of the countryside make it well-suited to pasture or forestry and relatively poorly suited to cultivated crops in many parts of the country. (Newsome et al., 2008; Landcare Research NZ Ltd, 1973). For every citizen there is 1.6 hectares of cropland but twice that in grazing land. In this context, applying a vegetarian diet to New Zealand might necessitate importing foods or dedicating more land to cultivation, which could well do damage to fragile hillsides. In essence, context is critical if the ecological footprint tool is to be used for policy or planning decisions.

If we apply the footprint concept to cities around New Zealand and abroad it reveals just how large a demand big cities like New York, Beijing and even Auckland place on their hinterlands (Figure 2.4).

Making a Difference

Michael Pollan, author of *The Omnivore's Dilemma* (2006), presents three strategies for a food-secure world: 'Eat (fresh) food. Mostly plants. Not too much.' These simple choices, in addition to 'waste not', 'local first' and 'fork to fork' can make a sizeable difference to our collective ecological foodprint either in isolation or in combination, as modelled below.

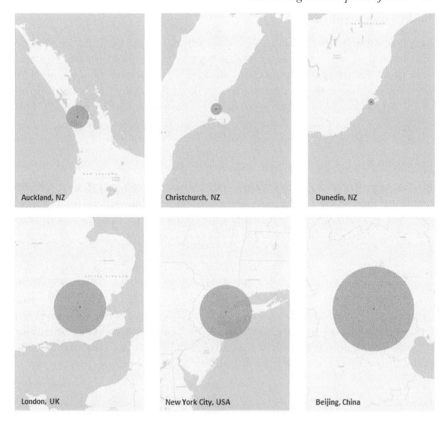

Figure 2.4 International foodprint comparisons for various cities. Assuming consumption and production values similar to New Zealand and ignoring what land is available for farmland or the needs of adjacent towns, foodprints of major world cities raise interesting questions about the future of urban growth, density and regional resilience. All images are shown at the same scale for 2011 city populations

Sources: Data from Groves *et al.* (2011); National Bureau of Statistics of People's Republic of China (2011); Office for National Statistics (2011); Statistics New Zealand (2011)

Eat (fresh) food

Eating fresh, lightly processed foods can help reduce the embodied processing that makes up 4 per cent of the conventional foodprint. Assuming Kiwis were to eat 25 per cent less processed food (or foods were processed 25 per cent less), the savings would amount to 0.004 hectares (40 square metres) per person – and food would probably taste a whole lot better, too.

Mostly plants

Eating less meat will make a huge difference. If New Zealanders were to shift from their current diet to a lacto-vegetarian diet inclusive of egg and milk products we could reduce our footprint per person by another 0.18 hectares (this assumes fish products were kept the same and consumption of all other food products increased to maintain calorie intake). This increases to 0.21 hectares if people were to go vegan.

Not too much

The average Kiwi is supplied with 23 per cent more calories than the 2,550 recommended daily intake. On top of previous changes, reducing food energy availability by 10 per cent can reduce the footprint to 0.157 hectares per capita – almost a third of what it was originally.

Waste not

While some wastage is inevitable, retail and household waste could easily come down. Canadian statistics suggest that 36 per cent of the food supplied is wasted. Reducing pre-consumer waste to 1 per cent and post-consumer waste to 28 per cent reduces the footprint to 0.139 hectares per capita while maintaining food energy availability above 2,550 kcal per person per day.

Local first

There are few better places in the world for growing food than New Zealand. While there may be good arguments for international food freight over local food in places like the UK (Saunders and Barber, 2007; Saunders *et al.*, 2006), this is not the case at home. If international transport were eliminated the footprint could be reduced to 0.131 hectares per capita. However, greater savings can be found at home: if we cut our driving trips to pick up food in half, the savings could amount to 0.008 hectares, reducing the foodprint to 0.123 hectares per person – 30 per cent of the current baseline.

Fork to fork

Cycling nutrients essential for plant growth, including phosphorus (P) and potassium (K), is essential for a sustainable food system. Composting vegetable wastes and even investing in composting toilet facilities will help keep these nutrients at home to support future generations of fruits and vegetables, as well as the people who enjoy them.

In closing, it may help to reflect on the light bulb analogy we began this chapter with. At a minimum, the New Zealand population is akin to 8 million 60-watt light bulbs burning continuously. However, to grow, fertilize,

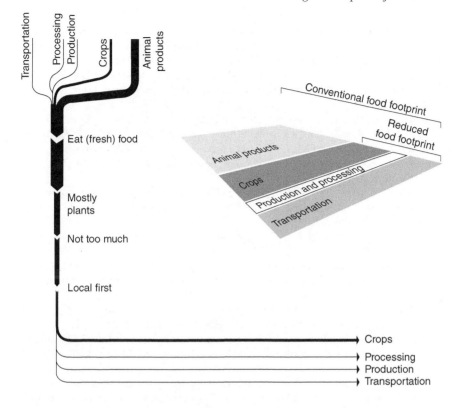

Figure 2.5 Reducing our ecological footprint. Applying the principles above, a Kiwi's foodprint shrinks from 0.382 to 0.123 hectares, a reduction of nearly 70 per cent

process and move the food for this population takes almost three times that energy input. This chapter shows that, even in New Zealand, with its skilled farmers and excellent quality of land, the crop and grazing land required to grow food and the energy land required to power the food system amounts to 0.382 hectares per capita or 0.71 gha. If the fair share footprint is around 1.8 gha per person, we need to adopt practices that will reduce our food footprint. However, this chapter also shows that changes in the way we grow, eat and move food can make a real difference, and that we could live within a fair share.

References

Andrew, R. and Forgie, V. (2008) 'A three-perspective view of greenhouse gas emission responsibilities in New Zealand', *Ecological Economics*, 68(1–2): 194–204.

AsureQuality New Zealand Ltd (2012) 'Stock Density Technical Information'. http://www.waikatoregion.govt.nz/Environment/Environmental-information/Environmental-indicators/Land-and-soil/Land/riv9-technical-information/#Bookmark_typical_stock_distrib (accessed 24 February 2012)

Baars, J.A. (1981) *Variation in Grassland Production in the North Island with Particular Reference to Taranaki*, Hamilton, New Zealand: Ruakura Soil and Plant Research Station, MAF.

Baisden, W.T., Keller, E., Timar, L., Smeaton, D., Clark, A., Ausseil, A. and Zhang, W. (2010) 'New Zealand's Pasture Production in 2020 and 2050', GNS Science Consultancy Report 2010/154, Lower Hutt, New Zealand: GNS Science.

Bascand, G. (2006) *Overseas Cargo Statistics: Year Ended June 2006*, Wellington, New Zealand: Statistics New Zealand.

Bascand, G. (2011) 'New Zealand Energy Use Survey'. http://www.stats.govt.nz/browse_for_stats/industry_sectors/Energy/EnergyUseSurvey_HOTP10.aspx (accessed 21 June 2012)

Biswas, A.K. and Biswas, M.R. (1976) 'Energy and food production', *Agro-Ecosystems*, 2(3): 195–210.

Buijs, S., Keeling, L., Rettenbacher, S., Poucke, E.V. and Tuyttens, F.A.M. (2009) 'Stocking density effects on broiler welfare: Identifying sensitive ranges for different indicators', *Poultry Science*, 88(8): 1536–1543.

Cardiff Council (2005) 'Cardiff's Ecological Footprint'. http://www.cardiff.gov.uk/content.asp?nav=2%2C2870%2C3148%2C4119 (accessed 21 May 2012)

Chilonda, P. and Otte, J. (2006) 'Indicators to monitor trends in livestock production at national, regional and international levels', Rome, Italy: Livestock Information and Sector Analysis and Policy Branch, FAO.

Christopher, T. (2010) 'Mapping Stock Unit Densities in New Zealand'. http://www.lincoln.ac.nz/research-themes/ecosystem-services/Research-Projects-and-Websites/Mapping-Stock-Unit-Densities-in-New-Zealand/ (accessed 12 February 2012)

Ewing, B., Moore, D., Goldfinger, S., Oursler, A., Reed, A. and Wackernagel, M. (2010) *Ecological Footprint Atlas 2010*, Oakland, CA: Global Footprint Network.

Ewing, B., Reed, A., Galli, A., Kitzes, J. and Wackernagel, M. (2010) *Calculation Methodology for the National Footprint Accounts, 2010 Edition*, Oakland, CA: Global Footprint Network.

Facanha, C. and Horvath, A. (2007) 'Evaluation of life-cycle air emission factors of freight transportation', *Environmental Science and Technology*, 41(20): 7138–7144.

FAO (2011a) 'Concepts and Definitions Used in Food Balance Sheets, 2012'. http://www.fao.org/DOCREP/003/X9892E/X9892e02.htm#P430_31575 (accessed 26 March 2012)

FAO (2011b) 'The State of Food Insecurity in the World'. http://www.fao.org/publications/sofi/en/ (accessed 26 March 2012)

FAO (2011c) 'Food Security Statistics'. http://www.fao.org/economic/ess/ess-fs/en/ (accessed 22 December 2011)

FAOSTAT (2010) 'FAOSTAT Production Balance Sheets'. http://faostat.fao.org/site/339/default.aspx (accessed 1 April 2012)

FAOSTAT (2011a) 'Commodity Balances 2006', FAO Statistical Division, United Nations. http://faostat.fao.org/site/616/default.aspx#ancor (accessed 22 December 2011)

FAOSTAT (2011b) 'Food Balance Sheet', FAO Statistical Division, United Nations. http://faostat.fao.org/site/368/default.aspx#ancor (accessed 19 July 2011)

Field, C. (2011) 'The Ecological Footprint of Wellingtonians in the 1950s', Master's in Building Science thesis, Victoria University of Wellington, Wellington, New Zealand.

Folke, C. (1988) 'Energy economy of salmon aquaculture in the Baltic sea', *Environmental Management*, 12(4): 525–537.

Folke, C., Kautsky, N., Berg, H., Jansson, A. and Troell, M. (1998) 'The ecological footprint concept for sustainable seafood production: A review', *Ecological Applications*, 8(1): S63–S71.

Getz, A. (1991) 'Urban foodsheds', *The Permaculture Activist*, 24(26).

Groves, R., Mesenbourg, T., Jackson, A., Hogan, H., Matos, M. and Weinberg, D. (2011) '2010 Census Redistricting Data (Public Law 940171) Summary File', US Department of Commerce. http://www.census.gov/prod/cen2010/doc/pl94-171.pdf (accessed 14 March 2012)

Gulland, J.A. (1971) 'The Fish Resources of the Ocean', Fishing News (for the FAO), West Byfleet, England.

Hedden, W.P. (1929) *How Great Cities Are Fed*, New York: Heath and Company.

Hollinger, D.Y., Maclaren, J.P., Beets, P.N. and Turland, J. (1993) 'Carbon sequestration by New Zealand's plantation forests', *New Zealand Journal of Forest Science*, 23(2): 194–208.

Honeyfield-Ross, M., Morel, P. and Visser, A. (2009) 'The growth potential of NZ pigs', in P.C.H. Morel and G. Pearson (eds) *Proceedings Massey University Advancing Pork Production Seminar Palmerston North 8 June 2009*, Palmerston North, New Zealand: Animal Nutrition Division, Massey University, pp. 13–16.

Isaacs, N., Camilleri, M., Burrough, L., Pollard, A., Saville-Smith, K., Fraser, R., Rossouw, P. and Jowett, J. (2010) 'Energy Use in New Zealand Households: Study Report SR221', Porirua, New Zealand: BRANZ.

Jansson, Å., Folke, C., Rockström, J. and Gordon, L. (1999) 'Linking freshwater flows and ecosystem services appropriated by people: The case of the Baltic Sea drainage basin', *Ecosystems*, 2(4): 351–366.

Landcare Research NZ Ltd (Cartographer) (1973) 'NZLRI Land Use Capability'. http://lris.scinfo.org.nz/layer/76-nzlri-land-use-capability/#/layer/76-nzlri-land-use-capability/metadata/ (accessed 14 January 2012)

Leach, G. (1975) 'Energy and food production', *Food Policy*, 1(1): 62–73.

Ministry of Agriculture and Forestry (2003) 'Animal Welfare (Broiler Chickens: Fully Housed) Code of Welfare 2003'. http://www.biosecurity.govt.nz/files/regs/animal-welfare/req/codes/broiler-chickens/broiler-chickens.pdf (accessed 12 January 2012)

Ministry of Economic Development (2011a) 'Energy Balance Sheets'. http://www.data.govt.nz/dataset/show/633 (accessed 4 February 2012)

Ministry of Economic Development (2011b) 'New Zealand Energy Data File, 2011', Wellington, New Zealand: MED.

Ministry of Fisheries (2012) 'New Zealand Fisheries at a Glance'. http://www.fish.govt.nz/en-nz/Fisheries+at+a+glance/default.htm (accessed 6 January 2012)

Ministry of Transport (2011a) 'New Zealand Household Travel Survey'. http://www.transport.govt.nz/research/travelsurvey/ (accessed 14 February 2012)

Ministry of Transport (2011b) 'New Zealand Vehicle Fleet Data Spreadsheet'. http://www.transport.govt.nz/research/newzealandvehiclefleetstatistics/ (accessed 14 February 2012)

National Animal Welfare Advisory Committee (2004) 'Animal Welfare (Layer Hen) Code of Welfare Report'. http://www.biosecurity.govt.nz/files/regs/animal-welfare/req/codes/layer-hens/lhc-report-ii.pdf (accessed 7 March 2011)

National Animal Welfare Advisory Committee (2010) 'Animal Welfare (Pigs) Code of Welfare 2010'. http://www.biosecurity.govt.nz/files/regs/animal-welfare/req/codes/pigs/pigs-code-of-welfare.pdf (accessed 12 January 2012)

National Bureau of Statistics of People's Republic of China (2011) 'Communique of National Bureau of Statistics of People's Republic of China on Major Figures of the 2010 Population Census [1] (No. 2)'. http://www.stats.gov.cn/english/newsandcomingevents/t20110429_402722515.htm (accessed 24 October 2011)

National Research Council (1989) *Recommended Dietary Allowances*, 10th Edition, Subcommittee on the Tenth Edition of the Recommended Dietary Allowances Food Nutrition Board Commission on Life Sciences, Washington, DC: National Academies Press.

Natural Resources Canada (2010) 'Comprehensive Energy Use Database', Ottawa: Natural Resources Canada. http://oee.nrcan.gc.ca/corporate/statistics/neud/dpa/tablestrends2/tran_ca_28_e_4.cfm?attr=0 (accessed 24 October 2011)

New Zealand Feed Manufacturers Association (2012) 'Feed Manufacturing in New Zealand'. http://www.nzfma.org.nz/Nzfm/nzfm.php (accessed 1 March 2012)

Newsome, P.F.J., Wilde, R.H., and Willoughby, E.J. (2008) *NZLRI Land Use Capability – LRIS Data Dictionary*. Lower Hutt: Landcare Research NZ Ltd, http://lris.scinfo.org.nz/layer/76-nzlri-land-use-capability/#/file/162-lris-data-dictionary-v3/ (accessed 3 November 2012)

Office for National Statistics (2011) 'Region and Country Profiles: Key Statistics 28th October, 2011'. http://www.ons.gov.uk (accessed 14 March 2012)

Parnell, W., Blakey, C., Gray, B., Fleming, L. and Walker, H. (2010) *New Zealand Adult Nutrition Survey*, Dunedin, New Zealand: LINZ Nutrition and Activity Research Unit, Otago University.

Pauly, D. and Christensen, V. (1995) 'Primary production required to sustain global fisheries', *Nature Chemical Biology*, 374: 255–257.

Peters, C.J., Bills, N.L., Lembo, A.J., Wilkins, J.L. and Fick, G.W. (2009) 'Mapping potential foodsheds in New York State: A spatial model for evaluating the capacity to localize food production', *Renewable Agriculture and Food Systems*, 24(1): 72–84.

Pollan, M. (2006) *The Omnivore's Dilemma: A Natural History of Four Meals*, New York: Penguin.

Ravindran, R. (undated) 'Misconceptions that Egg on Many Myths'. http://www.pianz.org.nz/pianz/wp-content/uploads/2010/10/Misconceptions-that-egg-on-Many-Myths.pdf (accessed 26 March 2012)

Richard Paling Consulting (2008) 'National Freight Demands Study'. http://www.transport.govt.nz/research/Documents/FreightStudyComplete.pdf (accessed 14 March 2012)

Richardson, J. (2010) 'Foodshed Vancouver', MASLA Master's thesis, University of British Columbia, Vancouver.

RuralFind (2011) 'Rural Livestock Units'. http://www.ruralfind.co.nz/livestock-units-data.html (accessed 3 January 2012)

Saunders, C. and Barber, A. (2007) *Comparative Energy and Greenhouse Gas Emissions of New Zealand's and the UK's Dairy Industry*, Christchurch, New Zealand: Lincoln University.

Saunders, C. and Barber, A. (2008) 'Carbon footprints, life cycle analysis, food miles: global trade trends and market issues', *Political Science*, 60(1): 73–88.

Saunders, C., Barber, A. and Taylor, T. (2006) *Food Miles: Comparative Energy / Emissions Performance of New Zealand's Agriculture Industry*, Christchurch, New Zealand: Lincoln University.

Stanhill, G. (1974) 'Energy and agriculture: A national case study', *Agro-Ecosystems*, 1: 205–217.

Statistics Canada (2002) 'Food Consumption in Canada', Ottawa: Government of Canada. http://dsp-psd.pwgsc.gc.ca/Collection-R/Statcan/32-229-XIB/0000232-229-XIB.pdf (accessed 14 March 2012)

Statistics New Zealand (2011) 'Subnational Population Estimates Tables'. http://www.stats.govt.nz (accessed 20 March 2012)

Vale, R. and Vale, B. (2009) *Time to Eat the Dog: The Real Guide to Sustainable Living*, London: Thames and Hudson.

Wells, C. (2001) *Total Energy Indicators of Agricultural Sustainability: Dairy Farming Case Study*, Dunedin, New Zealand: University of Otago.

3 Domestic Travel

Robert and Brenda Vale

Current wisdom suggests that the way to reduce the environmental impact of travel is to construct higher-density settlements so people walk rather than drive to the shops, school and work. This is based on evidence that cities with higher densities use less energy for transportation (Newman and Kenworthy, 1999: 69–72). However, even the higher density Asian cities achieve only relatively modest public transportation as a percentage of total transportation (for example, Bangkok 17 per cent, Seoul 18 per cent and Manila 23 per cent). In fact, Newman and Kenworthy (1999: 71) give the average for Asian cities of 11 per cent of the total of all transport energy being public transportation. This is only just a bit more than double the average for all European cities, of 5 per cent. This suggests that higher-density living does not have a huge impact on travelling, despite all the words in its favour (for example Jenks *et al.*, 1996; Ewing, 1997; de Roo and Miller, 2000).

Further evidence in support of the idea that density alone is not the answer to reducing the energy used for travelling comes from comparing the results of the Cardiff and Gwynedd Ecological Footprint (EF) studies (calculated for 2001). The Cardiff footprint study describes Cardiff as a city which is reasonably compact and well served by local buses and trains. In the 2001 census the Cardiff urban areas, including Penarth, had a total population of 327,706 (Office for National Statistics, 2004: 30) in an area of 7,572 hectares, giving a density of 43 persons per hectare (Pointer, 2005: 47). The EF for personal transport in Cardiff, including all public and private transport modes, is 0.99 hectares per person (WWF Cymru *et al.*, 2005: 27). A comparison can be made with Gwynedd, a large rural area with a scattered population of 116,843 persons in 254,800 hectares (Farrer and Nason, 2005: 1), at a density of 0.46 persons per hectare. Gwynedd is poorly served by public transport. The EF for all private transport in Gwynedd is 0.78 hectares per person (Farrer and Nason, 2005: 63), which is 21 per cent less than that of urbanised Cardiff. In fact people living in Cardiff, at one hundred times the density, as a whole travelled more in a year than the residents of Gwynedd. This runs counter to the whole philosophy of densification, which is based on the theory that a more urbanised and denser society will travel less.

Of more significance are the relative income levels of residents. In 2001 the mean gross annual earnings in Cardiff were £19,426, whereas in Gwynedd the mean was £15,656 (Office for National Statistics, undated). This is 19 per cent lower than Cardiff, close to the drop in transport EF of 21 per cent, suggesting that income might have as much influence on the footprint of travel as the built environment. In addition, although Cardiff is urban and dense, people do not always travel just locally. Not entirely surprisingly, given the cost of living in Cardiff, a study found that 52,000 people commuted to work in Cardiff from neighbouring local authorities and a further 17,000 came from further afield, including 2,000 from outside Wales. However, the same study found 23,000 people commuted out of Cardiff with 4,000 going outside Wales (Cardiff Research Centre, undated: 5–6).

This brief investigation of the EF of travel suggests that a simplistic solution, such as moving to higher-density living or the so-called Transport Oriented Development (TOD; Calthorpe, 1993), where living is within walking distance of work (and how do you make sure that people can find work within walking distance of their home?), may not be sufficient if a fair share travel footprint is the goal. At first sight, however, the results of a survey in Auckland, New Zealand would seem to support the thinking behind TODs (Auckland Regional Transport Authority, 2009). This found that people living in the central business district (CBD) are more likely to be single-car households and that 40 per cent walk to work and only 23 per cent drive. This looks promising, but the first thing to note is that high-density living does not remove car ownership. In addition, the drive to work is only 7 per cent of total annual distance travelled (Ministry of Transport, 2009a: 9). This means that, for high-density living to be successful, not only work but all shopping, medical, recreational and social journey goals need to be within the CBD for the low-emission transport modes of walking and cycling to be viable. This would involve a significant redesign of existing environments.

To try and understand what a fair earth share personal transport footprint would look like it is necessary to look in detail at what the current travel footprint is. The following study is based on typical travel in Wellington, the capital city of New Zealand. Table 3.1 shows the comparison between personal travel in Wellington and personal travel in Cardiff, the capital city of Wales. More people in Wellington use public transport for the journey to work (approximately 25 per cent of commuters) than in any other New Zealand city (Goodyear and Ralphs, c. 2006: 1). In this respect commuting in Wellington is more like commuting in a European city. However, this figure also means that three quarters of commuters are still driving to work.

Figures for flying are not easy to find, and the methods of calculation are explained below.

Table 3.1 Comparative annual travel distances per person in Cardiff and Wellington

Transport mode	Cardiff average distance (km/person/year)	Wellington Region average distance (km/person/year)*
Driving car or van	5,817 (including driving lorry)	7,600
Passenger in car or van	3,273 (including passenger in lorry)	3,710
Walking	264	320
Cycling	34	51
Bus	475**	290
Motorcycle	16	42
Train	334	no data
Other	203***	49****
Total land transport	10,418	12,062
Flying – domestic	163	414
Flying – international	11,205	5,381
Total	21,786	17,857

* Figures are measured from a bar chart

** 326 km/person/year local bus; 90 km/person/year non-local bus; 59 km/person/year private hire bus

*** 14 km/person/year invalid carriages, dormobiles, etc.; 147 km/person/year taxi; 2 km/person/year ferry; 40 km/person/year unknown

**** No breakdown available; assumed as train

Sources: Data for Cardiff from WWF Cymru et al. (2005: 81–82); data for Wellington from Ministry of Transport (2009a, 2009b)

- *Flying – domestic* For the year ending June 2011, 3,880,305 domestic visitors used air travel for domestic travel in New Zealand (MED, 2011). Huang (2011: 123) has worked out the average one-way domestic flight distance (i.e. trip distance) from Auckland to be 470 kilometres. The population of New Zealand as of 30 June 2011 was 4,405,300 (Statistics New Zealand, 2011a) giving an average domestic flying distance of 414 kilometres per person. This figure for the whole of New Zealand is used for Wellington in the absence of a more detailed breakdown.
- *Flying – international* For the year ended September 2011 there were 2.059 million short-term New Zealand resident traveller departures. Of these, roughly half flew to Australia and Fiji, a total of 1.067 million people (Statistics New Zealand, 2011b: 5). Huang (2011: 124) gives average distance to Oceania destinations (based on the proportion of flights departing Auckland International Airport) as 2,328 kilometres. This gives a total of 1.067 million passengers who each travelled 2,328 × 2 kilometres, or 4,656 kilometres. The remaining 992,000 passengers travelled an average of 9,442 kilometres (Huang, 2011: 125) × 2, or 18,884 kilometres. This gives an overall average for international travel

of 11,511 kilometres per passenger. This gives an average per capita figure for New Zealand international travel of 5,381 kilometres per year. These New Zealand average figures are used for air travel in Wellington.

A Fair Earth Share Travel Footprint in Wellington

In order to be able to work out the travel footprint the distances covered need to be converted to energy. Table 3.2 sets out the breakdown of average personal travel distance in Wellington converted to energy use. It also shows a possible figure for the transport part of a fair share footprint (the 'Reduced total'), assuming a two-thirds reduction.

The coefficients used in Table 3.2 are based on the energy for operating the vehicles and the energy embodied in producing the particular fuel but do not include the energy embodied in the vehicles (or shoes in the case of walking) or in the transport infrastructure. The latter will all add to the total but the energy in operation in all cases will be the major part (PTUA, 2009; Vale and Vale, 2009: 74–77). The inclusion of fuel for what are often regarded as totally environmentally benign forms of transport – walking and cycling – covers the impact of the food required for the particular activity described, assuming it is in addition to normal food intake and daily exercise regimes.

All energy values can be converted to land by using a value that either represents the sequestration rate of CO_2 or the land required to grow sustainable fuels such as wood or plant oils for bio-diesel. These values will therefore vary with climate and soil productivity. However, for a broad study such as this, where the focus is on how the domestic transport footprint can be reduced, using an accepted conversion rate of 100 gigajoules per hectare (Wackernagel and Rees, 1996: 72) will be sufficient for the purposes of comparison.

Scenario 1: A better car

It is clear from Table 3.2 that driving and flying are the main sources of transport impacts on the environment, making up 95 per cent of the total, so what possibilities are available for changing these? The first scenario for change involves no change in travel modes or infrastructure but only in technology. The typical 2.3-litre car, as found in New Zealand, is swapped for the most efficient car, in this case something like a Toyota Prius hybrid or a small diesel car, so that all driving is done with the best available proven technology. Assuming 5 litres per 100 kilometres average fuel consumption (56.5 miles per gallon in the imperial measure), the energy coefficient for driving, which makes up 81 per cent of all travel, will be halved to 2.14 megajoules per person-kilometre (MJ/person-km). The results of doing this are shown in Table 3.3.

Table 3.2 Annual energy use for personal travel in Wellington

Mode of travel	Average annual distance per person (km)	Assumptions	Coefficient (MJ/person-km)	Total energy (GJ)	Distance as a percentage of total	Energy as a percentage of total
Car	11,310*	2.3 litre car with one person	4.28**	48.41	63%	80%
Walking	320	Fuelled by bread	0.90	0.29	2%	0.5%
Cycle	51	Fast; fuelled by bread; shower	2.20	0.11	0.3%	0.2%
Motorcycle	42	Average	1.06***	0.04	0.2%	0.7%
Bus	290	50% diesel, 50% trolley-bus (renewable energy)	2.19	0.64	2%	1%
Train	49	Electric (renewable energy)	0.59	0.03	0.2%	0.05%
Flying – domestic	414	Average of three figures	3.64	1.51	2%	2.5%
Flying – international	5,381	Average international; based on a Boeing 747 at approx. 80% capacity	1.60	8.61	28%	12.8%
Total	17,857			Approx. 60		
Reduced total	17,857			Target 20		

* Total for being driver and passenger

** Average engine size for New Zealand light vehicle fleet in 2009 is approximately 2.3 litres (Ministry of Transport, 2010: 26). Average fuel consumption is around 10 litres per 100 km (Ministry of Transport, 2010: 9). Energy content of fuel is 34.2 MJ/litre (EPA Victoria, 2001). The embodied energy multiplier for petrol is 1.25 MJ/MJ (Alcorn, 2001: 9). So an average car with a single occupant uses (10 litres × 34.2 MJ/litre × 1.25 MJ/MJ) ÷ 100 = 4.28 MJ/person-km

*** Department for Environment, Food and Rural Affairs (2008: 22)

Source: Unless otherwise stated all figures are taken from Vale and Vale (2009: 122)

Table 3.3 Scenario 1: A better car

Mode of travel	Average annual distance per person (km)	Assumptions	Coefficient (MJ/person-km)	Total energy (GJ)
Car	11,310	A better car	2.14	24.20
Other	752	As before	n/a	1.10
Flying – domestic	414	As before	3.64	1.51
Flying – international	5,381	As before	1.60	8.61
Total	17,857			35.42 (Target 20)

If all driving is done efficiently, then a 41 per cent reduction in the overall energy footprint is achieved. The world would be a much better place if fuel-efficient cars were the only type ever manufactured. Having said that, it is the general change in cars – their increasing size, complexity and engine capacity – that is the problem, not just the lack of fuel-efficient technology. The fuel consumption of a modern hybrid, of 5 litres per 100 kilometres (56 miles per gallon) is no better than could be achieved by careful driving in a 1950s Morris Minor, quite simply because the four-seater Morris Minor was light and could not go very fast. However, changing car technology does not appear to be sufficient to achieve a fair share footprint.

Scenario 2: Conventional car, no flying

The second scenario is a change of behaviour rather than of technology, and simply involves no flying, but keeping everything else the same. Although the reduction in distance travelled is around 30 per cent, the energy footprint reduction is less than 20 per cent. The figures for Scenario 2 are shown in Table 3.4. For nearly all journeys there are substitutes for flying, even if some of these are much slower, such as going long distances by sea. In Europe the train is a lower-impact method of international travel, although often not the cheapest. This emphasises the fact that the current market-based economies do not allow for the impact on the environment of

Table 3.4 Scenario 2: Conventional car, no flying

Mode of travel	Average annual distance per person (km)	Assumptions	Coefficient (MJ/person-km)	Total energy (GJ)
Car	11,310	2.3 litre car with one person	4.28	48.41
Other	752	As before	n/a	1.10
Total	12,062	—	—	49.51 (Target 20)

what is being sold. Some governments are thinking forward, and China is planning a high-speed rail connection between Beijing and London within the next 10 years (Jones, 2011). However, if the impact of building new infrastructure is taken into account, this may not be much better than flying, although an electric train can use renewable electricity from wind power and other sources, whereas a plane cannot.

Scenario 3: A better car and no flying

Finally Scenario 3, as detailed in Table 3.5, shows the effect of combining the efficient car and no flying. Even changing the technology of the car and omitting flying do not quite do enough to reduce the transport footprint, although it gets close.

These results point out the problem of considering only efficiency, rather than effectiveness. Flying is in fact a more efficient means of transport in terms of fuel use per passenger kilometre travelled than driving in a typical car, as can be seen from the coefficients in Table 3.2, but is a large part of the total footprint because it involves long distances. Apart from dreamers like Frank Lloyd Wright in his design for Broadacre City (Wright, 1958: 81–82), few designers have proposed using any form of air transport for getting about the city. In Europe the extensive international high-speed rail network makes taking the train a viable alternative to flying for long-distance travel, but in New Zealand, with its limited passenger timetable and small number of long-distance rail travellers, there is little alternative to flying, even for domestic travel. Like the railway, long-distance coach services offer a very energy-efficient means of travel, but both these and rail travel have the problem of taking a long time. In New Zealand, flying makes it reasonable to go to Auckland for the day from any of the major cities, but a train trip even from Wellington to Auckland, both in the North Island, takes ten hours each way, without considering the further train ride and sea passage which would be required for travel to or from the South Island. But this is how people managed in the days before air travel was common, and business still got done.

Table 3.5 Scenario 3: A better car and no flying

Mode of travel	Average annual distance per person (km)	Assumptions	Coefficient (MJ/person-km)	Total energy (GJ)
Car	11,310	A better car, as before	2.14	24.20
Other modes	752	As before	n/a	1.10
Total	12,062			25.30 (Target 20)

These initial results suggest a compelling need to change modes of travel to those that have a lower energy footprint. But what journeys might be susceptible to change? Data from the New Zealand Ministry of Transport (2009b: 4) show that in 2005–2008, 94.2 per cent of the total distance travelled by New Zealanders on the country's roads was by car, either as driver or passenger. In the 'Travel to work' category in the Ministry of Transport's Household Travel Survey, 81 per cent of the distance travelled is by driving. However, it is trips other than the journey to work which make up the largest part of people's road travel. Travel to work is ranked only fourth for distance travelled. The category of travel for 'Shopping/personal business/medical reasons' is the largest travel category, and 70 per cent of the distance travelled for this category is driver travel. The second largest travel category in terms of distance is 'Travel to social destinations', where driving comprises 54 per cent of the distance travelled. This category covers holidays, visits to friends and family, entertainment trips, attending religious meetings and hobby-related trips. It should be noted that the reference to 'driving' is to the situation where a person is the driver of the vehicle. The figures show an additional percentage of the distance travelled in each category is also attributable to people travelling as passengers in a car (Ministry of Transport, 2009a: 9).

A Household Transport Footprint

In an attempt to see what life might be like and what might be possible in a more sustainable transport scenario, Table 3.6 sets out the authors' transport use for a year. This is a household of two people working full time. The house is in the low-density suburb of Khandallah, which is located fifteen minutes' ride by electric train or diesel bus from the Wellington city centre. The local village, which is about five minutes' walk from home, provides a supermarket, butcher, chemist, library, post office, doctor and dentist plus other shops and food outlets. The household has one car and car use in Table 3.6 is based on recorded driving over the past two years. The fuel consumption figure of 5 litres per 100 kilometres is based on the read-out from the car's in-built fuel consumption meter (4.7 litres per 100 kilometres over the three-year life of the car). All journeys to work are made using public transport (bus and train) and walking. Both members of the household are members of clubs and societies, and drive to meetings in the city and surrounding area once or twice a week, and they also drive for outings and holidays. However, public transport is also used for outings.

Table 3.6 Measured household transport footprint

Mode of travel	Annual distance per person (km)	Assumptions	Coefficient (MJ/person-km)	Total energy (GJ)
Car	6,000	Prius 1.5-litre hybrid; 5 litres per 100 km	2.14	12.8
Walking	1,500	Fuelled by bread	0.90	1.4
Bus	2,500	Diesel	1.01	2.5
Train	2,500	Electric (renewable energy)	0.59	1.5
Total land transport	12,500			18.2

The figures for walking and for public transport use are based on the actual situation, as follows:

Walking:
To bus stop in morning	5 min
To university from city bus stop	15 min
To railway station in evening	30 min
To home from local station	10 min
	60 min walking per day

Walking 5 days per week
If walking speed is 5 km/h, weekly distance is 5 km × 5 days, which is 1,300 km per year

Bus:
Assume 5 trips to work per week
Distance measured from map is 9 km
Total distance is 9 km × 5 trips × 52 weeks = 2,340 km

Train:
Assume 5 trips home per week
Distance measured from map is 8 km
Total distance is 8 km × 5 trips × 52 weeks = 2,080 km

Figures have been rounded in Table 3.6 to avoid unreasonable accuracy and to allow for uncounted trips. What Table 3.6 shows is that it is perfectly possible to have a low enough transport footprint provided that flying is eliminated. There can still be some car use, provided that the car is either small or has low fuel consumption. This modest car use, with public transport used for commuting, tends to bear out the authors' proposition that a 1950s lifestyle could have been possible for the whole world, albeit using Morris Minors rather than the latest hybrid technology (Vale and Vale, 2009: 37).

What would happen to these figures if all land transport were by bus instead of by car? Would that allow us to fly? If the amount of flying is the same as the Wellington average, yes, it would be possible to do both domestic and international trips, as shown in Table 3.7.

If all land transport currently done using a car were done by bus with no car use, the travel footprint could accommodate the average amount of flying currently done by a New Zealander. So there is a clear choice: you can drive a car, or you can fly, but it is not possible to do both, unless a way is found to operate aircraft from renewable energy. The US Department of Transportation (2001) found that 40.6 per cent of all air travel in 2001 was for business purposes. More recently the business reporter for Cleveland's newspaper *The Plain Dealer* stated that 80 per cent of US air travel is for leisure rather than business (Grant, 2011), while in the UK business passengers made up 30 per cent of all air travel in 2007 (CAA, 2011: 19). These figures suggest that business travel could continue by air, giving a 70 per cent reduction in the EF of air travel overall. However, there would be no flying for holidays and leisure, long-distance travel would be by rail where possible, or on passenger cargo liners as used in the 1950s (Vale and Vale, 2009: 119–120). Alternatively, if we gave up the use of cars and went over completely to public transport, walking and cycling, and if business made much greater use of electronic communication, it would be possible to make a long-haul flight (New Zealand to England) every five years. This would suggest that employment might need to contemplate some sort of sabbatical arrangement, so that if you were going to fly you could afford to be away long enough to get the most out of the travel.

The current fashionable rush to higher-density living, although good for developers since they can generally make more money out of putting more units on a given area of land, may not lead to lower transport footprints overall. Whether living in town or country, choosing not to fly is going to

Table 3.7 The effect of giving up the car to allow flying

Mode of travel	Annual distance per person (km)	Assumptions	Coefficient (MJ/person-km)	Total energy (GJ)
Bus to replace car	6,000	All land transport by car, 6,000 km, is replaced by bus; diesel	1.01	6.1
Walking, bus and train from Table 3.6	6,500	As before	n/a	5.4
Flying – domestic	414	As before	3.64	1.5
Flying – international	5,381	As before	1.60	8.6
Total				21.6 (Target 20)

Table 3.8 Land travel in urban and rural settlements in New Zealand

Mode of travel	Annual distance per person per year (km)	
	Main/secondary urban (population greater than 10,000)	Minor urban/rural (population less than 10,000)
Car/van – driver	6,421	9,405
Car/van – passenger	3,527	5,711
Walking	218	144
Train/bus/ferry	337	414
Cycling	72	63
Other, including motorcycle	203	169
Total	10,778	15,906

Source: Data from Ministry of Transport (2009a: 14)

make a big difference to the transport footprint. Denser living seems to affect mainly distance driven by car, as shown in Table 3.8. However, the rural population of New Zealand appear from these figures to use more public transport than their urban counterparts.

Table 3.8 shows that rural New Zealand residents travel 52 per cent further by car (adding driver and passenger figures). However, less car travel may not mean lower fuel consumption. It depends on the type of driving involved. Car fuel consumption in Europe is tested in both urban and open road situations by, among others, the UK Vehicle Certification Agency. The tests are carried out on a rolling road from a cold start when the engine has not run for several hours. The urban cycle consists of a series of accelerations, steady speeds, decelerations and idling. Maximum speed is 50 kilometres per hour, average speed is 19 kilometres per hour and the distance covered is 4 kilometres. This is immediately followed by the extra-urban cycle which consists of roughly half steady-speed driving and the remainder accelerations, decelerations, and some idling. Maximum speed is 120 kilometres per hour, average speed is 63 kilometres per hour and the distance covered is 7 kilometres. Table 3.9 shows the fuel consumption figures for three petrol cars chosen to represent a range of engine sizes.

Table 3.9 Car fuel consumption data for some typical vehicles

Car make and model	Urban (cold start; litres/100 km)	Extra-urban (litres/100 km)	Percentage difference
1.3-litre Toyota Yaris	6.2	4.5	38%
1.6-litre Volkswagen Golf	9.7	5.6	73%
2.4-litre Audi A6	14.1	7.1	99%

Source: Data from VCA (2010)

The figures for these three cars show that the urban fuel consumption is often significantly worse than the extra-urban. If rural driving is represented by the extra-urban cycle, it would seem very likely that rural drivers may not be experiencing significantly greater fuel consumption in spite of travelling longer distances.

In addition, Table 3.8, showing longer travel distances for rural residents, covers only land transport, which means that it may not be showing the whole picture. Table 3.10 shows the results of a study carried out comparing Cardiff and Gwynedd. It can be seen that in Gwynedd, which has a density of population only one hundredth of that of urban Cardiff, the overall transport footprint is significantly lower. This is not because the residents of Gwynedd do less driving – they drive nearly as much as Cardiff residents – but because they do less flying; as suggested earlier, this is probably linked to having less money.

Part of the key to reducing transport use might be to look more closely at why those who live in rural areas travel less. Aside from lower incomes, additional reasons may be because there is less choice in rural areas and so fewer things to travel for, or it may be that travel distances are long and people do not have time to make all the separate journeys the city dweller can, so they plan their trips to incorporate several activities. It may even be that rural areas offer a better quality of life so there is less need to travel in search of self-fulfilment.

Many writing in the 1970s, when the idea of sustainability first appeared (Goldsmith *et al.*, 1972), saw a future short of energy and resources as a rural future where food would be grown close to the place of consumption and where the shared interests of growing food would form part of social cohesion, as it has in rural communities for centuries. The high-density city will always have a high food footprint because it has to be fed from outside its boundaries (Vale and Vale, 2010: 19–26). A high density is also not going to reduce transport footprint if people continue to fly. This is underlined

Table 3.10 Annual average travel distances per capita (in modes of travel for which there is a difference) for urban Cardiff and rural Gwynedd

Mode of travel	Cardiff	Gwynedd
Air – international (km)	11,205	6,995
Air – domestic (km)	163	59
Air – total (km)	11,368	7,054
Car – driver (km)	5,733	5,626
Car – passenger (km)	3,239	3,178
Car – total (km)	8,972	8,804
Local bus (km)	326	176
Taxi/minicab (km)	147	51
Total transport EF (gha)	0.99	0.78

Source: Data from Farrer and Nason (2005: 62–63)

by a study of the ecological footprint of the Findhorn Foundation Community (FFC) in Scotland. Findhorn was founded as a commune in a rural area of Moray, Scotland in 1962 and has received a Best Practice designation from the UN Centre for Human Settlements. It contains 61 ecological buildings, built of natural and local materials such as local stone and straw bales and equipped with solar panels for hot water heating; these incorporate passive solar design, energy efficient appliances and an on-site sewage treatment plant; much of the community's electricity is supplied by four wind turbines; most food is provided by the Community Supported Agriculture Scheme, based on organic farming methods, and very little meat is eaten (Tinsley and George, 2006: 8); and a local economy is implemented using the Local Exchange Trading System which avoids the use of money in trade. Taken together, the ecological footprint of someone living in Findhorn is 2.56 gha (Tinsley and George, 2006: 34). One of the highest components of the EF turned out to be flying (Tinsley and George, 2006: 17). The annual EF for air travel was 0.25 gha per person, with 90 per cent of this being for leisure rather than Findhorn business. This is more than twice the average Scottish air travel EF (Tinsley and George, 2006: 18).

What this study shows is that it is not difficult to reduce the personal travel footprint and still have mobility, provided that either flying or the private car is eliminated. However unpalatable this may seem, it is the simplest solution, coupled with much more use of walking, cycling and public transport. In scenarios that do not eliminate flying it is hard to reach a fair earth share. Given modern levels of communication in the future, long-distance exchange is more likely to be via the Internet rather than in person. As the proverb states, 'If God had meant us to fly he would have given us wings.'

References

Alcorn, A. (2001) *Embodied Energy and CO_2 Coefficients for New Zealand Building Materials*, Wellington, New Zealand: Centre for Building Performance Research, Victoria University of Wellington.

Auckland Regional Transport Authority (2009) *Auckland Transport Facts*, http://www.arta.co.nz/what-we-do/auckland-transport-facts.html, accessed 14 July 2010.

CAA (2011) *CAP 796 Flying on Business: A Study of the UK Business Air Travel Market*, Regulatory Policy Group, November, London: Civil Aviation Authority.

Calthorpe, P. (1993) *The Next American Metropolis: Ecology, Community and the American Dream*, New York: Princeton Architectural Press.

Cardiff Research Centre (undated) *An Overview of Cardiff Employment and the Local Economy*, Cardiff, Wales: Cardiff Research Centre, Corporate Support, Cardiff Council.

Department for Environment, Food and Rural Affairs (Defra) (2008) *Guidelines to Defra's GHG Conversion Factors: Methodology Paper for Transport Emission Factors*, http://www.defra.gov.uk/environment/business/reporting/pdf/passenger-transport.pdf, accessed 24 May 2009.

de Roo, G. and Miller, D. (2000) *Compact Cities and Sustainable Urban Development*, Farnham, England: Ashgate.

EPA Victoria (2001) 'Which of the fossil fuels generates the most carbon dioxide?' http://www.epa.vic.gov.au/GreenhouseCalculator/calculator/reference/reference/greenhouse/gh07a.htm, accessed 12 August 2010.

Ewing, R. (1997) 'Is Los Angeles-style sprawl desirable?' *Journal of the American Planning Association*, 63(1), pp. 107–125.

Farrer, J. and Nason, J. (2005) *Reducing Gwynedd's Ecological Footprint: A Resource Accounting Tool for Sustainable Consumption*, Cardiff, Wales: WWF Cymru.

Goldsmith, E., Allen, R., Allaby, M., Davoll, J. and Lawrence, S. (1972) 'A blueprint for survival', *The Ecologist*, 2(1).

Goodyear, R. and Ralphs, M. (c. 2006) 'Car, bus, bike, or train: what were the main means of travel to work?, http://www.stats.govt.nz/browse_for_stats/people_and_communities/geographic-areas/commuting-patterns-in-nz-1996-2006.aspx, accessed 12 April 2011.

Grant, A. (2011) 'Air travel expected to increase this summer, but it will be more costly and planes will be packed', *The Plain Dealer*, 28 May, http://www.cleveland.com/business/index.ssf/2011/05/air_travel_expected_to_increas.html, accessed 21 October 2011.

Huang, N. (2011) 'A Modified Ecological Footprint Method for Assessing Sustainable Transport in the Auckland Region', PhD Thesis, University of Auckland, Auckland, New Zealand.

Jenks, M., Burton, E. and Williams, K. (eds) (1996) *The Compact City: A Sustainable Urban Form?* London: E. & F. N. Spon.

Jones, D. (2011) 'London to Beijing ... by rail?' *EU Infrastructure*, 6 April, http://www.euinfrastructure.com/news/china-europe-rail-link/, accessed 24 April 2011.

MED (2011) 'Transport types used by domestic visitors – current year end – trip taken', http://ocv.onlinedatacentre.com/mot/OAPAnalysis.html?c=DTS%20-%20Transport%20Used%20-%20Table.cub, accessed 25 October 2011.

Ministry of Transport (2009a) *Comparing Travel Modes: Household Travel Survey, v2 revised*, http://www.transport.govt.nz/research/Documents/Comparingtravelmodes_2009.pdf, accessed 16 August 2010.

Ministry of Transport (2009b) *How New Zealanders Travel*, http://www.transport.govt.nz/research/Documents/How%20New%20Zealanders%20travel%20web.pdf, accessed 18 May 2010.

Ministry of Transport (2010) *The New Zealand Vehicle Fleet: Annual Fleet Statistics 2009*, Wellington: Ministry of Transport.

Newman, P. and Kenworthy, J. (1999) *Sustainability and Cities: Overcoming Automobile Dependence*, Washington, DC: Island Press.

Office for National Statistics (undated) 'Annual Pay – Gross (£)', Table 7.7aa, http://www.statistics.gov.uk/downloads/theme_labour/ASHE_2001/2001_work_LA.pdf, accessed 24 April 2011.

Office for National Statistics (2004) *Census 2001: Key Statistics for Urban Areas in England and Wales*, London: HMSO.

Pointer, G. (2005) 'The UK's major urban areas', in Office of National Statistics, *Focus on People and Migration*, London: HMSO.

PTUA (2009) 'Common urban myths about transport', Public Transport Users' Association, www.ptua.org.au/myths/energy.shtml, accessed 12 August 2010.

Statistics New Zealand (2011a) *National Population Estimates: June 2011 Quarter*, Statistics New Zealand, http://www.stats.govt.nz/browse_for_stats/population/estimates_and_projections/NationalPopulationEstimates_HOTPJun11qtr/Commentary.aspx, accessed 25 October 2011.

Statistics New Zealand (2011b) *International Travel and Migration: September 2011*, 21 October, Wellington, New Zealand: Statistics New Zealand.

Tinsley, S. and George, H. (2006) *Ecological Footprint of the Findhorn Foundation and Community*, Moray, Scotland: SDRC.

US Department of Transportation (2001) *2001 National Household Travel Survey*, US Department of Transportation, http://ntl.custhelp.com/app/answers/detail/a_id/252/~/percentage-of-air-travel-for-business-vs-other-purposes, accessed 25 October 2011.

Vale, R. and Vale, B. (2009) *Time to Eat the Dog? The Real Guide to Sustainable Living*, London: Thames and Hudson.

Vale, B. and Vale, R. (2010) 'Is the high-density city the only option?' in Ng, E. (ed.) *Designing High-Density Cities for Social and Environmental Sustainability*, London: Earthscan.

VCA (2010) 'Fuel consumption testing scheme', VCA, http://www.vcacarfueldata.org.uk/information/fuel-consumption-testing-scheme.asp, accessed 18 August 2010.

Wackernagel, M. and Rees, W. (1996) *Our Ecological Footprint*, Gabriola Island, Canada: New Society Publishers.

Wright, F.L. (1958) *The Living City*, New York: Horizon Press.

WWF Cymru, Sustainable Development Unit, Cardiff Council and Centre for Business Relationships, Accountability, Sustainability and Society, Cardiff University (2005) *Reducing Cardiff's Ecological Footprint: A Resource Accounting Tool for Sustainable Consumption*, Cardiff, Wales: WWF Cymru.

4 Consumer Goods

Maggie Lawton

The second half of the twentieth century delivered a double blow to the sustainable use of resources, sending the global ecological footprint into overshoot in 1970 (Global Footprint Network, 2012a: 4). First, there has been a nearly threefold increase in global population since 1950 (population in 1950 was around 2.5 billion, and in 2012 was 7 billion; US Census Bureau, 2012), and, second, a dramatic increase in consumerism in developed and increasingly in developing countries, substantially on the back of fossil fuel-based products and derivatives. During the twentieth century the world increased its fossil fuel use by a factor of 12, whilst increasing its extraction rate of material resources by a factor of 34 (European Commission, 2011: 2). It appears that the concept of a finite planet and resource base, with an upper limit on the rate of natural resource regeneration, is still difficult for many to comprehend, despite being as absolute as any law of physics.

In our rapidly urbanising world, most of the 'things' we use and consume are purchased rather than grown or made by ourselves. Even prior to the advent of globalisation, a phenomenon that accelerated rapidly throughout the twentieth century, trade has always been a component of human lifestyles, as people came together in communities and shared their skills and produce. The category of 'consumer goods', which includes any goods bought for personal or household use that are the end result of production and manufacturing, makes up that proportion of each individual's ecological footprint over which they have most control. The level at which we consume is within our individual and collective capability to moderate and letting manufacturers worldwide realise that we will be more prudent and discerning in our shopping would change the face of the marketplace.

'Shop until we drop' has been a catch-cry for many in developed countries for a number of decades, shopping often being associated with fun, demonstrating social status or providing comfort and solace, satisfying a want rather than a need. As Bill Bryson said 'We used to build civilizations. Now we build shopping malls' (Bryson, 1992: 105). Oscar Wilde, known for his hedonistic tendencies, joined in with the refrain: 'anyone who lives within their means suffers from a lack of imagination' (undated). So it appears that the nascent desire to shop and liberally consume may long have been within us, fuelled more recently by the rise of the seductive combination of global connectivity, apparently abundant raw resources and cheap Asian labour.

The need to consume in excess has become a lynchpin of our economy, taking off after World War II as the United States in particular sought to maintain the upsurge in manufacturing that was reached during the war years. Victor Lebow, an economist and retail analyst, provided a synopsis of that development in a 1955 article: 'Our enormously productive economy demands that we make consumption our way of life, that we convert the buying and use of goods into rituals, that we seek our spiritual satisfactions, our ego satisfactions, in consumption' (Lebow, 1955).

On a similar theme was a comment made by Mayor Giuliani (not President Bush, to whom it is sometimes attributed) directly after the bombing of the World Trade Centre in 2001, when he told people to go about their normal day and, 'Show your confidence. Show you're not afraid. Go to restaurants. Go shopping' (Adler and Adler, 2002: 26). While Guiliani probably meant well, the comment could be considered a flippant response, out of keeping with the seriousness of the recent event, and therefore became the subject of some scrutiny.

Not only has the amount of consumer goods risen dramatically but there has also been a change in the composition of what we purchase. This includes technologies such as washing machines and microwave ovens that are alleged to free up time, although they have been shown to do no such thing (Bittman *et al.*, 2003). Once considered a luxury, they would now be considered by many to be necessities. Our perception of what is a necessity has changed, as new consumer goods have become available. Researchers at Oxford University have found that domestic electricity consumption from lights and appliances in the UK doubled from 1970 to 1995 (Boardman *et al.*, 1997: 8).

This rise in consumption of non-essential goods, however 'non-essential' is defined, sets wealthy and poor countries apart. This is well demonstrated by the correlation between income and ecological footprint (see Chapter 18, 'Footprints and Income'). There can be no doubt that many countries should have a more equitable access to the necessities of life and do not have currently their fair share of consumer goods, or even the basic requirements for life, food and shelter. Many of the social and economic intricacies of solving the problem of global equitable distribution of wealth are beyond the scope of this book, although we can state the obvious: in an equitable world with finite resources, there has to be a rebalancing of consumerism. That is, wealthier countries will have to use less, if poorer countries are to have their fair share. The focus of this chapter is therefore on countries and communities where incomes have given rise to consumer overload and how the consumer addiction can be moderated.

Why and how has this consumer society in developed countries arisen? The psychology of consumerism has many facets (Harre, 2011; Hamilton and Denniss, 2005): the social status associated with the latest and greatest new thing, the feel-good factor with shopping 'retail therapy', and even the fact that shopping has become a pastime. Look at many malls on a Sunday: families strolling, eating and purchasing incidental items they didn't know they needed. The advent of cheaper commodity goods has been a major fuel to this shopping bonanza; even those with little disposable income can find something in a two-dollar shop to satisfy the shopping urge. Online marketing bombards us with adverts and when we go looking for things through Internet shopping they are easy to find. A recent report from the New Zealand National Business Review (NBR, 2011) estimates that nearly 1.5 million New Zealanders over the age of 18 are shopping online, about two thirds of the adult population. Our access to shopping is now far greater than before, particularly through online shopping and seven-day trading in large malls.

The dictum is that 'you manage what you measure' and, unfortunately, that measurement is for most countries based on gross domestic product (GDP). Politicians, the media and economists concentrate on our GDP and worry about achieving sufficient growth. An increase in GDP is considered a good thing, even though it includes spending on goods that are often considered harmful, such as cigarettes, or support services required to counter a negative aspect of our society, such as prisons. Many problems are associated with GDP. In particular, it does not measure the externalities that are involved in the production and distribution of consumer goods. These include the environmental and social costs that are building up and contributing to, among other things, the overshoot in our ecological footprint, ecosystem degradation and social problems. While many acknowledge the shortcomings of GDP as a standalone indicator, we continue to use it and hence confuse spending power with quality of life.

Alternative or additional measures to GDP have been promoted, including the Genuine Progress Indicator (GPI; Anielski, 2001), which attempts to combine a country's economic fortunes with a measure of social well-being. Unlike GDP it differentiates what might be termed 'useful' growth from the negative aspects of growth. Despite considerable research and attempted applications, this and other related indicators, such as the National Happiness Index (United Nations, 2011), are still far from mainstream.

Along with greater consumerism has come the concept of planned obsolescence. Cheap goods do not generally last long, so they will need replacing. More expensive goods can be subject to perceived rather than actual obsolescence. This applies to major household items, clothes and electronic goods, in particular, which may still be functioning or functional but are no longer 'this year's thing', so they have to go. One possible positive development is the emergence of online second-hand shopping such as eBay or, in New Zealand, Trade Me. Unfortunately, along with the trade in second-hand goods this trend has also encouraged more importation of cheap goods by 'home-traders', so overall it may just have added to the sum total of consumer goods available.

The emergence of Asian countries as global economic bases has had much to contribute to the availability of consumer goods. There is no doubt that many consider these cheaper goods as a positive development, when a generation ago they were unable to afford so many 'luxury' items. The impact on developing countries is a mixed bag, as economic development in those countries has also brought environmental and social disruption. Even in western countries the lure of cheap overseas labour has created workplace disruption for many, as manufacturing has moved offshore. Similarly, the rise of cheap overseas goods has changed the face of retail, with high-street shops taking second place to the mega-stores. The environmental impact is evident in ecosystem damage the world over, caused by extraction and also by increases in pollution, including greenhouse gases.

While there are many sociological dissertations on the rise of Asian economies and their influence on consumer goods, it should be remembered that one class of resources more than any other has enabled that development: the advent of cheap carbon and hydrocarbon compounds, first coal, then oil and natural gas. This has contributed to all areas of the ecological footprint of developed countries – transport, infrastructure, housing, food and mass-produced consumer goods. Oil and its derivatives, in particular, are part of virtually every product or its supply chain worldwide. Its availability has dramatically changed consumer patterns and the nature of local economies.

Oil, transformed into plastic, is the basic material or a key ingredient in most of the non-food products we import, and oil is also critical to their national and international transport. Many other materials have also been extracted and incorporated into products. We now extract and use natural resource materials at a rate that exceeds their collective replenishment by approximately 35 per cent. The trend is marked each year by Earth Overshoot Day, which in 2011 fell on 27 September (Global Footprint Network, 2012b). To enable us to continue our consumerism we deplete stocks and accumulate waste that should be made available for future use, to maintain resource stocks in line with the replacement rate of the Earth's ecosystems, as a minimum. By failing to achieve that basic equation we are borrowing from the future and eventually, as with sub-prime mortgages in the recent economic crisis, this will become impossible to maintain.

Aided by oil, we now have a global economy designed to require growth to function satisfactorily and, to sustain that growth, we need to produce ever more consumer items. However, as continual growth is impossible in a finite world and with the predicted advent of less available and more expensive oil, key areas of consumerism will need to change. In doing so, they may help rebalance global resource-use distribution. Although 'peak oil' is now claimed by one source to have been postponed almost indefinitely (Maugeri, 2012), there is still the problem that, if all the oil is burned, the climate of the planet will be destroyed (Kharecha and Hansen, 2008).

There is no universal classification for consumer goods, but food, clothing, housing, household items and personal items form the major categories, which can then be broken down further. It is worthwhile examining each of the major classes, their emergence and how they are changing over time, also how that will impact on resource use and resulting pollution – the fundamental ingredients of ecological footprints.

Feeding Frenzy

Food is a subject expanded on elsewhere in this book (see Chapter 2), but, given its role as a central consumer good, it also gets a mention here. Our need for food is as basic as it gets. Without it we die. However, our need for food goes further than that – it is part of our culture and our social interaction, and increasingly it is a part of our consumer society. Food satisfies emotional as well as physical needs. What we eat goes well beyond what we need to survive and has, with growing wealth, given rise to a complex global system of production and distribution. While we enjoy greater access to a far larger variety of food, there are some significant downsides to these developments. In particular, the increase in obesity and health-related problems is one manifestation of an over-consumption of this essential consumer good (Frey and Barrett, 2007).

Similarly, our reliance on imported food has increased our vulnerability to shocks to the supply system. Food security is something that governments, even in developed countries, are starting to take seriously (Nord, 2009).

'Naked People Have Little or No Influence on Society' (Johnson, 1927)

Clothes make up a significant segment of consumer goods, with our buying behaviour having changed markedly over the last century. Like food, clothing is a necessity but often also a luxury. With our lack of body hair, nobody would claim clothing is optional, at least in most countries. However, it is so much more than just a need: clothing is the ultimate personal accessory. Whatever bodily flaws we may have can be hidden or at least softened by the right clothing. It is not surprising that Gok Wan's television programme, *How to Look Good Naked*, is so popular. The programme title is a bit of misnomer, as although people achieve the confidence to be naked in public, it is how they look in the clothes he suggests for them that gives them that confidence. There is pressure to stay current with clothes, maybe more so for women than for men. Designers ensure that shapes and styles, hem lengths and shoe heels keep changing, and marketers ensure that we feel pressure to be up with the latest trends. The young or young at heart are particularly targeted, with pre-teens now being a significant market segment.

It is not just the pace of change in fashion that has modified our clothes consumption; it is the availability of cheap clothes made from synthetic fabrics that flood the malls across the globe. Oil has again played a large part in the availability of these products, as oil-derived polymers are the basis of many forms of synthetic fibres. Synthetic clothing need not have a larger ecological footprint than clothes made from natural fibres (Vale and Vale, 2009: 193), nor is it necessarily more environmentally damaging. So much depends on the methods of farming, production and processing of the fibres, the place of manufacture and transport to market. An account of the complexity of this mass-production machine appears in the book *Where Underpants Come From* (Bennett, 2008), with similar tales of complexity for most of the consumer goods probable after considerable research; tracing these life cycles is no easy matter. Together with cheap labour, the increased availability of synthetic raw material means that clothes are now significantly more affordable than decades ago. While knitting and dressmaking, though often using imported materials, are still common interests, they are generally hobbies rather than necessities. Only the most determined will continue to make their own clothes from natural fibres, e.g. wool or hemp from their locality; for busy people the incentive to maintain that level of creativity is low.

Maybe we should be relearning some of those skills and thinking about where best to obtain the raw materials for clothes? We could certainly consider how to reduce the amount of clothing we buy. This is not just as an attempt to reduce our ecological footprint but because, as with most other consumer goods, our current reliance on oil may be short-lived. Falling oil supplies, whether due to climate considerations or price, will impact on the availability of synthetic materials made from oil, the ability to grow natural materials which require fertiliser made from natural gas, the processing and manufacture of garments and their transport globally from factory to customer. Clothes, as a consumer good, will need to undergo further evolution.

Plastic Fantastic

The remaining consumer goods will be clumped together. This category is mainly comprised of things that didn't exist – certainly not in their current form – before the industrial revolution and coal and oil supplies increased productivity and access to raw materials. The product revolution that is plastic has penetrated every aspect of our lives, and plastic is found somewhere in virtually all of these products. Into this group fall household furniture and goods, whiteware, electronic items, personal items such as cosmetics and hygiene products, a plethora of cheap plastic toys, and a bewildering array of other items. Catalogues tempt us with the weird and wonderful – just have a look at the Innovations New Year 2012 catalogue. Do you really want a 'Solar Powered Meerkat Crystal Ball Garden Light' (Innovations, 2012)? It's a rhetorical question – doesn't everyone? After all, it uses renewable energy and therefore must be good for the planet.

The digital age has added to the raft of goods available. Rather than reduce consumption, as previously envisaged, the resource requirements and waste generated from the digital age of electronic goods are a serious threat on many levels. The externalities, the social, health and environmental costs of the manufacture of electronic goods from the mining, manufacture and disposal of the vast amount metals and synthetic compounds involved, are well described by Elizabeth Grossman in *High Tech Trash* (2007). Trying to determine the ecological footprint associated with individual items such as a cell phones, iPods or other electronic goods is even more complicated than for clothes, but to maintain a flow of the raw materials for these products, as well as minimise their harm, major changes to production methods must occur.

We can take much more control over this area of consumer spending. Some basic rules for spending would be:

1. Don't impulse buy anything you don't know you need. If you really, really have an immediate overwhelming attraction to it, think about it for a while and then decide.
2. If you decide to buy it identify something you will go without to balance your purchases.
3. Buy to last and resist in-built or perceived obsolescence as much as possible.
4. Buy second-hand if practical.
5. Remove temptation wherever possible. If you can't find out how to stop being bombarded with ads on your computer screen or other media, learn to ignore them.

If this seems easy then you are either a saint or deluded. While our personas continue to be so much aligned to the goods that we display, it will continue to be difficult to comply with all these rules. However, any improvement will be welcome in contributing to the lowering of the world's ecological footprint. If, as is the norm with climate change, you are amongst the majority who believe that as an individual your actions are next to futile in making any difference, then in reducing resource use there are always the financial benefits to think about. Taking control over your spending habits and the amount of time you need to work to fulfil those purchasing desires will have a positive impact on your bank balance, your feeling of well-being and ultimately your quality of life. The correlation between happiness and material wealth is well documented and indicates that once we reach a certain level of wealth and consumption our satisfaction plateaus. Indeed in his book, *Affluenza*, Oliver James (2007) credits the consumer society with a raft of problems such as depression, substance abuse and personality disorder, conditions which seem to be heightened in affluent societies.

For those consumer items that we continue to purchase, the manufacturing process has to change. The current linear system of manufacturing is extraordinarily inefficient. Paul Hawken and colleagues in their seminal book, *Natural Capitalism*, note that only 1 per cent of the total North American materials flow ends up in and is still being used within products six months after their sale (1999: 81). Most industrial processes require high amounts of energy as well as raw materials, so any improvement in manufacturing systems would have a dramatic impact on the contribution that consumer goods make to our ecological footprint.

Another approach that has found favour is 'cradle to cradle' (McDonough and Braungart, 2002). The work of cradle to cradle practitioners is based on creating (and marketing) a 'total quality framework' which has been designed to support companies in creating products that are 'more good' rather than simply 'less bad' (Cradle to Cradle Products Innovation Institute, 2011). Similarly the sustainability principles of 'The Natural Step' organisation (undated) reflect concepts which if followed would lead to a serious reduction in the global ecological footprint over time, through systematic reduction in mineral extraction, pollution and biodegradation of the world's resources.

Robert Costanza and colleagues (1997) measured the value of all the services that nature provides to be an average of US\$33 trillion (US\$33 × 10^{12}) per year, compared with a global GDP of US\$18 trillion, demonstrating what we are at risk of losing if we continue to destroy ecosystems as we consume. Unsustainable consumption cannot, by definition, continue unabated. The proposed solutions are many but inertia, for a multitude of reasons, is hindering progress. There is no shortage of information, severe warnings or upbeat solutions but unless there is a major and immediate shift in consumption patterns, it is likely that the solution will be thrust upon us rather than planned for. Richard Heinberg in *Peak Everything* (2007) describes the impacts and possibilities brought about by the peak in availability of many of our essential minerals and materials and how that will impact on the economy.

There are clearly many ways in which this drastic waste of resources, born of our current consumer society and contributing much to our ecological footprint, can be reduced. This ranges from changing people's perception of the need for and value of goods and reducing their will to shop, to moderating the manufacturing processes, reducing transport for distribution, decreasing or eliminating packaging and reusing as much material as possible. Right now these seem like optional activities to many – sensible ones as, apart from anything else, waste costs money – but as we head further into resource overshoot we will be forced to make these changes. The resources we use will simply become hard to get. The trick for us, individually and collectively, is to put changes in place before we are forced to. We can all decrease our own footprints and probably gain economic and even health benefits in the process. Some communities are already tackling the projected reduction in availability of consumer goods. The Transition Towns movement is a prime example, reducing resource use and gaining social benefits along the way (Transition Network, 2012), as it promotes early adoption of behaviour that substantially reduces resource use and makes the transition smoother. It is in all our best interests to begin now.

Industry can greatly improve on the efficiency of those goods which we have now genuinely come to require. We have now extracted and wasted much of the world's mineral and natural wealth, which will be less and less available as the twenty-first century progresses. This component of our ecological footprint has to decrease; the only question is the extent to which we control the process and maintain or improve the quality of life for all the world's inhabitants as our global footprint shrinks to a sustainable level.

References

Adler, B. and Adler, B. (2002) *The Quotable Giuliani: The Mayor of America in His Own Words*, Simon & Schuster, New York

Anielski, M. (2001) 'Measuring the sustainability of nations: the genuine progress indicator system of sustainable well-being accounts', paper presented at *The Fourth Biennial Conference of the Canadian Society for Ecological Economics: Ecological Sustainability of the Global Market Place*, August, Montreal, Quebec, http://www.anielski.com/Documents/Sustainability%20of%20Nations.pdf, accessed 30 November 2012

Bennett, J. (2008) *Where Underpants Come From*, Simon & Schuster, London

Bittman, M., Rice, J.M. and Wajcman, J. (2003) 'Appliances and their impact: the ownership of domestic technology and time spent on household work', SPRC Discussion Paper No. 129, October, Social Policy Research Centre, University of New South Wales, Sydney

Boardman, B., Fawcett, T., Griffin, H., Hinnells, M., Lane, K. and Palmer, J. (1997) 'DECADE 2MtC', Energy and Environment Programme Environmental Change Unit, University of Oxford

Bryson, B. (1992) *Neither Here nor There: Travels in Europe*, Harper Collins Perennial, New York

Costanza, R., d'Arge, R., de Groot, R., Farberk, S., Grasso, M., Hannon, B., Limburg, K., Naeem, S., O'Neill, R., Paruelo, J., Raskin, R., Sutton, P. and van den Belt, M. (1997) 'The value of the world's ecosystem services and natural capital', *Nature*, 387, 15 May, pp. 253–260

Cradle to Cradle Products Innovation Institute (2011) 'A multi-attribute protocol', http://www.c2ccertified.org/index.php/product_certification/program_details, accessed 3 July 2012

European Commission (2011) *Communication from the Commission to the European Parliament, the Council, the European Economic and Social Committee and the Committee* of the Regions: Roadmap to a Resource Efficient Europe, com(2011) 571 final, 20 September, European Commission, Brussels

Frey, S. and Barrett, J. (2007) 'Our health, our environment: the Ecological Footprint of what we eat', paper A0001-33, presented at *Stepping Up the Pace: New Developments in Ecological Footprint Methodology, Applications,* International Ecological Footprint Conference, 8–10 May 2007, Cardiff

Global Footprint Network (2012a) *The National Footprint Accounts, 2011 edition,* Global Footprint Network, Oakland, CA

Global Footprint Network (2012b) 'Earth Overshoot Day', http://www.footprintnetwork.org/en/index.php/gfn/page/earth_overshoot_day/, accessed 3 July 2012

Grossman, E. (2007) *High Tech Trash: Digital Devices, Hidden Toxics, and Human Health,* Island Press/Shearwater Books, Washington, DC

Hamilton, C. and Denniss, R. (2005) *Affluenza: When Too Much is Never Enough,* Allen and Unwin, Sydney, Australia

Harre, N. (2011) 'Psychology for a better world: strategies to inspire sustainability', http://www.psych.auckland.ac.nz/uoa/home/about/our-staff/academic-staff/niki-harre/psychologyforabetterworld, accessed 14 July 2012

Hawken, P., Lovins, A. and Hunter Lovins, L. (1999) *Natural Capitalism: Creating the Next Industrial Revolution,* Little, Brown and Co., New York

Heinberg, R. (2007) *Peak Everything: Waking Up to the Century of Declines,* New Society Publishers, Gabriola Island, BC, Canada

Innovations (2012) 'Solar powered meerkat crystal ball – adorable garden light!', http://www.innovations.com.au/Product_Detail.aspx?ParentCategoryID=164&CategoryID=31&ProductID=111255&cm_mmc=pricecomp-_-myshop-_-Garden+%26+Outdoor+-%3e+Garden+Features+%26+Lighting-_-MKLIT&utm_source=myshopping&utm_medium=cpc&utm_campaign=Garden&utm_term=Meerkat+Solar+Ball+Light, accessed 3 July 2012

James, O. (2007) *Affluenza,* Vermilion, London

Johnson, M. (ed.) (1927) *More Maxims of Mark,* November, privately printed, New York

Kharecha, P.A. and Hansen, J.E. (2008) 'Implications of "peak oil" for atmospheric CO_2 and climate', *Global Biogeochemical Cycles,* 22, http://arxiv.org/ftp/arxiv/papers/0704/0704.2782.pdf, accessed 1 August 2012

Lebow, V. (1955) 'Price competition in 1955', *Journal of Retailing,* XXXI(1), Spring

McDonough, W. and Braungart, M. (2002) *Cradle to Cradle: Remaking the Way We Make Things,* North Point Press, New York

Maugeri, L. (2012) *Oil: The Next Revolution: The Unprecedented Upsurge of Oil Production Capacity and What It Means for the World,* Belfer Center for Science and International Affairs, Harvard Kennedy School, Cambridge, MA

NBR (2011) 'Group buying sites driving online shopping increase', *National Business Review,* 20 April, http://www.nbr.co.nz/article/group-buying-sites-driving-online-shopping-increase-hp-91264, accessed 26 January 2012

Nord, M. (2009) *Household Food Security in the United States, 2009,* Diane Publishing, Derby, PA

The Natural Step (undated) 'The four system conditions', http://www.naturalstep.org/the-system-conditions, accessed 3 July 2012

Transition Network (2012), 'TransitionNetwork.org', http://www.transitionnetwork.org/, accessed 3 July 2012

United Nations (2011) 'Gross National Happiness Index', http://www.uncsd2012.org/index.php?page=view&type=99&nr=266&menu=20, accessed 3 July 2012

US Census Bureau (2012) 'International Programs: International Data Base', Revised: April 30, 2012 Version: Data 12.0625, Code 12.0321, http://www.census.gov/population/international/data/idb/worldpoptotal.php, accessed 1 July 2012

Vale, R. and Vale, B. (2009) *Time to Eat the Dog? The Real Guide to Sustainable Living*, Thames and Hudson, London

Wilde, O. (undated) 'Oscare Wilde quotes', http://www.goodreads.com/author/quotes/3565.Oscar_Wilde, accessed 1 July 2012

5 The Dwelling

Nalanie Mithraratne

The ecological footprint of the dwelling is potentially quite a large contributor to the overall footprint. Based on existing studies the overall contribution from the dwelling and its use varies from as low as 8 per cent of the total (Close and Foran, 1998) to as high as 22 per cent (WWF Scotland, 2007) in different locations. Stechbart and Wilson (2010) in a recent study of Ontario, Canada estimated the dwelling contribution as 14 per cent of the total footprint, while Wiedmann *et al.*'s (2008) estimates for various locations in Australia range from 18 to 21 per cent. The lower estimate (8 per cent) by Close and Foran (1998) is based on 1993 dwelling occupancy in Canberra, Australia, which was 2.8 persons per dwelling. This is quite high compared to current estimates for average dwelling occupancy in the UK (2.3 in 2011; Department for Communities and Local Government, 2012: 2), Australia (2.5 in 2011; Department of Infrastructure and Transport, 2010) and in Scotland (2.2 in 2004; General Registrar Office for Scotland, 2006). The disparity in dwelling contribution to the total footprint may also be attributed to the different methodologies used by the authors, as there is no commonly agreed methodology for calculating the ecological footprint. While Close and Foran (1998) used process analysis, Wiedmann *et al.* (2008) used the economic input-output methodology to trace regional resource use. Both systems have inherent weaknesses, leading to inaccuracies which have been highlighted (Czamanski and Malizia, 1969; van den Bergh and Verbruggen, 1999; Lenzen and Murray, 2001). Despite these inaccuracies, ecological footprint is still a useful indicator of human impact. It could be argued that, globally, the dwelling contribution to the total footprint of wealthy nations is around 15 to 20 per cent of the total footprint if use of the dwelling is also included.

Apart from identifying the extent of the dwelling footprint it is also important to understand the factors contributing to it. If a life-cycle perspective is used, the relative contributions from construction, maintenance and demolition as well as the use of the dwelling are important to consider. In their detailed study, Close and Foran (1998) estimated the following contributions: construction, maintenance and demolition of dwelling, 31 per cent; use of the dwelling, 60 per cent; built-up land occupied by the dwelling and the surrounding garden, 10 per cent over a 50-year useful life for an Australian dwelling. Two recent studies of Cardiff and Aberdeen (WWF Cymru, 2005; WWF Scotland, 2007) estimated the use of the dwelling to contribute up to 89 per cent of the footprint of a dwelling over a 60-year period. The Cardiff study also estimated that the contribution to the total waste footprint from household renovation waste is negligible, at 0.3 per cent of the total. All this suggests that the use of the dwelling is the most significant phase of a dwelling's life and any measures to reduce the footprint need to target this phase.

Energy use is the main contributor to the use phase footprint of dwellings and both the type and the quantity of fuel used are important. In 1971, the average residential energy use in the UK was reported to be 21,770 kilowatt hours (kWh) per annum, which is very similar to the residential energy use in 2005, of 21,820 kWh per annum. The average residential energy use in the UK has declined since 2005, and in 2009 it was reported to be 19,020 kWh per annum (DECC and NSO, 2011: Table 3.4). Average residential energy use in Australia has remained constant, at 17 gigajoules (GJ) per person, since 1990 (DEWHA, 2008: ix) and the average residential occupancy is 2.6 persons per dwelling (NHSC, 2010: 139). Therefore the average residential energy use in Australia is 12,275 kWh per annum. Residential energy use in New Zealand is reported to be 11,410 kWh per annum (BRANZ, 2006: 17)

UK residential energy use seems to be about one-and-a-half times that of Australia and New Zealand. This could be due to the need for higher space conditioning energy due to colder and longer winters in the UK compared with shorter and milder winters in both Australia and New Zealand. On the other hand, New Zealand houses are well known for being underheated, cold and damp, with the mean living room temperature recorded as 17.8°C (BRANZ, 2006: 28). Therefore what is being compared is energy use to provide very different internal comfort conditions in the two locations. Cooking energy use in New Zealand is the highest of the three countries, which could be a result of different cooking and eating habits. The appliance energy use in Australia is similar to that of the UK. The energy use for refrigeration appliances is relatively high in the New Zealand house, at least partly due to continued use of faulty appliances. According to the *Household Energy End-use Project* (HEEP) study, 16 per cent of refrigeration appliances used in New Zealand houses are faulty (BRANZ, 2006: 62). Energy end-uses in British, Australian and New Zealand houses are shown in Table 5.1.

Reducing the Use-Related Dwelling Footprint by 50 Per Cent

What are the measures necessary to reduce the use-related dwelling footprint by 50 per cent in the UK, Australia and New Zealand? First of all, total residential energy use needs to be minimised as much as possible. In the three locations, energy uses other than cooking and lighting are

Table 5.1 Residential energy end-uses in UK, Australia and New Zealand

End-use category	Energy (kWh/year)					
	UK		Australia		New Zealand	
Space heating	11,602	61%	4,665	38%	3,879	34%
Water heating	3,424	18%	3,069	25%	3,309	29%
Cooking	571	3%	491	4%	685	6%
Lighting	752	4%	859	7%	913	8%
Refrigeration	691	4%	859	7%	1,141	10%
Appliances	1,980	10%	1,964	16%	1,483	13%
Standby			368	3%		
Total	19,020	100%	12,275	100%	11,410	100%

Sources: Data for the UK from DECC and NSO (2011); data for Australia from Department of Infrastructure and Transport (2010: 76); data for New Zealand from BRANZ (2006: 19)

responsible for close to 90 per cent of the total. Any attempts to reduce the cooking energy use, which is marginal at around 5 per cent, could simply displace the energy use from the residential sector to commercial catering services. It can be argued that, as a result of avoided waste heat, replacing incandescent lamps with energy efficient lamps can reduce the energy use by 75 per cent. However, this could also mean that the space heating energy requirement has to increase to compensate for the loss of free heat gains from lighting. Therefore the measures considered here are limited to energy uses other than cooking and lighting, such as space conditioning, water heating and appliances.

Reducing appliance energy use

Barrett *et al.* (2002: 98) estimated that 11 per cent of appliance energy use in the UK house is for standby, the energy used by appliances that are switched on but not in use. On this basis, annual standby energy use in the UK house is 218 kWh per annum. The data shown in Table 5.1 suggest that 3 per cent of the total energy use (368 kWh per annum) in Australian houses is for standby purposes. In New Zealand, an average standby power use of 61 watts was recorded by the HEEP study (BRANZ, 2006: iii). Therefore the annual standby energy use in New Zealand houses can be assumed to be 534 kWh per annum. Standby energy use can be avoided simply by switching off appliances at the wall when not in use.

The Energy Saving Trust (EST, undated: 13) in the UK identified that introducing new appliances can halve the electricity use for refrigeration appliances used in the UK house. Similarly the Energy Efficiency and Conservation Authority (EECA) in New Zealand estimates that replacing a 10-year-old refrigerator with a similar-capacity 3.5-star energy-rated model can reduce the electricity use for refrigeration in the New Zealand house by 50 per cent (EECA, 2012a). However, if these savings are to be realised it is essential that the old refrigeration appliances displaced by the new ones are disposed of rather than relegating them to garages and store rooms as a second refrigerator.

The energy use by the other household appliances can be reduced only by careful use and by limiting the number of appliances to those which are essential. But we keep buying more appliances, and new types that did not exist ten years ago. The Energy Saving Trust (EST, 2007: 3) in the UK has pointed out that by 2020 entertainment devices alone will use 45 per cent of electricity in the home.

Reducing water heating energy use

Water heating is a significant energy-use category in many households, irrespective of the geographic location. The hot water system consumes energy to heat water, maintain the storage temperature due to standing losses, and replace the hot water that has been used. A considerable amount of heat is lost through the hot water cylinder wall and distribution pipes. The higher the difference between the temperature of the hot water and the surrounding air the greater the losses. Standing losses from the typical hot water system in New Zealand houses have been estimated to be about 27 to 34 per cent of the total hot water energy consumption (Isaacs, 2004). Reardon *et al.* (2010: 200) estimated standing losses from a typical hot water system in an Australian house to be 30 per cent. According to DECC and NSO (2011: Table 3.15d), in 2009, 98 per cent of all residential hot water cylinders in the UK were insulated, although the level of insulation was not mentioned. Data on standing losses from the typical hot water system in the UK house were not available. Standing losses can be avoided by insulating the hot water distribution system and hot water cylinder, or replacing the existing cylinder with a factory-insulated cylinder with a high energy rating. This could reduce the water heating energy use by roughly 25 per cent.

Reducing the volume of water used can indirectly reduce the energy use for water heating. The Environment Agency (2008: 35) estimates that the UK household water use could be reduced by 28 per cent by introducing water metering, along with retrofitting existing WCs with variable flush devices and installing flow restrictors on taps. The Parliamentary Commissioner for the Environment in New Zealand suggests that technical measures such as dual-flush toilets, low-flow shower heads, front-loading washing machines and flow restrictors can reduce household water use per person by 50 per cent (PCE, 2001: 37). This reduction, however, has been calculated for a 5-person household, which represents only a small percentage of New Zealand houses and reflects both cold and hot water use. Mithraratne and Vale (2006) in their investigation of rain harvesting in New Zealand houses estimated a 15 per cent reduction in water demand due to technical measures. In Australia, the increasing water efficiency of clothes washers and shower heads coupled with declining household size is estimated to reduce the water heating energy use of an average Australian house by 9 per cent (DEWHA, 2008: 40, 50).

Another measure to reduce water heating energy is to use water at a lower temperature. Research by the Energy Saving Trust in the UK suggested that washing clothes at 40°C instead of 60°C could reduce the electricity use for clothes washing in a UK house by a third (EST, undated: 16). In 2009, UK household electricity use by washing machines and washer dryers was 262 kWh per annum (DECC and NSO, 2011: Table 3.10) Therefore the energy saved by using a lower temperature for clothes washing is 87 kWh per annum.

Cold water clothes washing is widespread in both Australia and New Zealand. In the Australian house on average 70 per cent of the total washing loads use the cold water cycle (DEWHA, 2008: 91). The average front-loading washing machine uses 55 per cent less energy compared to the average top loader, which is common in Australasia. However, at least half the front-loading clothes washers in the Australian market are reported to use a minimum temperature of 30°C in all wash cycles, using an in-built water heater. Therefore, with the increased use of front-loading washing machines in Australia, the energy use for clothes washing is likely to increase.

Table 5.2 shows the water heating energy savings that may be achieved in the three locations, if water savings are considered to be representative of energy savings.

According to EECA (2012b) solar water heating systems can supply 50 to 75 per cent of the total hot water requirement of a New Zealand house. In Australia, depending on the location of the house, 90 per cent of the hot water requirement can be supplied using solar water heating (Reardon *et al.*, 2010: 198). Ekins-Daukes (2009) estimated that at least 70 per cent of domestic hot water requirements can be supplied by solar water heating systems in the UK. However, the embodied energy of the solar water heating system will negate part of the energy savings achieved from employing solar water heating. Therefore, an efficient solar water heating system coupled with demand management measures and a well-insulated hot water distribution system can potentially eliminate much of the non-renewable energy demand for hot water in all houses.

Table 5.2 Water heating energy savings in UK, Australia and New Zealand

Strategy	Energy (kWh/year)					
	UK		Australia		New Zealand	
Average energy use	3,424	100%	3,069	100%	3,309	100%
Eliminate standing losses	—	—	−767	−25%	−827	−25%
Reduce water volume	−959	−28%	−276	−9%	−496	−15%
Lower water temperature	−87	−3%	—	—	—	—
Balance	2,378	69%	2,026	66%	1,986	60%

Reducing space heating energy use

Reducing the space heating energy use of New Zealand houses is problematic as houses in New Zealand are generally underheated. Any energy savings due to increased insulation levels and weather tightening could therefore be expected to be used as thermal comfort improvements by the occupants. In Australia, improvements to the building fabric in terms of ceiling insulation, double glazing, installing draught seals and shading devices are expected to reduce space heating energy use by 28 per cent (DEWHA, 2008: 42).

Yates (2006) reported that in the UK a 60 per cent reduction in energy use was achieved in tenement flats by an improvement package that includes insulating roof and walls, double glazing for windows, draught sealing for doors and gas central heating for space and water heating. It is also suggested that a 35 per cent reduction in space heating energy could be achieved in detached or semi-detached houses by improving insulation, installing new windows and wooden door frames and sealing suspended timber ground floors, coupled with repairing defects in plaster (Bell and Lowe, 2000). In 2005, detached and semi-detached houses and flats represented 53 per cent and 17 per cent, respectively, of the total housing stock in the UK (Ravetz 2008: 4464). The energy use of flats is likely to be lower than the average for the housing stock, while energy use of detached and semi-detached houses is likely to be higher than the average. Therefore the reduction in space heating energy use by upgrading the stock with similar measures is likely to be lower. If 60 per cent reduction in energy use from the average is assumed for the 17 per cent of the stock which are flats and a 35 per cent reduction in space heating energy use from the average is assumed for the rest of the stock, in the absence of better data, the space heating energy use is reduced to 7,048 kWh per annum.

Table 5.3 shows the impact of implementing the measures identified for houses in the three locations. If reduction in energy use is equated to reduction in ecological footprint these efficiency measures can reduce the dwelling use-related footprint by about 40 per cent in the locations considered.

However, in addition to the amount of energy, the fuel mix for generating energy also has an impact on the ecological footprint. The footprint of different domestic fuels according to Barrett *et al.* (2002) is shown in Table 5.4. Fossil fuels release different quantities of CO_2, and the ecological footprint is the area of land required to absorb the emission. The estimates shown in Table 5.4 are made on the basis that 5.2 hectares of newly planted forest can sequester 1 tonne of CO_2 per annum and that in terms of productivity a hectare of energy land is similar to 1.78 hectares of global land. Renewable energy sources are significantly lower in ecological footprint compared to fossil fuels.

Table 5.3 Residential energy end-uses that may be realised in UK, Australia and New Zealand

End-use category	Energy (kWh/year)		
	UK	Australia	New Zealand
Space heating	7,048	3,359	3,879
Water heating	—	—	—
Cooking	571	491	685
Lighting	752	859	913
Refrigeration	342	430	571
Appliances	1,762	1,964	949
Standby	—	—	—
Total (% reduction)	10,475 (45%)	7,103 (42%)	6,997 (39%)

Table 5.4 Ecological footprint of delivered energy from different fuel types. In the case of fuels used for combustion the ecological footprint will be affected by the efficiency of the heater

Fuel	CO_2 emissions (kg/100 kWh)	Ecological footprint	
		(gha/100 kWh)	(ha/GWh or m^2/100 kWh)
Non-renewable			
UK electricity	43	0.0147	147
Coal	30	0.0103–0.012	103–120
Oil	25	0.0086–0.01	86–100
Petrol	24	0.0082–0.01	82–100
Natural gas	19	0.0065–0.008	65–80
Renewable			
Solar photovoltaic	—	0.0024	24
Solid biomass	2	0.001	10
Wind	—	0.0006	6
Average hydro	—	0.00036	3.6

Sources: Data from Barrett et al. (2002: 46–47) and WWF Cymru (2005: 73); data for average hydro from Wackernagel and Rees (1996: 69)

Coal, which is the non-renewable fuel with the largest ecological footprint (excluding electricity), has a footprint around 30 times greater than hydro power generation, the renewable energy source with the smallest ecological footprint. Gas, which has the lowest ecological footprint of the fossil fuels, is the predominant fuel source used in the UK for domestic space heating. In 2009, gas supplied 83 per cent of UK domestic space heating, oil 9 per cent, electricity 5 per cent and solid fuel (including coal) only 3 per cent (DECC and NSO, 2011: Table 3.7) If gas were to be replaced by solid biomass the footprint related to space heating could be reduced to around a seventh of its original value.

How Can the Impact of Location on the Dwelling Footprint be Minimised?

Although home owners can reduce their footprint through efficiency measures such as insulation and solar water heating, a house connected to common services available at its location, such as electricity, water supply and waste water management, is responsible for its share of these services. If the grid electricity supply is considered, as an example of the impact, the footprint depends on the fuel mix used for generation. The footprint of electricity in the three countries considered here is shown in Table 5.5. The electricity generation mix varies between states in Australia, where electricity (46 per cent of the total) is a notable contributor to the domestic energy mix (DEWHA, 2008: ix). The footprint of Victorian grid electricity (almost all from coal) is four times that of Tasmanian grid electricity (mainly hydroelectric). According to Wiedmann *et al.* (2008), dwellings in Victoria contribute 21 per cent to the total footprint while dwellings in Australia on average contribute only 18 per cent to the total. This variation could partly be explained by the electricity generation mix, although behaviour and attitude of occupants and other, construction-related factors are also underlying causes.

As shown in Table 5.1, the current annual energy use of an Australian house is 12,275 kWh per annum, and 5,647 kWh per annum (46 per cent) of this is supplied by electricity. Depending on whether the house is located in Tasmania (0.01 global hectares [gha] per 100 kWh) or the State of Victoria (0.04 gha per 100 kWh), the footprint of electricity use varies from 0.58 gha per annum to 2.33 gha per annum, respectively. However, if the house generates all of its annual electricity needs on-site using solar photovoltaics, or if it obtains all of its electricity from grid-scale wind, the footprint of the electricity use would reduce to 0.034 gha per annum and 0.136 gha per annum, respectively, regardless of its location.

Table 5.5 Ecological footprint of electricity in the UK, Australia and New Zealand

Country	CO_2 emissions (kg/100 kWh)	Ecological footprint (gha/100 kWh)	(ha/GWh or m^2/100 kWh)
UK, average	43*	0.0147	147
Australia	30–121**	0.0103–0.0414	103–414
New Zealand, average	19.5***	0.0067	67

* Barrett *et al.* (2002: 46)
** DCCEE (2011: 20)
*** MfE (2009: 12)

A study by Mithraratne and Vale (2007) investigating the use of rain tanks in New Zealand houses reported the emissions associated with supplying 193 cubic metres of water a year as 34, 30 and 65 kilograms of CO_2 (kg CO_2) using mains supply, a concrete rain tank and a plastic rain tank, respectively. These values can be converted to footprint using the same method as for energy. However, the volume of water needed for a similar service from the mains supply and from a rain-tank system is different. In New Zealand, for each 100 cubic metres of water supplied using the mains network in the Auckland area, 12 cubic metres are lost due to network leakages and fire demand. When using rain-tank systems for total residential water supply, the supply volume is reduced, leading to further reductions in the water system operating requirements, i.e. electricity for pumping. Therefore using a concrete rain-tank system for water supply could reduce the footprint of water supply to 77 per cent of that of being connected to the mains supply. However, it is clear from the CO_2 emissions figures that the footprint of water supply, at least in a New Zealand city, is extremely low: using mains water the annual emission for a household is 34 kg CO_2 whereas for household electricity, even using New Zealand electricity which has a relatively low footprint, the annual emission is close to 2,000 kg CO_2. This applies irrespective of how the water is supplied, and therefore in this case water is not a priority for overall dwelling footprint reduction.

The Role of New Dwellings

Much of the focus of current efforts to reduce the environmental impacts of dwellings seems to be focused on employing more efficient construction materials and methods in new constructions. However, the dwelling statistics from the three countries considered shown in Table 5.6 suggest that new constructions represent less than 2 per cent of the total stock. This implies that the majority of dwelling stock already exists and the overall impact due to dwellings depends far more on the existing stock and how it is operated and maintained than on the new constructions, however efficient they may be.

The Energy Saving Trust in the UK (EST, undated: 14) has argued that new housing built to a SAP 80 rating ('SAP' is the Standard Assessment Procedure for energy performance approved by the UK government)

Table 5.6 Housing stock characteristics in the UK, Australia and New Zealand

Country	Existing stock (millions)	Annual additions	Additions as a percentage of the total
UK	21.87	174,900	0.80%
Australia	9.0	130,900	1.45%
New Zealand	1.6	20,000	1.25%

Sources: Data for the UK from Ravetz (2008); data for Australia from NHSC (2010); data for New Zealand from NZBCSD (2008)

instead of the average of 42 (in 2006) can reduce space heating energy use by 75 per cent. However, other data sources (DECC and NSO, 2011; Power, 2008; Ravetz, 2008) suggest that the average SAP rating of the housing stock in 2006 in the UK is 49. Therefore the actual savings would be lower than estimated. Even if the savings are assumed to be 75 per cent, based on the data shown in Table 5.6, this is only applicable to 0.8 per cent of the total stock and therefore highly marginal in its overall impact.

There is heightened awareness of the embodied energy associated with building materials, and choice of sustainable materials for new constructions is often discussed. Generally, the brick and tile houses with concrete floors common in the UK tend to have higher initial embodied energy compared with the timber-framed houses common in Australia and New Zealand. Mithraratne *et al.* (2007) estimated lifetime embodied energy of New Zealand houses to vary from 4.5 gigajoules per square metre (GJ/m^2) for light timber-framed, timber-clad houses with raised timber floors to 4.3 GJ/m^2 for light timber-framed brick veneer-clad houses with raised timber floors. The difference in the values is due to the lower maintenance needs of the brick veneer house. The lifetime space heating energy of the two houses, however, is estimated to be 12.4 GJ/m^2 and 11.8 GJ/m^2, respectively, which is roughly three times the embodied energy. If the energy used for water heating, cooking, lighting and appliances is added to the total, the embodied energy becomes an even smaller part of the whole.

In a study conducted in the UK (EHA, 2008) which considered carbon emissions rather than energy, it was also found that, if the total CO_2 emissions from construction, maintenance and operation are considered over a period of 50 years, embodied CO_2 emissions are only 475 kilograms per square metre (kg/m^2) or 28 per cent of the total of 1,700 kg/m^2. Therefore, design strategies such as better siting and higher levels of insulation to reduce the operating energy are much more influential on overall ecological footprint than the choice of construction materials. However, it also needs to be noted that, as dwellings become more energy efficient in their overall design, more of the responsibility for the size of the residential footprint is transferred from the dwelling and its design and construction to the users and their choices of appliances and lifestyle. The entertainment systems of a household with two teenagers, each with their own set of entertainment devices in their own bedrooms, could be using 12 gigajoules of electricity a year just for entertainment (Vale and Vale, 2009: 218–219), which is about 1,300 kilograms of CO_2 per year.

There is a global tendency for new houses to be larger in size, although the number of occupants continues to decline. In their investigation of house size using the most common construction type used in New Zealand, Mithraratne *et al.* (2007) found that, with 55 per cent and 100 per cent increases in floor area, the lifetime energy increased by 37 per cent and 79 per cent, respectively. However, this estimate included embodied energy and space heating energy only. In addition to increased space heating energy use, larger houses also tend to use more lighting, appliances, and so on, leading to further increases in lifetime energy. Having a smaller house could also be beneficial in increasing the area of the site available for growing food and generating energy on-site, which could further reduce the dwelling footprint, as discussed earlier. In terms of housing, small is definitely beautiful.

References

Barrett, J., Vallack, H., Jones, A. and Haq, G. (2002) *A Material Flow Analysis and Ecological Footprint of York, Technical Report*, Stockholm Environment Institute, York, England

Bell, M. and Lowe, R. (2000) 'Energy efficient modernisation of housing: a UK case study', *Energy and Buildings*, 32(3), pp. 267–280

BRANZ (2006) *Energy Use in New Zealand Households: Report on the Year 10 Analysis for the Household Energy End-use Project (HEEP)*, BRANZ Ltd, Porirua, New Zealand

Close, A. and Foran, B. (1998) *Canberra's Ecological Footprint Part 3*, CSIRO, Canberra, Australia

Czamanski, S. and Malizia, E. (1969) 'Applicability and limitations in the use of national input output tables for regional studies', *Papers of the Regional Science Association*, 23, pp. 65–77

DCCEE (2011) *National Greenhouse Account Factors*, Australian Government Department of Climate Change and Energy Efficiency, Canberra

DECC and NSO (2011) *Energy Consumption in the UK: Domestic Data Tables 2011 Update*, Department of Energy and Climate Change and National Statistics Office, London

Department for Communities and Local Government (2012) 'EHS Results Published', *EHS Bulletin*, 7, pp. 1–12

Department of Infrastructure and Transport (Australian Government) (2010) *State of Australian Cities 2010*, Major Cities Unit, Department of Infrastructure and Transport, Canberra

DEWHA (2008) *Energy Use in the Australian Residential Sector 1986–2020*, Department of the Environment, Water, Heritage and the Arts, Canberra

EECA (2012a) 'Fridges and freezers', http://www.energywise.govt.nz/node/18143, accessed 13 April 2012

EECA (2012b) 'Solar water heating', http://www.energywise.govt.nz/how-to-be-energy-efficient/your-house/hot-water/solar-water-heating, accessed 13 April 2012

EHA (2008) *New Tricks with Old Bricks*, Empty Homes Agency, London

Ekins-Daukes, N. (2009) *Solar Energy for Heat and Electricity: The Potential for Mitigating Climate Change*, Grantham Institute for Climate Change, Imperial College London, Briefing paper No. 1, pp. 1–12

Environment Agency (ed.) (2008) *Greenhouse Gas Emissions of Water Supply and Demand Management Options*, Environment Agency, Bristol, England

EST (undated) *The Rise of the Machines: A Review of Energy Using Products in the Home from the 1970s to Today*, Energy Saving Trust, London

EST (2007) *The Ampere Strikes Back: How Consumer Electronics are Taking Over the World*, Energy Saving Trust, London

General Register Office for Scotland (2006) 'News Release: average household size to fall below two people', http://www.gro-scotland.gov.uk/press/news2006/average-household-size-to-fall-below-two-people.html, accessed 7 July 2012

Isaacs, N. (2004) 'Supply requires demand: where does all of New Zealand's energy go?' in *Royal Society of New Zealand Conference*, Christchurch, New Zealand

Lenzen, M. and Murray, S. (2001) 'A modified ecological footprint method and its application to Australia', *Ecological Economics*, 37, pp. 229–255

MfE (2009) *Guidance for Voluntary Corporate Greenhouse Gas Reporting*, Ministry for the Environment, Wellington, New Zealand

Mithraratne, N. and Vale, R. (2006) 'Life-cycle impact of water supply system selection on typical New Zealand Houses', paper presented at the *5th Australian Life Cycle Assessment Conference*, 22–24 November, Melbourne

Mithraratne, N. and Vale, R. (2007) 'Sustainable choices for residential water supply in Auckland', paper presented at *Talking and Walking Sustainability, 2nd Conference of New Zealand Society for Sustainability Engineering and Science*, 20–23 February, Auckland

Mithraratne, N., Vale, B. and Vale, R. (2007) *Sustainable Living: The Role of Whole Life Costs and Values*, Butterworth-Heinemann, Oxford, England

NHSC (2010) *National Housing Supply Council: 2nd State of Supply Report*, National Housing Supply Council, Canberra

NZBCSD (2008) *Better Performing Homes for New Zealanders: Making It Happen*, New Zealand Business Council for Sustainable Development, Wellington

PCE (2001) *Ageing Pipes and Murky Waters: Urban Water System Issues for the 21st Century*, Parliamentary Commissioner for the Environment, Wellington, New Zealand

Power, A. (2008) 'Does demolition or refurbishment of old and inefficient homes help to increase our environmental, social and economic viability?' *Energy Policy*, 36, pp. 4487–4501

Ravetz, J. (2008) 'State of the stock—What do we know about existing buildings and their future prospects?' *Energy Policy*, 36, pp. 4462–4470

Reardon, C., Milne, G., McGee, C. and Downton, P. (2010) *Your Home Technical Manual*, Department of Climate Change and Energy Efficiency, Canberra

Stechbart, M. and Wilson, J. (2010) *Province of Ontario: Ecological Footprint and Biocapacity Analysis*, Global Footprint Network, Oakland, CA

Vale, R. and Vale, B. (2009) *Time to Eat the Dog? The Real Guide to Sustainable Living*, Thames and Hudson, London

van den Bergh, J. and Verbruggen, H. (1999) 'Spatial sustainability, trade and indicators: an evaluation of the "ecological footprint"', *Ecological Economics*, 29, pp. 61–72

Wackernagel, M. and Rees, W. (1996) *Our Ecological Footprint: Reducing Human Impact on the Earth*, New Society Publishers, Gabriola Island, BC, Canada

Wiedmann, T., Wood, R., Barrett, J., Lenzen, M. and Clay, R. (2008) *The Ecological Footprint of Consumption in Victoria*, Stockholm Environment Institute, York, England and Centre for Integrated Sustainability Analysis, Sydney, Australia

WWF Cymru (2005) *Reducing Cardiff's Ecological Footprint: A Resource Accounting Tool for Sustainable Consumption*, March, WWF Cymru, Cardiff, Wales

WWF Scotland (2007) *Scotland's Global Footprint: Reducing Our Environmental Impact, Final Report*, WWF Scotland, Dunkeld, Scotland

Yates, T. (2006) *Sustainable Refurbishment of Victorian Housing*, BRE Press, Building Research Establishment, Watford, England

6 Tourism

Abbas Mahravan

Nowadays, tourism is viewed as one of the most important forces influencing the environmental, social and economic development of its host destinations. As the United Nations World Tourism Organization (UNWTO) has determined, as an export category at the global scale, tourism ranks fourth after fuels, chemicals and automotive products. UNWTO also demonstrates a virtually uninterrupted growth in international tourist arrivals, from 25 million in 1950, to 277 million in 1980, to 435 million in 1990, to 675 million in 2000, and 940 million in 2011. It seems that over 13 per cent of the world's population are now international tourists. In addition the contribution to employment made by tourism is high, estimated to be in the order of 6 to 7 per cent of all available jobs worldwide (direct and indirect; UNWTO, 2011: 2).

The environmental, social and economic influences exerted by tourism on its host destinations can be categorized into their negative and positive aspects. The negative impacts relate to the uncontrolled development of tourism, especially its environmental impacts, whereas its positive influences can be a boost to the local economy and local development.

The negative impacts of uncontrolled development associated with the products and activities of tourism and tourists have been discussed by many researchers, for example Coppock (1982: 272), Cohen (1978: 220) and Gossling (1999: 310). These include threats to the survival of wildlife; environmental destruction; loss of and damage to habitat caused by the pressure of human feet or vehicles on soil and vegetation; damage to or destruction of flora and fauna by fire and pollution; disturbance to fauna, especially birds and mammals; and consumerism and extremely high per capita demand for resources. Impacts such as increasing the ecological footprint of the host destination because of the food consumed, transportation used, spread of built-up land, water consumed and sewage generated can be added to these.

However, researchers such as Weaver and Lawton (1999: 16) believe that some sectors of mass tourism, including alternative tourism and ecotourism, can be seen as legitimate forms of tourism, and the only ones that can be considered sustainable. They argue that special consideration should be given to ecotourism, which is widely defined as a variant of alternative tourism that puts primary emphasis on the natural environment to be used and protected as the basis for the attraction of a particular 'tourism product'. In addition, they see ecotourism as a positive interaction with the natural environment through appreciation of it or learning more about it, not just using it as a background for more conventional tourist activities such as sunbathing or white-water rafting and other adventure sports.

A Comprehensive Framework for Sustainable Development of Tourism

Under the influence of phenomena such as global warming and degradation of environmental resources and the resulting evolution of sustainability policies, strategies for tourism development have to become more comprehensive, to cover multi-dimensional aspects of sustainability. Mahravan (2012), through the investigation of many existing frameworks and agendas, has proposed a comprehensive framework for the sustainable development of ecotourism. This framework is based on the idea that the main environmental, socio-cultural and economic outcomes of sustainable development through ecotourism are productive activities. Ecological, cultural and social-ecological economic indicators can be used as tools to evaluate the sustainability of ecotourism and its related activities and products. Also proposed is a holistic strategy for tourism that can link in a sustainable way environmental conservation, the socio-cultural behaviour of both host people and visitors, and economic development. This strategy is the result of the proposed comprehensive framework and determines the interaction between socio-cultural products and activities (produced, consumed or conducted through ecotourism activities) and their environmental and economic influences on the host destinations.

The comprehensive framework is made up of three subsectors, comprising environmental, cultural and economic frameworks for sustainability through development of ecotourism. The integration of the three subsectors gives rise to a range of anticipated environmental, socio-cultural and economic outcomes for the sustainable development of ecotourism. It is thus possible to use the framework for evaluation of tourism and its related products and activities in terms of the extent to which they can be considered sustainable. The following discussion will explain each of the environmental, cultural and economic components of this comprehensive framework and the linkages between them.

Subsector 1: Ecological Framework

Table 6.1 shows the anticipated ecological outcomes which can be achieved from the sustainable development of tourism. These goals are suggested in

Table 6.1 Ecological outcomes of development of sustainable tourism

Ecological outcomes	Productive activities
Environmental awareness	Engagement of local and indigenous people as well as all related organizations (individuals, government and NGOs) in environmental education
Engagement of local and indigenous people in the conservation process	Involvement of local and indigenous participants in the development process as employed or volunteer staff
Attention to carrying capacity and sustainable yields	Striking a balance between carrying capacity, resource consumption and the ecological footprint of tourism activities, products and services
Protection of environmental resources and maintenance of biodiversity	Implementation of policies to reduce the ecological footprint of tourism activities, products and services
Use of renewable resources to generate energy used by tourism products and services	Enhancing use of green technologies based on renewable resource consumption and decreasing demands for fossil fuels to generate energy

Source: Mahravan (2012)

an attempt to progress the attitude of the tourism industry toward the environment from one which is merely economically exploitative to one of stewardship for natural and environmental conservation.

As shown in Table 6.1, awareness among local and indigenous people of the values attached to their environment, as well as their engagement in the natural conservation process, are considered goals which can be achieved through a community approach to tourism development. Furthermore, these two prime goals of awareness and engagement play a pivotal role as the basic principles for the realization of the other forecasted outcomes. In this approach, education as a potential main tourism activity has a close and effective relationship with all the other sustainable tourism activities shown in Table 6.1.

Ecological indicators

Ecological indicators have been identified as a means of assessing the condition of the environment. Current environmental problems are caused not only by using technologies dependent on fossil fuels but also by social-cultural patterns of consumption of environmental resources (see Chapter 1). The solutions to these environmental problems must therefore be considered not only as technological but also as social-cultural policies. As Azar *et al.* (1996: 89) reveal, most sets of ecological indicators proposed so

Table 6.2 Ecological indicators for evaluation of the environmental impacts of ecotourism development on host destinations

Ecological outcome	Indicator	Definition
Environmental awareness	1a	The number of local people who participate in the educational process
Engagement of local and indigenous people in the conservation process	1b	The number of people who engage in the environmental conservation process, including volunteer or employed participants
Attention to carrying capacity and sustainable yields	1e	The ecological footprint of tourism activities, products and services
Protection of environmental resources and maintenance of biodiversity	1e	The ecological footprint of tourism activities, products and services
Use of renewable resources to generate energy used by tourism products and services	1e	The ecological footprint of tourism activities, products and services

Source: Mahravan (2012)

far have focused on the state of the environment rather than on the interactions between society and ecosystems. Thus, ecological indicators need to focus on societal activities and interactions between nature and society through the use of materials and energies. This is where the ecological footprint (EF) is so valuable.

As shown in Table 6.2, the ecological footprint of tourism activities, products and services can be used as a main ecological indicator for evaluating the environmental impacts of tourism development on host destinations, as a means of achieving the ecological goals of tourism shown in Table 6.1. In Table 6.2, Indicators '1a' and '1b', which show the number of participants in environmental education and the conservation process, are also introduced as social-cultural indicators. These can be viewed as complementary to EF as indicators for evaluating the success of the social dimension of environmental conservation principles.

Subsector 2: Cultural Framework

Mahravan and Vale (2011) have introduced a cultural framework for ecotourism, together with the related indicators and required data. This framework can be used to evaluate the cultural footprint of ecotourism through sustainable development at a host destination.

As defined in the *Mexico City Declaration on Cultural Policies* in 1982 (UNESCO, 1982), 'culture' means the distinctive spiritual and physical, intellectual and affective traits characterizing a society or a social group. It is the sum total of the ways in which a group builds up a pattern for living that is transmitted from one generation to another. Cochrane (2006: 322) argues that culture comprises explicit and implicit patterns of behaviour that are passed on by symbols, constituting the distinctive achievement of human groups, including their embodiments in artifacts. The essential core of culture consists of traditional (i.e. historically derived and selected) ideas, and especially their attached values. Cultural systems may on the one hand be considered products of action, and on the other as conditioning elements of further action.

From an ecological perspective, culture is a system which, as Rapoport (1969) demonstrates, has an interconnection with the environment. Culture can be manifested in the religious beliefs, intellectual and spiritual engagement, materials and products (such as art, architecture, food and textiles) of a given group or society. If a society is viewed as an organism that lives in an environment, culture can be defined as the way in which a society physically and spiritually makes a linkage between itself and its surroundings through using materials and resources or conducting activities for living.

Through investigation of the 2009 *UNESCO Framework for Cultural Statistics* (FCS; cited in UIS, 2010), the *Cultural Indicators for New Zealand* (Ministry for Culture and Heritage, 2009) and other proposed cultural frameworks such as Choi and Sirakaya's (2006) model, Mahravan (2012) proposed a cultural framework that can act as a component of a comprehensive framework for the sustainable development of tourism. This framework also determines the main cultural outcomes for tourism, the productive activities that can contribute to achieving these outcomes and a set of related cultural indicators that can be used as a tool to evaluate the development of culturally sustainable tourism (Tables 6.3 and 6.4). In this view, as defined by UIS (2010), cultural tourism is a customized excursion into other cultures and places to learn about the host society in terms of its environmental and cultural heritage and products, lifestyle, historical context and socio-cultural values. Accordingly, this definition can also cover spiritual tourism or ecological tourism (ecotourism).

This definition shows that learning about culture is a core activity for cultural tourism, and all cultural activities and related goods and services can be used in the process of learning or education through the development of tourism. On the other hand, education can be conducted in different ways, for example by learning about culture through face-to-face contact with local people, visiting museums or direct interaction with architectural spaces, or by using photography as a tool to record cultural events. These characteristics of education allow it to be viewed as part of a comprehensive framework which can be used to define cultural indicators and evaluate tourism and its sub-segments, such as ecotourism, as culturally sustainable.

Table 6.3 Cultural outcomes of sustainable development of tourism based on cultural education

---

Awareness of local participants of their existing cultural heritage and methods of protecting this through the educational process

Participation of local people in the cultural development process (programming, management and monitoring for the production, consumption and presentation of cultural products)

Democratized environment for participation of people in cultural development, with equal access to cultural sources

Culture-based development of economic systems that guarantee equal distribution of cultural capital and income among local people

Combination of ordinary tourism activities with compatible cultural activities

Protection and restoration of tangible cultural heritage

Protection and restoration of intangible cultural heritage (oral traditions and expressions, rituals, languages, social practices)

Tourist experience of (preferably) authentic culture or (alternatively) staged authentic cultural heritage of host destinations

---

Sources: Mahravan and Vale (2011); Mahravan (2012)

Cultural indicators

Table 6.4 presents the cultural indicators for the development of sustainable tourism. Factors that form a framework for choosing these indicators are data availability, measurability and international comparability (Mahravan, 2012).

The suggested cultural indicators can be categorized into the three themes of:

a   The *number* of local and indigenous people who participate in the development process, including through education, management, monitoring, protection of cultural heritage and economic activities (1a, 2a, 3a)
b   The *quantity* of tools, goods, services and places which are used or protected during the development process (1b, 1c, 3c, 4a, 5a, 6a, 7a)
c   The *economic benefits* which participants earn from tourism development based on cultural education (3b).

As shown in Table 6.4, the proposed cultural indicators attempt to assess quantitatively the engagement of local and indigenous people in the sustainable development process through assessing cultural activities and products related to cultural education-related tourism. The integrated findings of these cultural indicators create a base for evaluating whether

Table 6.4 Proposed cultural indicators for development of sustainable tourism based on cultural education

Cultural outcome	Indicator	Title	Definition
Awareness of local participants of their existing cultural heritage and methods of protecting this through the educational process	1a	Cultural education	The number of local people who participate in the educational process
	1b	Educational places	The areas, buildings and related infrastructure which are used for education
	1c	Educational tools and equipment	The tools and equipment that are used for education
Participation of local people in the cultural development process (programming, management and monitoring for the production, consumption and presentation of cultural products)	2a	Participation in the cultural development process	The number of people who engage in the cultural development process including volunteer or employed participants
Culture-based development of economic systems that guarantee equal distribution of cultural capital and income among local people	3a	Employment of local people	The number of local people employed in the culture-based economic system
	3b	GDPs*	Sustainable portion of GDP
	3c	Restored and protected cultural heritage (this indicator covers indicators 5a and 6a)	The quantity of the intangible and tangible cultural heritage that is protected or restored through development of tourism
Combination of ordinary tourism activities with compatible cultural activities	4a	Local, national or international cultural products	The quantity of local products which are combined with national or international products
Protection and restoration of tangible cultural heritage	5a	Restored tangible cultural heritage	The quantity of tangible cultural heritage that is protected or restored through development of tourism
Protection and restoration of intangible cultural heritage (oral traditions and expressions, rituals, languages, social practices)	6a	Products related to intangible cultural heritage	Cultural goods and services that contribute to restoration of intangible heritage
Tourist experience of (preferably) authentic culture or (alternatively) staged authentic cultural heritage of host destinations	7a	Authenticity	Original cultural products that are labelled as local products

* GDPs – sustainable Gross Domestic Product

policies and practices can be considered culturally appropriate through collecting and analysing data related to each indicator.

Subsector 3: Economic Framework

As Herman Daly, a former Senior Economist at the World Bank, has said, the economy is totally supported by the biosphere (Daly, 2005). The economy needs to have a social and ecological interaction with society and its surrounding environment. This model considers such an economy to be sensitive to the social-ecological footprint of its product and outcomes in terms of the degree to which they are socially appropriate and environmentally friendly. Following this social-ecological perspective, the United Nations Environment Programme (UNEP; 2011: 1) defines the main outcomes of a 'green' economy as improving human well-being and social equality while significantly reducing ecological destruction and scarcities. In this definition, the prefix 'green' does not merely put emphasis on being ecologically friendly, and it can be assumed to be the same as the prefix 'sustainable' for an economy. Relying on these definitions of a green or sustainable economy, the main outcomes of sustainable economic development can be categorized into the three areas of social-cultural, ecological and economic outcomes, which are linked together in a sustainable way. The first two of these are shown in Table 6.5, and the third, the economic outcome, is discussed separately.

Social-cultural outcomes

The main social-cultural outcomes of sustainable economic development, as pointed out by researchers such as Costanza (2009: 20), are opportunities for the involvement of local participants in activities that achieve economic growth and equal distribution of capitals among all components of a society. This can be viewed as a strategy that contributes to horizontal economic development through tourism.

Table 6.5 Anticipated social-cultural and environmental outcomes from sustainable economic development

Social-cultural outcomes	Environmental outcome
Engagement of local and indigenous people in the economic development process	Conservation of environmental resources through a social-economic educational process
Development: arts, culture and heritage make a growing contribution to the economy	

Source: Mahravan (2012)

The second cultural outcome for sustainable economic development relies on the strategy that results from the conservation of cultural heritage, which provides the people engaged with an opportunity to present their cultural products and capitals. This has been conceptualized in Table 6.5 as 'Development: arts, culture and heritage make a growing contribution to the economy'.

Environmental outcomes

A sustainable economic strategy relies on the idea that conventional economic systems are the cause of environmental problems for people through degradation of natural resources. However, these problems can be solved through changing social behaviours in terms of patterns of resource consumption. As the Organisation for Economic Co-operation and Development (OECD; 2008) demonstrates, the world's problems cannot be solved through economic growth alone, and all economic, social and ecological dimensions of any activity or product are interconnected. Considering only one of these aspects at a time means ignoring other aspects of the sustainable environmental outcomes of an economic system. From an ecological viewpoint, the main anticipated environmental outcomes of sustainable economic development can be summarized as 'Conservation of environmental resources through a social-economic educational process' (Table 6.5).

Sustainable portion of GDP (GDPs) as an ecological-economic indicator

As described, the main outcomes of sustainable economic development can be categorized into the three areas of socio-cultural, ecological and economic outcomes. Table 6.5 shows the social-cultural and environmental outcomes, but the economic outcome is more difficult to measure. The shortage of appropriate economic indicators can be seen as one of the critical issues for measurement of whether an activity (such as ecotourism) can be considered to be economically sustainable. As many researchers and related organizations, such as the New Zealand Council of Trade Unions (NZCTU; 2010: 22) and the United Nations Environment Programme (UNEP; 2011: 26), have shown, the use of conventional economic indicators such as GDP and other macroeconomic aggregates can lead to a distorted picture of economic performance, particularly because such measures do not reflect the extent to which production and consumption activities may be drawing down natural capital.

Many attempts have also been made to introduce economic indicators that include the social and ecological dimensions of sustainable economic development, but the results contain potential weaknesses. For instance, NZCTU (2010: 21) proposes taxation as a strategy for monitoring and controlling the environmental degradation caused by economic development. It suggests that polluters should face taxes on their emissions, including greenhouse gas emissions, with the aim of paying the costs of all significant 'externalities' (side-effects such as pollution and global warming). How the level of environmental pollution or degradation can be measured and costed is a question still to be answered.

Measurement of the sustainable portion of GDP

Mahravan (2012) attempted to develop an economic indicator that shows economic progress and provides a comprehensive measure of well-being and environmental sustainability. To achieve this aim, an economic indicator is needed which is as clear and appealing as GDP but which is more inclusive of other dimensions of progress, in particular its environmental and social aspects.

Development of local domestic products can be viewed as a fundamental social-cultural priority in a sustainable development strategy. This activity can be measured economically using GDP. To measure the environmental impacts of these products through GDP, it is essential to integrate both conventional gross domestic product and the cost that must be spent to conserve or restore the environment which is damaged in the process of development. Mahravan (2012) proposed a method of calculating the sustainable portion of GDP (called 'GDPs') to form an economic indicator which integrates the social-economic benefits of local products and their environmental impacts. The method of measuring GDPs involves the following five factors:

a   EF (gha) – the ecological footprint of an activity or a product
b   GDP ($) – earned GDP from this activity or product
c   EF1 (gha) – the overshoot portion of the EF of the product or activity relative to the available or target biocapacity
d   E1 (GJ) – the overshoot portion of the life-cycle energy use of the product or activity related to the biocapacity needed to absorb its carbon dioxide emissions
e   CE1 ($) – the cost of generating through renewable resources the overshoot portion of the life-cycle energy used.

(a) Ecological footprint (EF)

The ecological footprint (EF) measures the extent to which humanity is using nature's resources faster than they can regenerate. EFs are usually presented together with biocapacities, which measure the bioproductive supply. The EF of a given product or activity is equal to the area (in global hectares, gha) needed to absorb the CO_2 emissions generated through using fossil fuel to produce its life-cycle energy use (in gigajoules, GJ).

(b) Gross Domestic Product (GDP)

The GDP of an activity is all the money produced by that product or service.

(c) Ecological footprint overshoot (EF1)

If an EF is larger than the available biocapacity (BC) for a selected time period, EF/BC resource accounting produces a deficit or overshoot (EF1). A deficit occurs where human resource extraction and waste generation exceed an ecosystem's ability to regenerate the extracted resources and absorb the waste generated.

(d) Energy overshoot (E1)

E1 is the overshoot portion of the life-cycle energy use of the product or activity such that its CO_2 emissions are more than the available biocapacity for their absorption.

(e) Cost of generating E1 from renewable resources (CE1)

CE1 is the money needed to generate the overshoot portion of the life-cycle energy used (E1) through the use of renewable resources. Its value depends on the level, type and cost of the available technology and renewable resources. As a result, CE1 can be different from one location to another, from region to region, or from country to country.

Once these five factors have been calculated, the following equation can be used to evaluate the sustainable portion of GDP (GDPs):

$$GDPs = GDP - CE1$$

Basically, the closer GDPs is to GDP, the more sustainable the activity.

Cultural Footprint of Tourism (CF)

Figure 6.1 illustrates how the anticipated cultural outcomes of the sustainable development of tourism – its cultural footprint (CF) – can be conceptualized as the influences exerted by tourism on a given host society that change its attitudes to existing cultural and natural heritage and capital, and that develop cultural communication between the host society and its visitors. Since the economic development of tourism is introduced as a social-economic activity that is also environmentally sensitive, the cultural changes of the host society can direct the economic system toward more sustainable economic strategies.

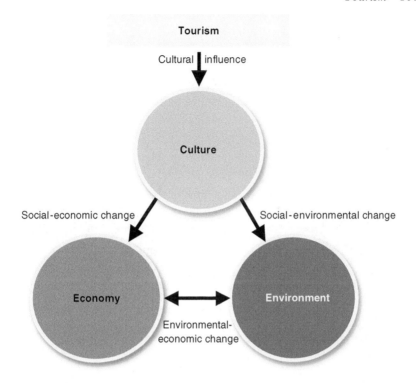

Figure 6.1 Relationship between tourism, the culture of the host society, the environment and the economic system

Figure 6.2 shows a proposed model that demonstrates the quantitative interaction between the cultural, environmental and economic changes caused by tourism in its host societies. In this model, the cultural, environmental and economic indicators (proposed in the comprehensive framework) are each set on one of the apexes. The cultural indicators determine and measure the quantity of the tourism products and activities (for example producing local foodstuffs) that are anticipated as cultural productive activities.

In this model, the environmental impacts of the cultural products and activities are calculated by using related ecological indicators such as EF. The EF of the cultural products and activities related to tourism determines whether these products and activities are environmentally sustainable or not. Evaluation of the EF can be conducted through comparison between the measured EF and the fair-share EF (as explained in Chapter 1) for each product and activity. Likewise two different products – for example organic and conventional foods – or activities – for example walking and golf – that are offered by a tourism venture can be ecologically compared with each other by making comparison between their EFs.

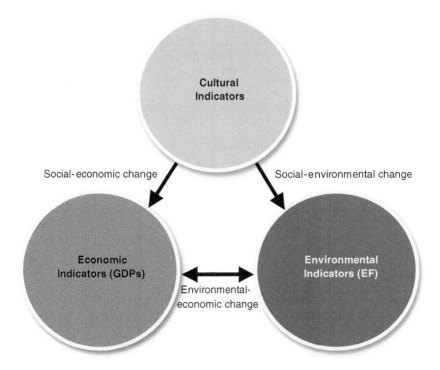

Figure 6.2 A model to evaluate interaction between cultural, environmental and economic changes caused by ecotourism development

The social-economic footprint of the cultural products and activities can be measured by using economic indicators such as GDPs. Since the GDPs of each product or activity is influenced by its EF, if a cultural product or an activity contributes to reducing the EF of tourism, this activity contributes to increasing GDPs.

A Case Study: The Otago Central Rail Trail

The Otago Central Rail Trail (OCRT), stated by the New Zealand Tourism Strategy 2015 (Tourism Industry Association *et al.*, 2007: 62) to be a successful ecotourism project, has been used as a case study to investigate its environmental, cultural and economic influences on its host destination through the comprehensive framework discussed above (Mahravan, 2012). Data from that case study are used here as an illustration.

OCRT at the regional scale

Table 6.6 demonstrates that the 2011 EF of the OCRT is equivalent to 1,617 gha, which derives from the integration of the calculated EFs of

Table 6.6 EF of the OCRT in 2011 – total and per visitor. Total OCRT visitor numbers were 11,788

Category	EF (gha)	Percentage	EF (gha/person)
Transportation	1,167	72.2%	0.1
Food	401.2	24.8%	0.034
Accommodation	42.4	2.6%	0.0036
Activities	6.6	0.4%	0.00056
Total	1,617	100%	0.138

transportation, food, accommodation services and activities used by 11,788 OCRT visitors that year. As shown in the table, transportation has by far the largest footprint, at 1,167 gha (72.2 per cent) of the total EF, followed by 401.2 gha (24.8 per cent) for food, 42.4 gha (2.6 per cent) for accommodation services and 6.6 gha (0.4 per cent) for visitor activities. The total EF of transportation can be further divided, giving 38.2 per cent of total EF as international transportation and 34 per cent as domestic transportation (Mahravan, 2012). This shows the importance of travel to a tourist destination for its overall environmental impact.

The sustainable EF of holidays is calculated to be 0.03 gha per person (Vale and Vale, 2009: 358) using the data in the report *Reducing Cardiff's Ecological Footprint* (WWF Cymru, 2005: 97) to calculate tourism's share of a fair-share EF, and this includes the EFs of transportation, food, accommodation and waste water that are consumed, used or produced by visitors. Mahravan's (2012) study used this sustainable EF of holidays (0.03 gha per person) as the goal to be achieved through the sustainable development of the OCRT. Comparing its present EF (0.138 gha per person) with its target EF (0.03 gha per person) gives the overshoot portion of the current EF, EF1, which is 0.108 gha per person. The total overshoot portion of the EF of 11,788 OCRT visitors in 2011 can then be calculated as 0.108 gha per person × 11,788 visitors = 1,273 gha.

As EF (gha) = energy used (GJ) ÷ 100 (GJ per gha – the carrying capacity of land relative to CO_2 absorptions; Wackernagel and Rees, 1996: 69), energy used (GJ) = EF (gha) × 100 (GJ/gha). As shown in Table 6.6, the EF of the OCRT is 0.138 gha per person, so its energy use is equivalent to:

$$0.138 \text{ gha/person} \times 100 \text{ GJ/gha} = 13.8 \text{ GJ/person}$$

However, its sustainable EF is 0.3 gha per person, so its sustainable energy use is:

$$0.03 \text{ gha/person} \times 100 \text{ GJ/gha} = 3.0 \text{ GJ/person}$$

Therefore the overshoot portion of OCRT energy use is:

$$13.8 \text{ GJ/person} - 3.0 \text{ GJ/person} = 10.8 \text{ GJ/person}$$

110 *Abbas Mahravan*

Table 6.7 The energy use of OCRT sustainable tourism in 2011

Current energy use (GJ/person/year)	Goal energy use (GJ/person/year)	Overshoot energy use (GJ/person/year)
13.8	3	10.8

This is set out in Table 6.7 above.

As a result the total yearly overshoot energy use of the 11,788 OCRT visitors is equivalent to:

10.8 GJ/person × 11,788 visitors = 127,310.4 GJ

OCRT GDP and GDPs

According to Central Otago District Council (CODC; 2011: 23), in 2011 the OCRT's total contribution to GDP was NZ$6,245,289. To calculate the sustainable portion of GDP (GDPs), Mahravan (2012) established that the cost of generating 1 GJ of overshoot energy use through using renewable resources, including wind and solar, is NZ$19.8 per GJ per year, allowing for a tenfold decrease in solar costs as predicted by the National Renewable Energy Laboratory in the United States (US Department of Energy, 2011: 60). Thus the portion of OCRT GDP that must be spent to generate the 127,310 GJ of overshoot energy is equivalent to 19.8 × 127,310 = NZ$2,520,738. Consequently the sustainable portion of OCRT GDP is equivalent to NZ$6,245,289 (GDP) − NZ$2,520,738 (the cost of the overshoot energy) = NZ$3,724,551. The sustainable GDP of the OCRT is 40.4 per cent (NZ$2,520,738) less than its conventional GDP. This means 40.4 per cent of the total current OCRT GDP would have to be spent to reduce its environmental impacts to the level of fair-share living, related to the fair-share EF of holidays as discussed above. The assumed mix, and the assumed cost, of renewable energy make a considerable difference to the GDPs calculation; using the current cost of solar technology, the OCRT is in GDP overshoot, with GDPs greater than GDP.

Using local products such as home-prepared food instead of conventional commercial products, using refurbished buildings instead of new buildings for accommodation, using open air spaces instead of indoor spaces as part of architecture, and conducting outdoor activities with lower EFs such as walking, sightseeing and horse riding can all contribute to reducing the total EF of the OCRT and therefore to increasing its total GDPs. For example, in 2011, producing 17 per cent of the total of 65,165 kilograms of food consumed by OCRT visitors locally would reduce its total EF (1,617 gha) by 1.4 per cent (23.2 gha). Looked at in another way, producing 1 kilogram of home-cooked food instead of 1 kilogram of conventional food can reduce the total EF of the OCRT by 143 square metres. In addition, producing home-cooked food (as a cultural product) contributes to increasing income by NZ$3.80 per kilogram. Using local products in ecotourism development is not only culturally appropriate, but also more environmentally friendly and economically viable.

Tourism and Transportation

As shown in Table 6.6, transportation forms by far the largest part of the EF of the OCRT. This might be less the case for a less isolated destination than New Zealand. In terms of the EF of transportation, in 2011 the EF of the OCRT's 2,593 international visitors was 617 gha (0.24 gha per person) and the EF of the 9,195 domestic visitors was 552 gha (0.06 gha per person). Comparison between the two EFs shows that the EF of an international visitor is four times that of a domestic visitor. Changing from international visitors to domestic visitors could contribute to reducing the EF of the OCRT by 0.18 gha per person (0.24 − 0.06). If all the OCRT visitors came from New Zealand, the total EF would fall from 1,617 gha to 1,157 gha, a reduction of 28 per cent. So it may be important in the future to promote local rather than international tourism.

To return to the international tourism figures from the World Trade Organization, what level of international tourism might it be possible to support in a fair-share economy? Until the 1960s and the growth of affordable international air travel, much international travel used 'cargo-passenger liners' – cargo ships which carried a limited number of passengers. It has been shown that this is a very low-impact way of travelling, as the passengers are almost incidental to the energy of transporting the cargo, taking about 100 times less energy than flying (Vale and Vale, 2009: 118–122). The current world shipping fleet includes 16,224 general cargo ships and 4,831 container ships (International Chamber of Shipping, 2012), both of which could carry passengers. So there are roughly 21,000 ships available. If the average voyage is two weeks and each ship can take 200 passengers (a typical figure for cargo-passenger ships in the 1950s), there could be just over 100 million 'international tourist arrivals' in a year (this crude calculation assumes that all international air travel is replaced by cargo-passenger ships). This would suggest that, in a fair-share society, international travel would be at about 10 per cent of its current level.

References

Azar, C., Holmberg, J. and Lindgren, K. (1996) 'Social-ecological indicators for sustainability', *Ecological Economics*, vol. 12, pp. 89–112

Central Otago District Council (CODC) (2011) *OCRT User Survey Gross Results*, CODC, Alexandra, Otago, New Zealand

Choi, H.C. and Sirakaya, E. (2006) 'Sustainability indicators for managing community tourism', *Tourism Management*, no. 27, pp. 1274–1289

Cochrane, P. (2006) 'Exploring cultural capital and its importance in sustainable development', *Ecological Economics*, vol. 57, pp. 318–330

Cohen, E. (1978) 'The impact of tourism on the physical environment', *Annals of Tourism Research*, vol. 5, no. 2, pp. 215–237

Coppock, J.T. (1982) 'Tourism and conservation', *Tourism Management*, vol. 3, no. 4, pp. 270–276

Costanza, R. (2009) 'Toward a new sustainable economy', *Real World Economics Review*, no. 49, pp. 19–21

Daly, H. (2005) 'Economics in a full world', *Scientific American*, September, vol. 293, no. 3, pp. 100–107

Gossling, S. (1999) 'Ecotourism: a means to safeguard biodiversity and ecosystem functions?' *Ecological Economics*, vol. 29, no. 2, pp. 303–320

International Chamber of Shipping (2012) 'Shipping facts – Shipping and world trade: Number of ships (by total and trade)', http://www.marisec.org/shippingfacts/worldtrade/number-of-ships.php, accessed 5 July 2012

Mahravan, A. (2012) 'The Main Characteristics of an Architecture that Will Contribute to Sustainable Development through Eco-tourism', PhD thesis submitted 5 April 2012, Victoria University of Wellington, New Zealand

Mahravan, A. and Vale, B. (2011) 'Cultural framework and related indicators for evaluation of sustainability through development of tourism', paper presented and published in *Shanghai International Conference on Social Science (SICSS 2011)*, Shanghai, China

Ministry for Culture and Heritage (2009) *Cultural Indicators for New Zealand*, Ministry for Culture and Heritage, Wellington, New Zealand, http://www.mch.govt.nz/files/CulturalIndicatorsReport.pdf, accessed 31 July 2012

New Zealand Council of Trade Unions (NZCTU) (2010) *Alternative Economic Strategy*, http://union.org.nz/sites/union/files/NZCTU%20Alternative%20Economic%20Strategy.pdf, accessed 4 July 2012

Organisation for Economic Co-operation and Development (OECD) (2008) *Key Environmental Indicators*, OECD, Paris, France, http://www.oecd.org/dataoecd/20/40/37551205.pdf, accessed 14 July 2012

Rapoport, A. (1969) *House, Form and Culture*, Prentice-Hall, Inc., Englewood Cliffs, NJ

Tourism Industry Association, New Zealand Ministry of Tourism and Tourism New Zealand (2007) *New Zealand Tourism Strategy 2015*, Tourism Industry Association, Wellington, New Zealand

UIS (2010) 'The 2009 UNESCO Framework for Cultural Statistics (FCS)', Statistical Commission forty-first session, item 4(e) of the provisional agenda, items for information: culture statistics background document, Institute for Statistics of the United Nations Education Scientific and Cultural Organization.

UNESCO (1982) *Mexico City Declaration on Cultural Policies*, UNESCO, Paris, http://portal.unesco.org/pv_obj_cache/pv_obj_id_A274FC8367592F6CEEDB92E91A93C7AC61740000/filename/mexico_en.pdf, accessed 1 November 2012

United Nations Environment Programme (UNEP) (2011) *Towards a Green Economy*, UNEP, Nairobi, Kenya

UNWTO (2011) *UNWTO Tourism Highlights: 2011 Edition*, http://mkt.unwto.org/sites/all/files/docpdf/unwtohighlights11enlr.pdf, accessed 31 July 2012

US Department of Energy (2011) *2010 Solar Technologies Market Report*, D DOE/GO-102011-3318, November, NREL, Golden, CO

Vale, B. and Vale, R. (2009) *Time to Eat the Dog? The Real Guide to Sustainable Living*, Thames and Hudson, London

Wackernagel, M. and Rees, W. (1996) *Our Ecological Footprint: Reducing Human Impact on the Earth*, New Society Publishers, Gabriola Island, British Columbia, Canada

Weaver, D. and Lawton, L. (1999) *Sustainable Tourism: A Critical Analysis*, Cooperative Research Center for Sustainable Tourism research report series, Griffith University, Queensland, Australia

WWF Cymru (2005) *Reducing Cardiff's Ecological Footprint*, WWF Cymru, Cardiff, Wales

Part II.II
Collective Footprint

7 Infrastructure

Ning Huang

Introduction

The ecological footprint (EF) related to the provision of infrastructure will account for a share of the total EF of transport activities, but what is the size of this share? As shown in Figure 1.5 in Chapter 1, the share is small for roads, but is it the same for other modes of transport? The EF of transport infrastructure has three parts: the energy EF for constructing the infrastructure, the energy EF for operating the infrastructure, and the built EF of the physical land area occupied by the infrastructure. The calculations cannot be made in the abstract, so this chapter will estimate the EF of transport infrastructure using Auckland, New Zealand as the basis for data collection and analysis. The infrastructure investigated here focuses on roads, railway tracks, airports, related official buildings and other facilities. The calculations are explained in some detail to give an idea of the complexity of the process.

Energy EF for Constructing Transport Infrastructure in Auckland

Energy EF for constructing transport infrastructure means the EF related to embodied CO_2 emissions from road network construction, rail network construction and airport construction.

Some attempts have already been made to calculate embodied CO_2 emissions from transport infrastructure construction. For example, Bennett (2008) worked out the embodied CO_2 emissions of road pavements by analysing two kinds of road surfacing commonly used in New Zealand: unbound granular with chipseal surfacing and structural asphalt with asphaltic concrete surfacing. Crawford (2009) investigated the life-cycle greenhouse gas emissions associated with timber (river red gum) and reinforced concrete railway sleepers in Australia. Kiani *et al.* (2008) analysed the embodied CO_2 emissions from ballast railway track beds and concrete slab track beds in the UK. Lenzen *et al.* (2003) have given data for greenhouse gas emissions of the Second Sydney Airport by life-cycle input-output analysis, including initial embodied energy and maintenance input energy.

Road network

In order to estimate the embodied CO_2 emissions for road network construction the total quantity of roads in Auckland must be known. Data from Transfund NZ (2004) provide the lengths of different kinds of roads in the Auckland Region in 2004, as shown in Table 7.1.

There are three main types of surface used on New Zealand's roads and highways (NZTA, 2009): asphalt, chip-seal and unsealed surfaces.

As a country with a large area relative to its population, New Zealand has a significant amount of relatively cheap gravel, or 'unsealed' roads. Table 7.2 shows an assumed allocation of the three kinds of surfacing to all roads in Auckland (Huang, 2011: 306).

A recent research report lists embodied CO_2 emissions from materials and emitted CO_2 emissions from construction and maintenance for the two main kinds of roads in New Zealand: unbound granular with chip-seal surfacing and structural asphalt with asphaltic concrete surfacing (Bennett, 2008: 7). The analysis in this report is based on the construction and maintenance over a 30-year lifetime. The result shows that total embodied CO_2 emissions for the two kinds of roads are 127 tonnes and 284 tonnes respectively, on the basis of a unit with 10-metre width (3.75 metres for each lane × 2 lanes + 1.25m wide shoulders on each side) and 1-kilometre length. Thus the embodied CO_2 emissions of 1 square metre will be:

Unbound granular with chip-seal surfacing:
$$127 \text{ t} \div (10 \text{ m} \times 1000 \text{ m}) = 0.0127 \text{ t/m}^2$$

Structural asphalt with asphaltic concrete surfacing:
$$284 \text{ t} \div (10 \text{ m} \times 1000 \text{ m}) = 0.0284 \text{ t/m}^2$$

Huang (2011: 310) calculated that the embodied CO_2 emissions for a gravel road ('unbound granular without surfacing') are 0.0072 tonnes per square metre. Therefore the total embodied CO_2 emissions for the road network in Auckland are as shown in Table 7.3.

Table 7.1 Road lengths in the Auckland Region in 2004

Road type	Length (km)		
	Total	Sealed	Unsealed
State highways	326.3 (105.8 km motorway)	326.3	—
Local urban roads	4,037.3	4,008.0	29.3
Local rural roads	3,585.9	2,389.3	1,196.6

Table 7.2 Surface area of roads with different surfacing types in Auckland. Sealed roads with other surfacing are considered as 'Structural asphalt with asphaltic concrete surfacing' for the embodied CO_2 emissions calculation because detailed data for their construction are not available. 'Unbound granular without surfacing' is the technical description of a gravel road

Surfacing types	Unbound granular with chip-seal surfacing (m²)	Structural asphalt with asphaltic concrete surfacing (m²)	Unbound granular without surfacing (m²)
Sealed local roads with chip-seal surfacing	54,916,000	0	0
Sealed local roads with asphaltic concrete surfacing	0	7,854,000	0
Sealed local roads with other surfacing	0	802,000	0
Motorway	0	2,856,600	0
Expressways	3,748,500	0	0
Unsealed local roads	0	0	12,259,000
Total	58,664,500	11,512,600	12,259,000

Table 7.3 Embodied CO_2 emissions for road construction (30-year lifetime)

Categories	Area (m²)	Embodied CO_2 emissions (t/m²)	Total embodied CO_2 emissions (t)
Unbound granular with chip-seal surfacing	58,664,500	0.0127	745,039
Structural asphalt with asphaltic concrete surfacing	11,512,600	0.0284	326,958
Unbound granular without surfacing	12,259,000	0.0072	88,265
Total embodied CO_2 emissions	82,436,100	—	1,160,262 (over 30-year lifespan) 38,675 (for 1 year)

Table 7.3 shows that total CO_2 emissions from road construction and maintenance in Auckland in one year are 38,675 tonnes. McDonald and Patterson (2003) give the CO_2 absorption factor for New Zealand as 13.2 tCO_2/ha·yr. So total energy EF related to embodied CO_2 emissions from road network construction per year is 2,930 hectares. From the figures in Table 7.1, the total length of the road network in Auckland is 7,949.5 kilometres. Thus, per kilometre of the whole road system, the embodied

CO_2 emissions and embodied energy EF respectively are 4.9 tonnes per kilometre and 0.4 hectares per kilometre, respectively.

Railway network

The railway network includes the construction of tracks, rail bridges and tunnels, railway stations and other associated buildings and infrastructure. In this section only the embodied CO_2 emissions from railway track construction are calculated, due to the unavailability of data for the other factors. Railway tracks are usually comprised of steel rails, sleepers (also known as 'ties') and ballast. The calculation is more complex than that for roads, where there were pre-existing data that could be used.

Steel rails

In different countries, the weight of steel rail varies. Typical figures are in the range of 75–76.9 pounds per yard in America, 30–68 kilograms per metre in Australia, 65–75 kilograms per metre in Russia, and 40–60 kilograms per metre in Europe (Singh, 2009). In New Zealand the track uses a gauge of 3 feet 6 inches, and traditional track construction – flat-bottom rail on wooden sleepers – weighs 35 kilograms per metre (70 pounds per yard; Taieri Gorge Railway, undated). It is assumed here that the typical weight of steel rail in New Zealand is now 40 kilograms per metre. The total length of railway track in Auckland is 188 kilometres, and each track contains two rails. So the total weight of steel rail in Auckland is:

$$188,000 \text{ m} \times 2 \times 40 \text{ kg/m} = 15,040,000 \text{ kg}$$

Here it is assumed that the rails imported into New Zealand some years ago are mostly from its nearest neighbour, Australia. The embodied CO_2 emissions factor of steel in Australia is 5.85 $kgCO_2/kg$ (Crawford, 2009: Table 7), so the embodied CO_2 emissions from the material of the steel rails will be:

$$\text{Total: } 15,040,000 \text{ kg} \times 5.85 \text{ kgCO}_2/\text{kg} = 87,984,000 \text{ kg} = 87,984 \text{ t}$$
$$\text{Per kilometre: } 87,984 \text{ t} \div 188 \text{ km} = 468 \text{ t/km}$$

This result is for the material only. The embodied emissions from the processing, capital equipment and other goods and services should also be taken into account. Research by Mithraratne shows that the quantity of CO_2 emissions for converting 1 kilogram of steel into 1 kilogram of steel product is 1.28 kilograms. These emissions are due to use of fuel and electricity in the metal factory, waste steel, and capital items such as machinery and factory use. Mithraratne's research focuses on the detailed conditions of New Zealand and adopts the SimaPro LCA software modified with material, energy and fuel data for New Zealand (Mithraratne, 2010). She also states that the emissions due to laying steel rails in place are likely to be neglected because they are very insignificant when compared with the value of the total. According to her study, the embodied emissions from the processing, capital equipment and other goods and services account for 51 per cent (1.28 ÷ [1.28 + 1.242]) of the total emissions. Using this figure of 51 per cent for the next estimation, the CO_2 emissions from steel rails will be:

Total: 87,984 t ÷ (100% − 51%) = 179,559 t
Per kilometre: 468 t/km ÷ (100% − 51%) = 955 t/km

Here, the average life of steel rail in New Zealand is assumed to be 60 years. Thus, the embodied CO_2 emissions for the steel rail are:

Overall: 179,559 t ÷ 60 yr = 2,993 t/yr
Per kilometre: 955 t ÷ 60 yr = 15.9 t/km/yr.

Sleepers

When estimating the embodied CO_2 emissions from railway sleeper manufacture and maintenance, data for Australia are used as shown in Table 7.4 below.

Crawford (2009: Table 8) gives the initial embodied CO_2 emissions factor per sleeper as 234 kilograms for reinforced concrete and 556 kilograms for timber. This is on the basis of using an Australian hardwood, river red gum, which has a CO_2 intensity of 3.27 $kgCO_2$/kg. The CO_2 intensities of the concrete (50 MPa) and the steel are 0.225 $kgCO_2$/kg and 5.85 $kgCO_2$/kg, respectively (Crawford, 2009: Table 7).

Table 7.4 Initial and recurring embodied CO_2 emissions of reinforced concrete and timber sleepers, per kilometre of track. Recurring embodied emissions shows the embodied emissions for replacement of sleepers over a 100-year period

Categories	Initial embodied emissions (tCO_2/km)	Recurring embodied emissions (tCO_2/km)*		
		20-year service life	30-year service life	50-year service life
Reinforced concrete	328	n/a	984	328
Timber, virgin fastenings	812	3,249	2,244	n/a

Source: Crawford (2009: Table 6)

Table 7.5 Comparisons of emissions intensity between Australia and New Zealand

Material	Australia	New Zealand
Timber	3.27 kgCO$_2$/kg (or 2.94 tCO$_2$/m^3) for air-dried red gum	−1.662 kgCO$_2$/kg (or −698 kgCO$_2$/m^3) for air-dried pine (assumes carbon lock-up)
Reinforced concrete	0.225 kgCO$_2$/kg (or 0.54 tCO$_2$/m^3) for 50 MPa concrete	0.189 kgCO$_2$/kg for 40 MPa concrete 0.159 kgCO$_2$/kg for 30 MPa concrete
Steel	5.85 kgCO$_2$/kg (or 45.9 tCO$_2$/m^3)	1.242 kgCO$_2$/kg

Sources: Data for Australia from Crawford (2009); data for New Zealand from Alcorn (2003)

Because this result is worked out on the basis of Australian data, it needs to be converted on the assumption that sleepers are made in New Zealand and not imported. Table 7.5 shows some comparisons of emissions intensity for Australia and New Zealand.

The assumptions made in these kinds of life-cycle calculations can make a considerable difference. Table 7.5 shows the embodied CO$_2$ emissions factor for timber in Australia is 3.27 kgCO$_2$/kg or 2.94 tCO$_2$/m^3. In contrast, the New Zealand figure is negative, at −1.662 kgCO$_2$/kg or −698 kgCO$_2$/m^3, because it takes account of the ability of timber to lock up carbon, as explained by Alcorn (2003: 10). The quantity per sleeper is 0.09 m^3 of timber (Crawford, 2009: Table 1). The CO$_2$ emissions per timber sleeper in Australia and New Zealand, respectively, will be:

$$0.09 \text{ m}^3 \times 2940 \text{ kg/m}^3 = 264.6 \text{ kg}$$
$$0.09 \text{ m}^3 \times (-698 \text{ kg/m}^3) = -62.82 \text{ kg}$$

However, Crawford (2009) specifically makes the point that he assumes that all the carbon in a sleeper will be emitted as it decays over a 100-year life, whereas Alcorn's (2003) data assume that the carbon remains.

It is assumed that CO$_2$ emissions for aspects other than material (for example, emissions from fabrication, engineering and assembly, emissions from capital goods and transport, and emissions from service) are similar in the two countries. Thus the initial embodied CO$_2$ emissions per timber sleeper in New Zealand could be:

$$556 \text{ kg} - 264.6 \text{ kg} + (-62.82 \text{ kg}) = 228.58 \text{ kg}$$

The figures in Table 7.5 suggest the CO$_2$ intensity of 50 MPa concrete in New Zealand should be approximately similar to that in Australia. For steel, there is a distinct difference between the emissions factors of the two countries due to the electricity consumed for making steel. The factor in Australia is nearly five times that in New Zealand. However, due to the small quantity of steel used in reinforced concrete sleepers the result based on the Australian CO_2 emissions factor for steel will be adopted for New Zealand. Thus, the factor of 234 kg is also used for New Zealand concrete sleepers.

Based on the previous explanation the CO_2 emissions factors of the concrete and timber sleepers in New Zealand are 234 and 229 kilograms, respectively. Currently in New Zealand there is a mix of approximately 5.86 million timber and concrete sleepers on the network, at 0.65 metre spacing (PricewaterhouseCoopers, 2004: 183). So there are 1,538 sleepers per kilometre (1000 ÷ 0.65) in New Zealand, and this analysis assumes there is an equal split between the two materials. The initial embodied CO_2 emissions for sleepers per kilometre of railway track in New Zealand are:

Reinforced concrete: 234 kg/sleeper × 1,538 sleeper/km = 360 t/km
Timber: 229 kg/sleeper × 1,538 sleeper/km = 352 t/km

A report focusing on the railway stock asset replacement says that the average life of a timber sleeper in New Zealand is approximately 25 to 30 years, compared to 50 years for a concrete sleeper (PricewaterhouseCoopers, 2004: 183). Using the 50- and 25-year service lives, the initial embodied CO_2 emissions for sleepers per kilometre of railway track in New Zealand for a counted period of 100 years are:

Reinforced concrete: 360 t/km + 360 t/km = 720 t/km
Timber, virgin fastenings: 352 t/km + 352 t/km + 352 t/km + 352 t/km = 1,408 t/km

The average embodied CO_2 emissions of sleepers per kilometre over a 100-year period are:

(720 t/km + 1,408 t/km) ÷ 2 = 1,064 t/km

The total embodied CO_2 emissions for the sleepers of 94 kilometres of double-tracked railway (188 kilometres) in Auckland are:

1,064 t/km × 188 km ÷ 100 yr = 2,000 t/yr

Emissions per kilometre will be 10.6 tonnes annually.

Ballast

Track ballast is another key component of the railway network. No embodied CO_2 emissions factors for track ballast are available for New Zealand, so research from the UK is used for this investigation. Kiani *et al.* (2008) analysed the embodied CO_2 emissions for UK ballast track beds, as shown in Table 7.6 below.

Table 7.6 Specifications for track ballast in the UK

Category	Specification
Mass of single track	Aggregate for sub-base: 3,600 kg/m
	Ballast: 2,652 kg/m
Life expectancy	Ballast track bed: 20~30 yr
	Ballast cleaning: 10~15 yr
	Ballast tamping: 1~2 yr
Embodied energy (aggregate and ballast)	0.02~0.10 GJ/t
Energy source	Fuel oil: 77%
	Electricity: 23%

The embodied energy intensity shown in Table 7.6 ranges from 0.02GJ per tonne to 0.10 GJ per tonne due to different scenarios for maintenance and life expectancy. An average value of 0.06 GJ per tonne is adopted for the next calculation.

Embodied energy from fuel oil: $(3,600 + 2,652)$ t/km × 0.06 GJ/t × 77% = 289 GJ/km
Embodied energy from electricity: $(3,600 + 2,652)$ t/km × 0.06 GJ/t × 23% = 86 GJ/km

It is assumed that the fuel oil used is diesel. In New Zealand the CO_2 emissions factors of diesel and electricity are 69.5 kt/PJ and 0.18 t/MWh, respectively (MED, 2007), which convert to 69.5 kg/GJ and 50 kg/GJ (1 MWh = 3.6 GJ). If it is assumed that the energy consumption and its composition are similar in New Zealand and in the UK (the electricity mix in the UK is different but because the figure is small this has not been accounted for here), the embodied CO_2 emissions from the ballast of railway track, in tonnes per kilometre, are:

289 GJ/km × 69.5 kg/GJ + 86 GJ/km × 50 kg/GJ = 24.386 t/km

The total embodied CO_2 emissions of ballast in Auckland therefore become:

188 km × 24.386 t/km = 4,585 t

The lifetime of ballast is about 25 years (as shown in Table 7.6), so the embodied CO_2 emissions from ballast for the railway track in Auckland are 4,585 tonnes ÷ 25 years = 183 tonnes per year. The figure per kilometre of track is 1.0 tonne annually.

Table 7.7 shows the embodied CO_2 emissions from railway track in Auckland.

The total embodied CO_2 emissions from the railway track are 5,176 tonnes, so the EF is 5,176 tonnes ÷ 13.2 tonnes per hectare = 392 hectares.

Table 7.7 Annual embodied CO_2 emissions from railway track in Auckland

Category	Lifetime assumption (yr)	Total embodied CO_2 emissions (t/yr)	Embodied CO_2 emissions per km (t/yr)	Percentage of total	Initial embodied CO_2 emissions (t/km)
Steel rails	60	2,993	15.9	57.8%	955
Sleepers (timber and reinforced concrete)	Timber sleeper: 25 Reinforced concrete sleeper: 50	2,000	10.6	38.6%	Timber sleepers: 352 (for 25 years) Reinforced concrete sleepers: 360 (for 50 years)
Ballast	25	183	1.0	3.5%	24 (for 20–30 years)
Total	—	5,176	27.5	100%	

The embodied CO_2 emissions and embodied energy EF annually are respectively 27.5 tonnes and 2.1 hectares per kilometre of railway track, with the majority of this being due to the steel rails.

Airport construction

The constructed infrastructure of an airport should include not only the runways but also the construction of office and service buildings, and relevant car parking and access ways. However, the main component of an airport is the runway. Only this part is considered next, as few data for the other components are available.

There are two runways at Auckland International Airport: a main runway with dimensions 3,635 × 45 metres and a main taxiway/runway which is 2,910 × 45 metres (WCC, 2010: Annex 1).

Since there are no available data for embodied CO_2 emissions for airport construction in New Zealand, some overseas figures will be used for the estimation. Table 7.8 shows some physical comparisons between the Second Sydney Airport (proposed in 1997) and Auckland Airport. Because this second Sydney Airport has a comparable scale to Auckland Airport, its data will be used for Auckland Airport.

The emissions factor of 70.4 $kgCO_2$/GJ is the factor for automotive diesel oil (ADO) used for the Sydney airport construction. The figure for ADO in New Zealand is 69.5 $kgCO_2$/GJ. If the result in Table 7.8 is used to represent New Zealand, the total CO_2 emissions for the construction of Auckland Airport will be:

$$69.5 \text{ kg/GJ} \div 70.4 \text{ kg/GJ} \times 0.24 \text{ Mt} = 236{,}932 \text{ t}$$

Table 7.8 Comparison between Second Sydney Airport and Auckland Airport

Scale category	Second Sydney Airport*	Auckland Airport
Runways	Two 4,000-m long parallel runways	One 3,635-m long runway and another 2,910-m long main taxiway/runway in parallel
Passenger capacity	30 million passengers a year	An anticipated 24 million passengers a year by 2025** Projected capacity: 30 million passengers/40,000 aircraft movements per year***
Energy consumption (PJ)	3.5	No data
Greenhouse gas emissions (Mt)	0.24****	No data

* Lenzen *et al.* (2003)
** AIAL (2009)
*** WCC (2010)
**** Emissions factor is 0.0704 $MtCO_2$/PJ (70.4 $kgCO_2$/GJ)

In total, 236,932 tonnes of CO_2 emissions would be related to the airport construction if it were in Auckland rather than in Sydney. Because the physical scale of the Second Sydney Airport is bigger than that of Auckland Airport (as shown in Table 7.8), a ratio of 80 per cent is adopted. So the estimated CO_2 emissions from the construction of Auckland Airport can be assumed to be:

$$236,932 \text{ t} \times 80\% = 189,546 \text{ t}$$

The designed lifespans for the parts of an airport are shown in Table 7.9.

Table 7.9 Designed lifespan of airport constructions

Category	Lifespan (years)
Runways, taxiways and aprons	100
Terminal buildings, pier and satellite structures	50
Tunnels, bridges and subways	50
Terminal fixtures and fittings	20
Transit system	20–50
Plant and equipment (runway lighting and bulking plant)	5–20
Motor vehicles	4–8
Retail units, bars and restaurants	3–5
Office equipment	5–10

Source: Edwards (2005: 46)

If the average lifespan is assumed to be 50 years, the annual embodied CO_2 emissions from Auckland Airport will be:

$$189{,}546 \text{ t} \div 50 = 3{,}791 \text{ t}$$

The embodied energy EF is:

$$3{,}791 \text{ t} \div 13.2 \text{ t/ha} = 287 \text{ ha}$$

Energy EF for Operating Infrastructure

Roads: EF from infrastructure operation

The CO_2 emissions from the infrastructure operation for roads will come from the routine energy consumption (fuel and electricity) for such things as associated office buildings, car park buildings, bus depots, traffic signals and street lights. Previous studies on the EF of transport modes have excluded the EF calculation of infrastructure operation for roads (Chi and Brain, 2005; Barrett and Scott, 2003; Wood, 2003). A possible reason is that this EF will be very small when compared to other EFs relating to roads, for instance the fuel to run vehicles and the energy for their manufacture. This can be seen in Figure 1.5 in Chapter 1, in which the total embodied energy of the roads in Vancouver is miniscule in comparison with the embodied energy of the vehicles that run on those roads, and even smaller when compared to the energy used for operation of the vehicles.

Trains: EF from infrastructure operation

Ontrack (2008: 32) gives nationwide data for the greenhouse emissions relating to annual railway operation in New Zealand, apart from those involved in powering locomotives and multiple units. Table 7.10 shows detailed figures.

These figures are for New Zealand, not just the Auckland Region. As Auckland has 94 kilometres of railway tracks and New Zealand has 4,000 kilometres in total (KiwiRail, 2008), this ratio will be selected for distributing the CO_2 emissions shown in Table 7.10. Thus, the CO_2 emissions from the railway infrastructure operation in Auckland are 315 tonnes and the EF is 315 tonnes ÷ 13.2 tonnes per hectare = 24 hectares.

Aviation: EF from infrastructure operation

Auckland Airport produced its first company emissions profile for the 2006 financial year. Total CO_2 equivalent emissions for 2006 to 2008 are shown in Table 7.11.

Table 7.10 CO_2 emissions from railway operation in New Zealand

Category	Quantity	CO_2 emissions (tCO_2)
Motor vehicle travel	1,529,889 litres	3,653
Rental vehicles	297,971 km	71
Work trains	413,361 litres	1,120
Service freight	203,370 litres	551
Mobile plant and equipment	161,917 litres	729
Scope 1 direct emissions		*6,124*
Electricity	10,466,196 kWh	6,541
Scope 2 indirect emissions		*6,541*
International air travel	418,536 km	46
Domestic air travel	1,950,947 km	272
Embodied emissions	1,287 t	422
Scope 3 indirect emissions		*740*
Total direct and indirect emissions		13,405

Table 7.11 Auckland Airport emissions profile for years 2006 to 2008

Source	FY 2006 (tCO_2)	FY 2007 (tCO_2)	FY 2008 (tCO_2)
Fuel			
Natural gas	1,025	1,752	1,846
Petrol 91	218	207	192
Petrol 96	44	42	29
Diesel	320	314	294
Avgas	2	2	0
Jet A1	77	48	77
Electricity	6,555	6,585	6,833
Air travel type			
Air travel short haul, < 500 km	41	24	24
Air travel medium haul, 500–1,600 km	1	8	12
Air travel long haul, > 1,600 km	130	73	78
Other			
Construction	2,240	1,903	1,922
Total	10,654	10,958	11,306

Source: AIAL (2010a)

The outcome based on 2006 data is used here, and the total EF is 807 hectares.

Built EF of physical land area for infrastructure

The physical land areas occupied by roads, railway tracks and the airport need to be investigated to give the most complete picture possible of the footprint of infrastructure.

Roads

The calculation of the built land used for roads is derived from multiplying the length of different kinds of roads by their respective widths. The roads in the Auckland Region can be divided into two main categories: state highways and local roads.

Table 7.1 in this chapter lists the lengths of different kinds of roads in the Auckland Region. The length of motorways, expressways (the other part of the state highway network excluding motorways), local urban roads and local rural roads are 105.8, 220.5, 4,037.3 and 3,585.9 kilometres, respectively. The widths of motorways, rural expressways and urban expressways in New Zealand are 31.77, 29.06 and 28.54 metres, respectively (Vessey, 2008). The width of motorway is for one direction.

Using these figures, the total area of the motorway and the expressways in the Auckland Region can be estimated as shown below.

Motorway land area: 105,800 m × 32 m × 2 = 6,771,200 m^2
Expressway land area: 220,500 m × 29 m = 6,394,500 m^2

The total land area of the state highways in the Auckland Region is therefore 13,165,700 square metres, or 1,317 hectares.

It is reasonable to assume that the average road reserve of all local roads is 20 metres. Hence, the total area of these in the Auckland Region can be calculated as shown below.

Local road land area (including urban and rural roads):
(4,037,300 m + 3,585,900 m) × 20 m = 152,464,000 m^2 = 15,246 ha

So the total road area (the EF of built land for roads) in the Auckland Region is:

1,317 ha + 15,246 ha = 16,563 ha

130 *Ning Huang*

Rail track

The land area of rail track can be derived from multiplying the total length of rail track by its width.

In 2004, the total route length of rail in the Auckland Region was approximately 94 kilometres (ARTA, 2006: 23). It is assumed that the average width of the railway track is 20 metres (New Zealand South, undated). This means the total land area of the railway track will be:

$$94{,}000 \text{ m} \times 20 \text{ m} = 1{,}880{,}000 \text{ m}^2 = 188 \text{ ha}$$

Airport

Data from the AIAL show that the total freehold land area of Auckland International Airport is 1,500 hectares (AIAL, 2010b). This means the total EF of built land for aviation in the Auckland Region is taken as 1,500 hectares.

Some Comparisons of Three Transport Modes

Huang (2011: 345–347) has calculated all the other parts of the EF related to the three transport modes dealt with above. A comparison between the EF of infrastructure and the total EF for each mode can be seen in Table 7.12.

These detailed findings on infrastructure and its role as part of the total footprint reinforce the comments made in Chapter 1. Infrastructure has a relatively small impact because of its relative durability. In the case studied here, of transport infrastructure, infrastructure generally lasts far longer than the vehicles that use it. In a similar way, the embodied energy of buildings is a small part of the total footprint of buildings when compared

Table 7.12 Comparison between infrastructure-related EF and total EF for Auckland Region road, rail and aviation, domestic and international, including both passengers and freight. The EF of Auckland Region road, rail and air transport is 560,333 ha/yr; the total area of Auckland Region is 500,000 ha

	Mode		
Categories	*Road*	*Rail*	*Aviation*
Infrastructure EF (ha/yr)	19,493	604	2,593
Embodied energy EF (ha/yr)	2,930	392	286
Operational EF (ha/yr)	n/a	24	807
Physical EF (ha/yr)	16,563	188	1,500
Total EF (ha/yr)	401,535	3,296	155,502
Infrastructure EF as percentage of total EF	4.9%	18.3%	1.7%

to the energy used to operate them; again, this is a function of the relative longevity of buildings.

This study of infrastructure is of a very particular place, the region of Auckland in New Zealand. Locality is of great importance. Mithraratne (2011: 63), in a study of land transport throughout New Zealand, found that primary and secondary railway tracks had construction energy demands similar to motorways and urban local roads respectively. When total energy was considered (fuel for vehicles, embodied energy of vehicles, infrastructure and maintenance), she found that, while Auckland's diesel suburban trains had similar life-cycle overall energy use per passenger-kilometre to that of private cars, Wellington's mostly electric suburban trains used less than a third of the energy of private cars, overall, as did long-distance passenger trains, while rail freight in general used only a third of the energy of the best road freight (Mithraratne, 2011: 90).

To put this discussion of infrastructure into perspective, Table 7.13 makes clear the changes in Auckland's transport over the years.

In the nearly fifty years between the two sets of data in Table 7.13, the population of the Auckland Region has risen by 264 per cent, but the total footprint for ground transport has risen by a shocking 865 per cent. This is not due to increases in infrastructure. Some of the increase is due to the increased number of trips taken per person per year, which is up by 55 per cent, but the biggest difference seems to be in mode, with public transport's share of total trips falling from 59 per cent in 1955 to only 5 per cent in 2004, to be replaced by private cars. In the last fifty years it seems that the biggest increase in footprint has come from the change from public to private modes of transport.

People were able to make quite a few trips in 1955. Probably many trips then made by bicycle (which are not shown in Table 7.13, as it covers only motorized transport) are now made by car, so it is possible that people were

Table 7.13 Comparison of Auckland Region motorized ground transport (car, bus, tram, trolley bus, ferry) in 1955 and 2004

Date	Population, number (percentage change)	Trips made per person per year, number and percentage share (percentage change to 2004)			Transport EF, ha/person/year (percentage change to 2004)	Total transport EF, ha/year (percentage change to 2004)
1955	361,600	Car: Public transport: Total:	207 303 510	41% 59%	0.069	24,975
2004	1,316,700 (+ 264%)	Car: Public transport: Total:	749 39 788	95% 5% (+ 55%)	0.183 (+ 165%)	240,958 (+ 865%)

just as able to get around in 1955 as they are now. This is discussed further in Chapter 10, which looks in greater detail at life in the 1950s.

The role of the infrastructure of transport is small in the overall picture of transport's footprint. The meaning of this for the bigger picture is that, in general, things which have a long life (roads, railway tracks, bridges, tunnels, water pipes, sewers, buildings, etc.) generally have a small EF in relation to their construction. The things to worry about in terms of EF are either those which have a short life, such as motor vehicles, meaning that their embodied energy is spread over a short period, or those which have an ongoing EF demand in the form of fuel or food.

References

AIAL (2009) *About Us*. Auckland International Airport Ltd, New Zealand

AIAL (2010a) *Climate Change*. Auckland International Airport Ltd, New Zealand

AIAL (2010b) *Facts & Figures: Auckland Airport*. Auckland International Airport Ltd, New Zealand

Alcorn, A. (2003) *Embodied Energy and CO_2 Coefficients for NZ Building Materials*. Centre for Building Performance Research, Victoria University of Wellington, New Zealand

ARTA (2006) *A Step-change for Auckland: Rail Development Plan 2006*. Auckland Regional Transport Authority, Auckland, New Zealand

Barrett, J. and Scott, A. (2003) 'The application of the ecological footprint: a case of passenger transport in Merseyside', *Local Environment*, vol. 8, no. 2, pp. 167–183

Bennett, N. (2008) 'Evaluating economic and environmental sustainability of road projects', in *Proceedings of NZ Transport Agency and NZIH 9th Annual Conference*. Napier, New Zealand

Chi, G. Q. and Brain, S. J. (2005) 'Sustainable transport planning: estimating the ecological footprint of vehicle travel in future years', *Journal of Urban Planning and Development*, vol. 131, no. 3, pp. 170–180

Crawford, R. H. (2009) 'Using life cycle assessment to inform infrastructure decisions: the case of railway sleepers', in *Proceedings of Australian Life Cycle Assessment Conference: Sustainability Tools for a New Climate*. Australian Life Cycle Assessment Society, Melbourne, Australia

Edwards, B. (2005) *The Modern Airport Terminal: New Approaches to Airport Architecture* (2nd edition). Spon Press, New York

Huang, N. (2011) *A Modified Ecological Footprint Method for Assessing Sustainable Transport in the Auckland Region*. PhD Thesis, University of Auckland, New Zealand

Kiani, M., Parry, T. and Ceney, H. (2008) 'Environmental life-cycle assessment of railway track beds', *Proceedings of the ICE – Engineering Sustainability*, vol. 161, no. 2, pp. 135–142

KiwiRail (2008) *New Zealand's Integrated Transport and Infrastructure Network: Business Overview and Review of Strategic Issues for Shareholding Ministers*. KiwiRail Group, New Zealand

Lenzen, M., Murray, S., Korte, B. and Dey, C. (2003) 'Environmental impact assessment including indirect effects: a case study using input-output analysis', *Environmental Impact Assessment Review*, vol. 23, pp. 263–282

McDonald, G. W. and Patterson, M. G. (2003) *Ecological Footprints of New Zealand and Its Regions*. Ministry for the Environment, Wellington, New Zealand

MED (2007) *New Zealand Energy Greenhouse Gas Emissions 1990–2006*. Ministry of Economic Development, Wellington, New Zealand

Mithraratne, N. (2010) personal email communication, 6 January 2010

Mithraratne, N. (2011) *Lifetime Liabilities of Land Transport Using Road and Rail Infrastructure*. NZ Transport Agency research report 462, Land Transport New Zealand, Wellington

New Zealand South (undated) *Otago Central Rail Trail*, http://www.wildflowerwalks.co.nz/wildflowerwalks/railtrail.html, accessed 17 August 2012

NZTA (2009) *The Official New Zealand Road Code*. New Zealand Transport Agency, Wellington, New Zealand

Ontrack (2008) *Ontrack 2008 Annual Report*. Ontrack, Wellington, New Zealand

PricewaterhouseCoopers (2004) *Ministry of Economic Development Infrastructure Stocktake: Infrastructure Audit*. PricewaterhouseCoopers New Zealand, Auckland

Singh, S. (2009) 'Rail manufacturing plant in the Kingdom of Saudi Arabia', presentation at the SBB Steel Markets Middle East Conference, 29 September 2009. Boulder Steel Limited, North Ryde, NSW, Australia

Taieri Gorge Railway (undated) *Technical Information*, http://www.taieri.co.nz/technical_info.htm, accessed 25 Nov 2009

Transfund NZ (2004) *Roading Statistics 2003/2004*. Transfund New Zealand, Wellington

Vessey, J. (2008) personal email communication, OPUS Ltd, New Zealand, 3 October 2008

WCC (2010) *Whenuapai Airbase: Impact of Closure on Civil Defence Emergency Management in the Auckland Region*. Waitakere City Council, New Zealand

Wood, G. (2003) 'Modeling the ecological footprint of green travel plans using GIS and network analysis: from metaphor to management tool?' *Environment and Planning B: Planning and Design*, vol. 30, pp. 523–540

8 Government

Jeremy Gabe and Rebecca Gentry

The central theme of this book is the recognition of limits to consumption behaviour and how these limits can be established using a combination of science and ethics to arrive at a 'fair earth share'. This chapter examines the advantages and disadvantages of three governance frameworks that exhibit a potential to help society live within its fair earth share: command-and-control, community-scale management, and market-based.

The types of natural resources described in the ecological footprint used to calculate fair earth share – energy, materials, food and the capacity for ecosystem services such as greenhouse gas sequestration – are often referred to by economists as 'natural capital'. Thus, a fair earth share could also be considered as a per capita ration of natural capital, presumably within a framework where the stock of natural capital is non-declining (a sustainable yield). Economists often measure natural capital by calculating its value in currency; this implies that natural capital is substitutable. For example, if today's global market price of US$100 is used as a proxy for the value of a barrel of oil, then one can assume that a barrel of oil contributes as much to human well-being as 15 minutes' worth of a US$400-per-hour lawyer's knowledge.

Recent research by the United Nations into the capital stocks of nations consistently shows a decline in the value of natural capital as a result of consumption, with subsequent growth in the value of human capital (education, population, labour) and manufactured capital (buildings, machinery, etc.; UNU-IHDP and United Nations Environment Programme, 2012). The growth in the monetary value of human and manufactured capital often, but not always, exceeds the decline in the value of natural capital. Hence proponents of substitutability between capitals – 'weak sustainability' – justify the sustainability of such transfers by citing the net gain in capital, measured in currency value.

But by *not* measuring natural capital in units of currency value, the question at hand in this chapter is what form of government is necessary to maintain strong sustainability, or the recognition that many natural capital stocks are 'critical' – non-substitutable – and must be managed separately from other forms of capital. The key justification for this accounting framework is that natural capital is highly vulnerable to market failures that result in the market price being artificially low. For example, in most markets, the cost of coal does not include a contribution by the consumer to the damage done by global warming, which is largely a result of fossil fuel combustion. Similarly, the cost of petrol does not include a contribution for public health costs that result from highway crashes or urban air pollution. If such 'externalities' are included in the market price through government taxation, for example, the costs of natural capital depletion increase.

The depletion and degradation of natural capital can also be a result of the 'tragedy of the commons'. Natural resources, particularly renewable resources, are often common property with open access. For example, in most countries, the fish in the sea were traditionally free to anyone with the skill to catch them. Because there is a limited supply at any time, there is an incentive to catch as many as possible, otherwise someone else will catch them. For a long time, technological limitations prevented humanity from depleting fish stocks (fishing beyond the sustainable yield), but modern fishing methods and population growth have led in many places to overfishing and destruction of fish habitat. In response, governments have intervened to prevent this tragedy of the commons.

Government intervention is critical to the preservation of natural capital, as such examples of market failure attest. But what kind of government is likely to succeed?

Command-and-Control

One option is command-and-control, where a government mandates a limit (the command), such as a fair earth share, and enforces it (the control). If the government commands the economy must remain within a fair earth share *and* it is able to enforce these limits, this method can be successful at preserving natural capital. Command-and-control is the most common mode of governance around the world when it comes to environmental regulation and has been credited for numerous environmental improvements over the past half-century (Davies and Mazurek, 1998).

This book argues that there is little choice in what society must command: a fair earth share limit. How society empowers its government to command – via democratic agreement or autocratic decree, for example – is complex, but beyond the scope of this essay. For discussion, we make the necessary assumption that society must command that consumption remains within a fair earth share.

How a government controls its command is more instructive, because unwanted side-effects of effective control policies, such as human rights violations and economic inefficiency, have encouraged the development of alternative governance strategies. One of the often-cited examples of command-and-control being used for environmental preservation is the contrast between the barren deforestation of Haiti and the lush, heavily forested Dominican Republic in the mid-twentieth century. Sharing the same Caribbean island, both countries were under autocratic rule for most of this period. However, the leadership of the Dominican Republic took the control of their policy of sustaining forest resources very seriously, going as far as army raids on illegal logging camps with the aim of executing the loggers (Diamond, 2005). On the other hand, Haitian leadership did not control its resources and allowed foreign governments to control agricultural markets, leading to severe depletion of natural capital (Lawless, 1992; Schwartz, 2008).

When viewed through the lens of ethical human rights, the violent control policies of the Dominican Republic are not considered acceptable. Thus, most countries use much less harsh control methods – often financial penalties or temporary incarceration. However, these less harsh penalties may actually induce non-compliance, as fines can be perceived as an alternative behaviour choice that is socially acceptable (Gneezy and Rustichini, 2000). If a fine is simply a price, then the control mechanism becomes quasi-market-based; if paying the fine generates greater profits for a firm than complying with the commanded limit, the firm will not comply. So these penalties may not be enough to ensure successful government control.

The bulk of this chapter explores two potential solutions to problems with top-down direct control in the conventional command-and-control model. Small communities have shown the ability to be self-controlling through social bonds, so we discuss community management as a solution to the tragedy of the commons. Following this, we turn to the potential for fixing market failures associated with natural capital and creating a more effective market-based governance regime.

Community Management

A very different strategy relies on communities, rather than central government controllers, to enforce environmental limits. In this model, the ownership of natural capital is shared within a small community. Members are assumed to be familiar with each other, so consumption beyond a fair earth share is restricted by social norms within the community. It is difficult to hide consumption beyond what is considered the social norm in such small communities, so members face a form of self-limitation that goes beyond the traditional economic theory of consumption (which states that individuals consume to increase their well-being, limited mainly by their resource budget, not social norms).

Nobel prize-winning political economist Elinor Ostrom argues that community governance is critical in examples of societies that have avoided the tragedy of the commons. The guiding principles in her work are that successful community resource management has clearly defined boundaries which exclude outsiders, a shared sense of resource entitlement, effective monitoring, recognition of community self-determination and appropriate sanctions for those that violate social norms (Ostrom, 1990).

One example of successful community natural capital management comes from remote Pacific island societies. A few centuries ago, these communities faced a dilemma over limited natural capital similar to the one now faced by the world as a whole. Trade between very remote islands was severely limited by the hazards of making sea crossings in canoes, so there were effective limits to natural capital stocks. Using this model, social scientist Jared Diamond has reviewed a body of anthropological research associated with island societies. His review suggests that Tikopia, a small island of five square kilometres in the Solomon chain, is an archetype, as an example of a society that was able to preserve both its natural capital base and a thriving community for over 3000 years (Diamond, 2005).

Diamond argues that strong community governance and social norms of natural capital conservation enabled Tikopian society to thrive. The island's population was kept stable at around 1,200 inhabitants, which was small enough that each islander was familiar with all the others. And the islanders made difficult decisions; a commonly cited example is their decision to exterminate their imported pig populations, which provided food but also consumed food and caused erosion. It was decided that fishing caused less environmental damage and less resource depletion. Underlying population stability was a mix of contraception, abortion, infanticide and 'self-expulsion suicide' (paddling out to sea with no intention of returning).

When compared with Ostrom's principles for effective common-pool resource management, the geographic context of historic Tikopia – a remote island in the Pacific – enabled it to succeed, but it still required some effort. Its isolation meant that it was easy to exclude outsiders and recognise community self-determination. But Diamond reminds us that many other small remote Pacific island societies – Easter Island, a larger society that erupted into civil war following deforestation, for example – did not manage to conserve their natural capital. So the ability of Tikopia to conserve its natural capital base largely rested on the ability of the community to develop social norms regarding entitlement to resources, to monitor consumption, and to sanction violators. While Diamond is not explicit about how Tikopia managed these three outcomes, the implied hypothesis is that familiarity within the small island population led to both self-monitoring and self-sanction, while experiential learning led the islanders to conclude collectively that a population of around 1,200 was the limit to enable them to survive frequent tropical cyclone damage (Diamond, 2005).

Modern technology and globalisation mean that the geographical advantage enjoyed by Tikopians, enabling them to exclude outsiders, establish a clear resource entitlement and recognise self-determination, is less possible today. Communities and resources are much more mobile. Existing forms of state governance are common. In this modern context, fisheries and housing development provide two examples of how collective action within small communities can develop and conserve natural capital.

The idea of co-management, or the sharing of management responsibilities between communities and existing government, has been widely implemented as a strategy for sustainably managing fisheries. Fisheries co-management can take a wide variety of different forms, but is often based on the premise that small communities, like Tikopia, can successfully self-manage their natural resources. Several studies have identified key social, governmental and ecological attributes of successful community management of fisheries (Pomeroy *et al.*, 2001, Gutierrez *et al.*, 2011). Strong cohesion and leadership within the community as well as clearly defined boundaries, both in terms of membership of the community of fishers and clear authority over the space or species to be managed (spatial use rights or stock quotas) seem particularly important for successful outcomes. On an island such as Tikopia, the community and resources are easy to delineate, but at a more global scale there is a clear role for existing governments to define and empower communities. In addition, larger-scale governments are likely to be needed to allocate natural capital stock rights to these communities and perhaps to manage failed communities.

Balance between resource mobility and community scale is also a very important consideration for community management of fisheries. One of the studies cited above found that multi-species fisheries were less likely to be successfully managed under co-management (Gutierrez *et al.*, 2011). It is suggested that failure is often a result of mismatch between the distribution and mobility of target species and the scale at which they are being managed. Intuitively this makes sense – community management can be successful when the community has the ability to exert full control over the resource. If a fish population moves beyond the area in which governance is in place to manage it, then it becomes difficult to exclude outsiders from depleting the resource. This is a problem for highly migratory species, such as the southern bluefin tuna, which move between the jurisdictions of multiple countries and the ungoverned high seas. Despite significant and sustained efforts to reach effective multinational agreements to rebuild southern bluefin tuna populations, the tragedy of the commons has proven difficult to overcome and overfishing continues to threaten the species.

It is not just resources that can be mobile. Small communities can also have geographically mobile populations. One of the present authors' work on sustainable housing development illustrates how communities within a highly mobile society must reassess boundaries to develop the social bonds necessary for effective governance over natural capital (Gabe *et al.*, 2009). A small Maori community planning a housing development via numerous *hui* (tribal meetings) recognised that, in order to achieve their shared environmental values, it was necessary to take responsibility for actions outside what could be achieved through co-location and neighbourhood design. Importantly, their values and social responsibility did not stop at the boundaries of the development. In another project, market-driven developers of a master-planned 'sustainable' community drew clear boundaries around their development and showed little willingness to assume responsibility for managing environmental damage that was not directly related to the construction of the 'hard' infrastructure (buildings, roads, etc.) they were hired to develop.

In these two cases, the Maori development serves as an example of how a small community can nest within a larger population. The community established itself through shared ancestry and cultural history, which provided clear boundaries excluding outsiders, and it clearly showed the desire for self-determination and institutions to support collective action (the *hui*). However, the geographical property asset that the community controlled was not the appropriate boundary for natural capital management, since this modern community is much more mobile than, for example, the Tikopians described previously. Instead, the behaviour of the community – both inside and outside its physical location – is the boundary necessary to preserve natural capital.

Although the market-driven developers were 'building a community', they lacked the development of social bonds, collective institutions and other 'soft' infrastructure. One could infer from their prospectus that a cohesive community would organically develop for no other reason than co-location in a collection of houses built by the same developer that were branded using the adjective 'sustainable'. But new occupants are free to exclude themselves from taking part in forming the soft infrastructure of the community, while still consuming the community's resource base. Hence this model does not appear likely to establish a shared sense of resource entitlement and 'outsiders' can exist within the community's geographical boundaries.

In isolation, community governance looks like a promising solution that may enable society to live within a fair earth share. Importantly, communities must be small enough that the social bonds and norms created within them can limit natural capital consumption without excessive monitoring costs. If the Tikopian example of a population of around 1,200 is an optimum size for self-monitoring and enforcement, then the current global population of 7 billion will need to form approximately 6 million small communities. Such an outcome could produce a remarkable burst of innovation in resource management strategy, as communities fail and succeed over time, sharing experiential knowledge.

But is small scale a limiting factor? The organisational feat of building 6 million small communities would be incredibly challenging in a mobile, globalised society. And as the fisheries example attests, when a community is established an overarching government is still needed, to define resource boundaries, enforce rules for inter-community relations and manage the fallout from failed communities. So community governance cannot be a solution on its own. In addition, the two modern examples in this chapter show that natural resource mobility and population mobility add some vexing challenges to community governance as a solution which ensures the world remains within its fair earth share.

Global population growth also challenges community governance. Returning to Tikopia, the modern island still maintains its population at approximately 1,200 inhabitants. But instead of infanticide and suicide, emigration to other Solomon Islands is the modern, more ethically pleasing method of exclusion. So what becomes of these emigrants in a global system of community governance? In order to maintain the hypothetical global network of small communities, global population growth would need to be at most zero, such that emigrants from overpopulated communities could find space in underpopulated communities. Otherwise communities would grow beyond their optimum number, weakening the social bonds that ensure natural capital sustainability. And there would be no resource base to establish a new community.

Market-Based Governance

Both community management and command-and-control exhibit scale limitations. Community governance must be small enough that resource management becomes self-enforcing through social bonds, while the idea of an environmentally benevolent global dictator or global technocrat able to enforce environmental limits seems more suited to utopian fiction (or perhaps dystopian fiction, if the control methods are harsh). Economists often remind society that the free market is a type of natural governance system. So could the 'invisible hand' of the global market for goods and services be the force that prevents a global tragedy of the commons?

As this book makes clear, business as usual is not conserving global natural capital within a single earth share, so this question is effectively asking whether the system that created part of the problem can be part of the solution. But there is some logic to this irony. The introduction to this chapter acknowledged that market failures, such as artificially low prices because of externalities, are one of the main reasons why natural capital must be accounted for separately. Supply and demand models tell us that, if prices are too low, consumption will be too high. A logical conclusion is that if these market failures are fixed, then the market could conserve natural capital. So market-based governance strives to 'get the prices right' through regulatory policy that removes market failures.

Fixing these market failures is of great interest to economists, but, as Robert Stavins has reminded the American Economic Association, throughout the 100-year existence of the society, the economic problem of the commons has remained unsettled (Stavins, 2011). He goes on to argue that one of the greatest contributions to market-based governance was the recognition that incomplete property rights are responsible for many market failures associated with natural capital. Clarifying poorly defined property rights, particularly the right to damage the environment (Coase, 1960), is the theory behind 'cap and trade' systems that have been credited with reducing sulphur dioxide air pollution in the United States (Stavins, 1998).

However, the example of climate change mitigation has shown that clarifying rights to damage the environment has limitations. Global warming is, according to Stavins (2011), 'the ultimate commons problem'. The limitation is not that assigning property rights to damage the environment can fail once implemented, but that an agreement to assign property rights at the global scale is needed to fix a global market failure. In the absence of a command-and-control power to structure the global market, it appears necessary to seek a comprehensive multinational consensus to limit – and therefore price – greenhouse gas emissions. But attempts to reach a consensus on scientific and ethical grounds (the approach of this book) have failed, along with attempts at reduction based on the traditional economic logic of intergenerational efficiency (Bosetti and Frankel, 2012).

Lost in the scepticism over climate change agreements and general pessimism with the apparent inability to govern the global commons is the recent emergence of an alternative approach to market-based governance – fixing information asymmetry. Producers know much more about the production process for their outputs than consumers, including how much natural capital was consumed in production. This information asymmetry is important because consumers might change their behaviour if they knew as much as the producer. One example is the social issue of child labour; manufacturers can reduce their production cost by hiring child labourers in some countries, but their consumer base generally finds this socially repulsive and can choose to boycott products made with child labour. If the boycott is large enough, the reduced demand for the product made with child labour could put the producer out of business or force him to cease child labour production. In regard to natural capital conservation, consumers rarely know the ecological footprint of the products they consume. So if it became clear that similar services consumed different amounts of natural capital, would supply and demand alter to favour the one with less environmental impact?

To provide this additional information to consumers, a range of environmental product certifications has emerged. The dominant model is for a manufacturer voluntarily to hire a third-party – not the consumer or producer – to audit their production process and provide a certificate of eco-friendly approval. The building industry provides a useful example. New buildings, existing buildings, tenancy fit-outs and even entire neighbourhoods can receive certification from third-party non-governmental organisations such as the United States Green Building Council, which administers the LEED brand, and the Green Building Council of Australia, which administers the Green Star brand. For a detailed history of green building certification tools in these countries, which are representative of most green building certification efforts, readers can consult Cole (1999, 2005) on North American markets and Mitchell (2010) on Australian markets.

LEED and Green Star certification are based on expectations of greenness, measured by proxy by building standards and simulations, so their certifications are most applicable to new buildings or major renovations. There are very few mandatory actions in obtaining certification; prospective LEED designers must choose to comply with at least 40 per cent of the optional standards for certification. One criticism has been mixed results in building operation that may reduce credibility in the ratings, including numerous buildings that failed to meet their producers' lofty green expectations (see Gabe, 2011). Despite this uncertainty, econometric studies have put forward a strong argument that green-certified office buildings have altered supply and demand by offering their owners higher sales prices, higher rents and higher occupancy rates than comparable uncertified properties (Miller *et al.*, 2008; Eichholtz *et al.*, 2010; Fuerst and McAllister, 2011; Newell *et al.*, 2011).

Voluntary certification has an unfortunate limitation: environmental benefits become correlated with the participation rate (Borck and Coglianese, 2009). Hence most of the research and effort in certification is designed to increase participation rates in the schemes, even if, ironically, the standards are reduced to meet industry demands, such as a lower cost of certification (transaction cost). For example, one of the present authors worked with a national green building council to improve their building energy consumption standards, which typically require building designers to simulate the annual energy consumption of the proposed building in use. With numerous variables associated with operating a building – maintenance, occupant demands, hours of occupancy, climate, etc. – this task requires a lot of effort on the part of designers, so it is expensive to get a reasonably accurate simulation. In this case, it cost NZ$20,000 to run a single simulation using generic values for the operating variables, or NZ$50,000 to run multiple simulations that would predict the actual consumption within a 5 per cent margin of error. Needless to say, the industry firmly rejected the cost increase, despite the fact that it appears trivial in the context of multi-million-dollar construction projects and leads to more buildings with significant deviation in use from their 'potential' energy consumption.

Recent experience in Australia shows that government can use certification schemes to overcome most of the problems associated with voluntary certification by introducing mandatory disclosure of in-use performance. The Australian Federal Government passed the Building Energy Efficiency Disclosure Act 2010, which requires all large existing office buildings to obtain an annual 'NABERS' certificate of their in-use energy-related greenhouse gas emissions, and, importantly, to market the rating in all advertising for sale or lease. In targeting only existing buildings, the government avoided the high transaction costs and uncertainty of simulation. The new regulation uses easily verified environmental indicators, such as energy consumption (converted to units of greenhouse gas emissions), which building managers receive on their monthly utility invoices. Second, the marketing requirement introduces social enforcement and social incentives on a much larger scale than Tikopia. Prospective tenants in the Sydney office market are now accustomed to seeing the NABERS star rating on lease advertising.

The standard economic framework of energy efficiency in Australian office buildings argues that it is a poor investment in the larger context of reducing the costs of office work. In Sydney, a range of annual data can be compiled to show that, in 2010,

- office energy costs were $17 per square metre
- net office lease costs were $542 per square metre
- salaries in the financial services and insurance sectors (the most typical Sydney office tenants) were approximately $4,700 per square metre.

(Energy and lease cost data here are taken from the Property Council of Australia [2010]. The labour rate is calculated by taking the average March 2011 quarter salary for Sydney in 'Banking and Financial Services' [A$97,113] and 'Insurance and Superannuation' [A$87,780] from the My Career website, which publishes average salaries based on advertisements placed for each sector [My Career, 2011]. The midpoint between these two salary averages is then divided by the average Australian CBD office worker density of 19.5 square metres of net lettable area (NLA) per employee [Warren, 2003].)

Hence, improving labour productivity is the better investment (green building firms claim to do this, too, but the data are less certain than energy conservation). Yet despite the triviality of energy costs, a number of large office building owners entered the NABERS certification scheme voluntarily, before mandatory disclosure, and proactively invested in energy efficiency. Recent evidence across all of Australia suggests that owners receiving multiple certificates have reduced their greenhouse gas emissions from energy consumption, on average, by over 10 per cent in the first two years (Gabe, forthcoming). Owners that re-certified five years in a row reduced greenhouse gas emissions by 25 per cent. This five-year reduction is equal to the amount suggested by climate change mitigation experts as a reasonable 50-year target for the entire building sector (see Pacala and Socolow, 2004). So, by adding social incentives, the Australian government's mandatory disclosure scheme has rapidly accelerated private investment in natural capital conservation.

Current research is looking for the limits to this bottom-up combination of market governance and community management. In theory, there is equilibrium between society's desire to invest in economically efficient ways and its desire to conserve the environment, so this governance model may not reach the fair earth share because other market failures exist. It may also only work for indicators that are cheap and easy to measure accurately, such as energy consumption.

But the Australian example demonstrates that such a simple regulation can ignite human innovation and private investment in conserving natural capital, with results that exceed expectations. So if there is a single generic recommendation in this chapter, it is that this new model of mandatory disclosure should be experimented with in resource conservation policy. And, unlike assigning property rights which require a command-and-control style initiation to market-based governance, fixing information asymmetry can begin without a global agreement because consumers 'price' social responsibility, the limiting force that drives success in community management. In this context, non-disclosure also conveys information to consumers, so governments that do not adopt it could pay a price in reduced demand.

Conclusion

Historical examples give salience to the argument that command-and-control and community management regimes can conserve natural capital within a fair earth share, but neither appears – in isolation – to answer the question of which governance structure will protect the planet. The success of modern environmental regulation in capital-intensive countries, along with environmentally benevolent dictatorial rule in the Dominican Republic, demonstrates that command-and-control governance can preserve natural capital. But the side-effects – perceived economic inefficiency and violent human rights violations – have led society to look for alternatives. The community management regime in historic Tikopia represents one potential alternative, in which social bonds and collective action limit resource consumption. However, successful examples involve clear geographical and resource boundaries. Community governance faces major challenges when resources and communities are mobile. In addition, both community management and command-and-control suffer from a scale limitation. Larger-scale governance is needed to define the boundaries of community management, global agreements on environmental limits are rare, and global dictators able and willing to enforce environmental limits have never emerged in human history.

Within the context of modern globalisation, where transport and relocation are cheap, market-based governance may be a solution at the global scale. But since 'business as usual' is largely responsible for undervaluing natural capital, the global market needs regulatory changes that remove market failures associated with undervalued natural capital. Fifty years ago there was a perceived breakthrough, and it was thought that defining incomplete property rights could fix these market failures. But recent experience has shown that the global governance needed to comprehensively assign property rights is lacking. Instead, there are hints that market governance could help break the scale limitations of community management by introducing social responsibility through improved information symmetry between producers and consumers. There is some evidence that mandatory disclosure schemes with low transaction costs, where producers must advertise their environmental credentials, can direct innovation and investment towards natural capital conservation.

So, as Elinor Ostrom often said, there is no panacea with regard to governance models for a fair earth share. But there are common themes. Successful community management and mandatory disclosure schemes both rely on limitations provided by societal responsibility, not financial budgets. Current governments wishing to conserve natural capital must experiment with the best ways to introduce societal responsibility, whether it is defining community resource management schemes, when feasible, or improving the information provided to consumers on producer resource management.

References

Borck, J. C. and Coglianese, C. (2009) 'Voluntary Environmental Programs: Assessing their Effectiveness.' *Annual Review of Environment and Resources*, 34, pp. 305–324

Bosetti, V. and Frankel, J. (2012) 'Politically Feasible Emissions Targets to Attain 460 ppm CO_2 Concentrations.' *Review of Environmental Economics and Policy*, 6, pp. 86–109

Coase, R. (1960) 'The Problem of Social Cost.' *Journal of Law and Economics*, 3, pp. 1–44

Cole, R. J. (1999) 'Building Environmental Assessment Methods: Clarifying Intentions.' *Building Research & Information*, 27, pp. 230–246

Cole, R. J. (2005) 'Building Environmental Assessment Methods: Redefining Intentions and Roles.' *Building Research & Information*, 33, pp. 455–467

Davies, J. C. and Mazurek, J. (1998) *Pollution Control in the United States: Evaluating the System*, Resources for the Future, Washington, DC

Diamond, J. M. (2005) *Collapse: How Societies Choose to Fail or Succeed.* Viking Penguin, New York

Eichholtz, P., Kok, N. and Quigley, J. M. (2010) 'Doing Well by Doing Good? Green Office Buildings.' *American Economic Review*, 100, pp. 2492–2509

Fuerst, F. and McAllister, P. (2011) 'Green Noise or Green Value? Measuring the Effects of Environmental Certification on Office Values.' *Real Estate Economics*, 39, pp. 45–69

Gabe, J. (2011) 'Market Implications of Operational Performance Variability in Certified Green Buildings.' In *SB11 Helsinki World Sustainable Building Conference*, 18–21 October 2011, Finnish Association of Civil Engineers & VTT Technical Research Centre of Finland, Helsinki

Gabe, J. (forthcoming) 'The Environmental Performance of Green Buildings in Australia'

Gabe, J., Trowsdale, S. and Vale, R. (2009) 'Achieving Integrated Urban Water Management: Planning Top-Down or Bottom-Up?' *Water Science and Technology*, 59, pp. 1999–2008

Gneezy, U. and Rustichini, A. (2000) 'A Fine is a Price.' *The Journal of Legal Studies*, 29, pp. 1–17

Gutierrez, N. L., Hilborn, R. and Defeo, O. (2011) 'Leadership, Social Capital and Incentives Promote Successful Fisheries.' *Nature*, 470, pp. 386–389

Lawless, R. (1992) *Haiti's Bad Press: Origins, Development, and Consequences*. Schenkman Books, Rochester, VT

Miller, N., Spivey, J. and Florance, A. (2008) 'Does Green Pay Off?' *Journal of Real Estate Portfolio Management*, 14, pp. 385–399

Mitchell, L. M. (2010) 'Green Star and NABERS: Learning from the Australian Experience with Green Building Rating Tools.' In Bose, R. K. (ed.) *Energy Efficient Cities: Assessment Tools and Benchmarking Practices*. World Bank Publications, Herndon, VA

My Career (2011) 'Salary Centre.' http://content.mycareer.com.au/salary-centre/, accessed 14 June 2011

Newell, G., Macfarlane, J. and Kok, N. (2011) *Building Better Returns: A Study of the Financial Performance of Green Office Buildings in Australia*. Australian Property Institute, Sydney

Ostrom, E. (1990) *Governing the Commons: The Evolution of Institutions for Collective Action*. Cambridge University Press, Cambridge

Pacala, S. and Socolow, R. (2004) 'Stabilization Wedges: Solving the Climate Problem for the Next 50 Years with Current Technologies.' *Science*, 305, pp. 968–972

Pomeroy, R. S., Katon, B. M. and Harkes, I. (2001) 'Conditions Affecting the Success of Fisheries Co-management: Lessons from Asia.' *Marine Policy*, 25, pp. 197–208

Property Council of Australia (2010) *Benchmarks 2010 Survey of Operating Costs: Sydney Office Buildings*. Property Council of Australia, Sydney

Schwartz, T. T. (2008) *Travesty in Haiti: A True Account of Christian Missions, Orphanages, Fraud, Food Aid and Drug Trafficking*. Booksurge Publishing, Charleston, SC

Stavins, R. N. (1998) 'What Can We Learn from the Grand Policy Experiment? Lessons from SO_2 Allowance Trading.' *Journal of Economic Perspectives*, 12, pp. 69–88

Stavins, R. N. (2011) 'The Problem of the Commons: Still Unsettled after 100 Years.' *American Economic Review*, 101, pp. 81–108

UNU-IHDP and United Nations Environment Programme (2012) *Inclusive Wealth Report 2012: Measuring Progress Toward Sustainability*. Cambridge University Press, Cambridge

Warren, C. (2003) 'New Working Practice and Office Space Density: A Comparison of Australia and the UK.' *Facilities*, 21, pp. 306–314

9 Services

Soo Ryu

The 'business as usual' approach long employed by industrialized societies is in need of a restructure. It results in unequal distribution of depleting resources, with UN statistics showing that the 20 per cent of the world's population that lives in rich countries consumes up to 80 per cent of the world's resources (Wackernagel and Rees 1996: 155). This shows the problem of excessive consumption by a minority whose ethical responsibility should be finding ways to reduce their ecological footprint while maintaining a good standard of living without compromising the prospects for growth and development of those whose basic standard of living is poor (Wackernagel and Rees 1996: 154). This requires the will to change both current lifestyles and the systems on which these lifestyles are based. It is not only individuals but also larger groups, such as governments, policy makers, institutions and corporations that can set an example and achieve reductions in their ecological footprint. Sustainability needs to be more than just environmental – it needs to embrace ethical, political, social and economic concerns (Dos Santos Martins, 2009: 3).

As a response to these pressing issues, there is an increasing participation and involvement from universities around the world in the practice of sustainability. Universities have always had an important leadership role in society in demonstrating the types of changes that need to occur with respect to the prime issues of the time. An academic institution's role of providing leadership in its community can be utilized to encourage and influence both more sustainable living (Mardon, 2007: 29) and the application of technological advancement through research. Therefore institutions are critical place for changes to occur, as they act as a bridge between the living, working and learning environments. But what does it really mean to be sustainable?

A Case Study: Victoria University of Wellington School of Architecture and Design, New Zealand

Many of the critical services people depend on to run a country include education, infrastructure, health, welfare and law and order services. Most of these services take place in or operate from office-type buildings. As an example of what can be achieved in a services-type situation, a tertiary education institution is chosen here as a case study. The data presented are from a much more detailed study which looked at the practical implications of trying to reduce carbon emissions (Ryu, 2010).

In 2008, Victoria University of Wellington (VUW) School of Architecture and Design (SoAD) became the world's first school of architecture and design to become certified carbon neutral through purchased and donated carbon credits from the renewable electricity supplier Meridian Energy (Faculty of Architecture and Design, 2008: 4; carboNZero, 2009: 2). SoAD's greenhouse gas (GHG) footprint for the year 2007 was reduced from 335 tonnes to zero by purchasing an extra 135 tonnes (40 per cent) of credits to add to the 200 tonnes (60 per cent) donated. The calculation took into consideration emissions from gas for heating, waste to landfill, transport by air for staff travel and other miscellaneous factors.

How many factors should an institution bear in mind when considering carbon neutrality? A lack of records and difficulty in attaining accurate data diminished the ability to take into consideration all the externalities, including all goods purchased for operation of the school, food consumed during operation hours and transport for travel to the school by students and staff (Faculty of Architecture and Design, 2008: 2).

The carbon offsetting approach could be a future problem as purchasing offsets in a free market is not always reliable and the scientific foundation for offsetting is complicated to calculate, often not precise and does not guarantee the capture of carbon forever (Smith, 2007: 6). Resorting to carbon offsets is a global phenomenon not just among tertiary institutions but also among businesses and government enterprises as well. Offsets do not provide incentives for individuals and institutions to greatly change consumption patterns or internal systems and it could be argued that the promotion of offsets indicates a flaw in our current economic system that gives people a loophole.

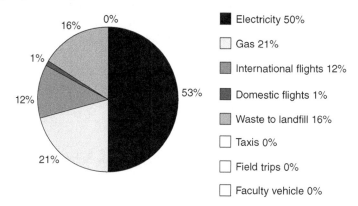

Figure 9.1 GHG emissions (tonnes CO_2e) by emission source for SoAD

Source: Data from Faculty of Architecture and Design (2008: 4)

Goals for Institutions to Set as Targets

In order to analyse a range of implications and solutions, three reduction standards are explored and established here. The lowest reduction goal for this study is based on the New Zealand Ministry of Environment's proposal, under the Kyoto Protocol, to reduce New Zealand's greenhouse gas emissions to 1990 levels. The Ministry stated: 'New Zealand's emissions are increasing, with emissions in 2005 about 25 per cent higher than they were in 1990' (Ministry for the Environment, 2008), meaning that a 25 per cent reduction is needed.

A medium-range reduction goal is derived from the NZ Green Building Council (Fernandez, 2008: 64), which suggests through its Green Star rating tool an approximately 50 per cent reduction from SoAD's current energy use.

The highest level of reduction is based on Monbiot's assertion that developed countries need to reduce emissions by at least 90 per cent to avoid excessive global warming (Monbiot, 2006: 16). This has been set as the highest target for this study.

It is clear that currently there is no consensus over which reduction target should be achieved – all are voluntary. This poses a problem, as voluntary targets mean that there is no incentive for anyone to make reductions, which may well be achieved only at an economic loss. It is clear that in the long term New Zealand's responsibility is a 90 per cent reduction, a target that might seem unrealistic at present. Given that the reductions in ecological footprint (EF) needed for developed countries to achieve a fair share are at least 50 per cent of current footprints, the three carbon reduction ranges – 25, 50 and 90 per cent – are here studied in detail, to show the likely effects of each level of reduction on an institution. Carbon emissions are used here as a proxy for EF reductions. This is a reasonable approach, as much of the EF uses energy and emissions as the basis for calculation.

Solutions: What Institutions Need to Do to Be within Their Footprint

This section explores various solutions for reducing the three major contributors to SoAD's current carbon footprint, which are energy, transport and waste. It looks at the difference between behavioural changes (low cost) and technological investment (high cost) in order for SoAD to reduce its footprint to meet the three levels of reduction targets. This research can help in understanding the implications of attempting to reduce consumption patterns of office-type buildings with similar activities and behavioural patterns. The purpose of this example is to show what levels of reduction are possible with different types of behaviour and technical change (Tables 9.1, 9.2 and 9.3).

150 *Soo Ryu*

Table 9.1 25, 50 and 90 per cent reduction schemes for energy use

Reduction target	Solutions for reducing electricity and gas use
25% reduction	• *Lighting* Up to 32 per cent lighting electricity reduction achieved by de-lamping and installing lower wattage lamps (and/or dimmers), due to existing lighting being above recommended levels. • *Equipment/computers* Switch off 50 per cent of student computers currently left on after closing hours (11:30 pm to 7:30 am) during main opening period throughout the year (March to November). Switch off 50 per cent of student computers after class hours (5:30 pm onwards) and switch all staff computers to go into sleep mode after 20 minutes of inactivity. The common accepted behaviour is that many computers are left on even when not in use. • *VAC (ventilation and air conditioning)* Switch off 50 per cent of labs with air conditioning after class hours during the main operational period in the year. Close 100 per cent of computer labs with air conditioning during summer school period (November to February), when the building is mainly vacant of students. In Wellington, New Zealand air conditioning is not really needed due to the temperate climate (the mean temperature in Wellington during the hottest month of January is 17°C). Reduce excessive fresh air intake to reduce air handling unit load during the main opening period throughout the year. • *Heating* Reduce the heating period to five months (May to September) by eliminating heating in summer months and not heating spaces that are intermittently used, such as corridors and the atrium.
50% reduction	• *Lighting* Turn off 50 per cent of remaining lights left on after school closing hours. Close down at least 50 per cent of the 'student services' (studios and lab spaces) part of the building during the summer school period. Close off 50 per cent of the teaching areas (lecture theatres and seminar rooms) after class hours during the main school opening period throughout the year. Install motion sensors in intermittently used spaces (corridors, staircases, toilets, kitchens, etc.). • *Equipment/computers* All computers need to be set to sleep mode, hibernating after 20 minutes of inactivity during operation hours. Or all staff computers need be replaced with the equivalent laptop and 70 per cent of student computers need to be replaced with laptops to achieve a combined 50 per cent saving in electricity. • *VAC* Achieve 50 per cent or greater reduction in ventilation and air conditioning load by changing the computers to laptops (with similar specifications) in computer labs. The power rating of the computers is an indication of heat output. Laptop electricity use is 65 to 80 per cent lower than that of current desktop computers. • *Heating* Replace existing single-glazed aluminium windows with argon-filled, low-emissivity double glazing with thermally broken frames. Increase roof and wall insulation and improve the building airtightness to reduce heating load during winter months.

Reduction target	Solutions for reducing electricity and gas use
90% reduction	- *Lighting* Savings require a combination of switching from T8 Fluorescent to LED equivalent lighting and lowering existing lighting levels. LED power consumption is 60 to 80 per cent lower but around 10 times more expensive than an equivalent fluorescent light; however, lifespan is five times longer, reducing the price difference to half. - *Equipment/computers* Switch all standard computers to laptops with equivalent specification (79 per cent savings) and reduce the number of computers available. 11 per cent savings made by reducing the number of student PCs by 63 per cent. - *VAC* Eliminate air conditioning through passive cooling methods and changing the occupants' cooling expectations. Retrofit the building so that it does not need any active cooling system. 70 per cent VAC electricity reduction can be made by eliminating air conditioning throughout the whole year. 11 per cent VAC electricity reduction can be made by reducing fresh air intake during non-cooling periods (nine months of the year) to Building Code requirement. This gives a total of 81 per cent reduction. The remaining 9 per cent saving would need to be met via renewable electricity generation, as minimum fresh air intake cannot be further compromised as passive ventilation to all spaces is improbable. - *Heating* In addition to the 50 per cent reduction solutions, further reductions are possible by changing the heating period during the heating season from all-day heating to only morning and evening (7:00 am to 9:00 am and 5:00 pm to 11:00 pm). Change the volume of the space to be heated by lowering the current ceiling height from 4 to 2.5 metres. This can be done by installing false ceilings in offices. Increase occupancy in the staff offices from one person to two by allowing hot-desking and allowing researchers to work at home. Reduce the heating level from 21°C or higher to the World Health Organization minimum heating standard of 18°C. Increase wall insulation, approximately doubling it to achieve an R-value of 4.8.

Table 9.2 25, 50 and 90 per cent reduction schemes for transport

Reduction targets	Solutions for reducing transport use
25% reduction	• *Land* 40 per cent of staff who drive their cars to work (one person per car) must switch to public transport (bus or train) or, for those who live further away, carpool in a car of four. Existing bus riders need to walk or cycle to work or use electric trolley buses, which in Wellington are running on renewable electricity. • *Air, domestic* Around 30 per cent of the average staff domestic air travel distance of approximately 600 kilometres per year (Auckland to Wellington) needs to be replaced by train, bus, or a car with four or more passengers, all of which have much lower CO_2 emissions. However, in the current market, travelling by air can sometimes be cheaper than train, and takes one tenth of the time. Alternatively, 25 per cent of current staff travel needs to be replaced by tele- or video-conferencing. • *Air, international* Staff air travel to the Asia Pacific region (particularly to Australia), which makes up around 25 per cent of all international flights taken, needs to be replaced with tele- or video-conferencing, as the geographical and cultural distances are not great.
50% reduction	• *Land* In conjunction with the 25 per cent savings made by staff, half the students that live within 5 kilometres of the school should cycle, and those within 2 kilometres should walk. These schemes need to consider incentives for staff and students to choose alternative forms of transport. Alternatively, the school needs to reduce the number of times the students need to come into school by hosting lectures online, utilizing online submissions and scheduling timetables more efficiently so students have a full, productive day. • *Air, domestic* Around 50 per cent of staff domestic air travel over a distance of approximately 600 kilometres needs to be replaced with train or bus. Alternatively, 50 per cent of current staff domestic travel needs to be replaced by tele- or video-conferencing. • *Air, international* The choice is between either replacing half of international flights taken with tele- or video-conferencing or using an alternative transport method. This could be achieved by using a cargo passenger ship, which would use 99 per cent less energy per kilometre than flying. To travel to Hong Kong, the journey would take 16 days with full board on such a ship costing about the same as a premium economy air fare. This would call for staff to be away longer when they travel, although during this time they can still work, as Internet access is usually available. However, there are no regular timetables and places on such ships are limited.

Reduction targets	Solutions for reducing transport use
90% reduction	- *Land* All staff and students would need to move closer to the city to allow them to walk or cycle, or they would need to travel on transport powered by renewable electricity, such as electric buses or trains. If they choose to live further away they would need to work from home and come to school only once a week. - *Air, domestic* All travel in the North Island needs to be by train. Travel is allowed only within the island, as taking the ferry to cross to South Island results in higher emissions. Alternatively, 90 per cent of current staff travel needs to be replaced by tele- or video-conferencing. - *Air, international* Unfortunately there are currently no alternatives to air travel from New Zealand other than by sea. All major trade routes to Australia, Asia, the Americas and Europe need to be replaced with passenger cargo ships as the main form of travel. Staff have the choice to travel this way, use web technology to enable video-conferencing, or simply to reduce travel.

Table 9.3 25, 50 and 90 per cent reduction schemes for waste

Reduction targets	Solutions for reducing waste
25% reduction	- *Paper* Replace all virgin paper with 100 per cent recycled paper. Currently the majority of the paper used is virgin. - *Recyclables* Reduce all glass bottle consumption because glass has a higher emission impact than plastic due to its heavy mass for transportation. This can be achieved by reusing glass bottles. - *Organics* First point of reduction should be reducing food scraps, of which more than half could have been eaten. This may be linked with the fact that the average New Zealander is overeating and wasting food. The use of worm farm compost bins to convert organic waste into compost requires money, maintenance and appropriate land space. It is also important to note that worms do not eat meat, some dairy and certain type of strong-smelling vegetables and fruits. However, this solution is much more economical than paying for rubbish collection, and compost can either be sold or reused for the gardens.

Reduction targets	Solutions for reducing waste
25% reduction (cont'd)	• *Non-recyclables* A lot of takeaway meals are consumed, creating huge packaging waste. This can all be eliminated by students and staff eating in local restaurants and cafés. Ideally, the school could operate a canteen serving cheap, healthy meals to students and staff using ingredients from local markets. Paper towels can be replaced with reusable cotton towels. The relatively large amount of cigarette butts generated touches on the sensitive topic of whether smoking should be tolerated. A society that markets and sells poison which eventually leads to death and which puts strain on the health system indicates the present situation, in which global society works for financial rewards even at the cost of many lives. Dealing with waste to landfill is a government problem and regulations are required to ban materials that cannot be recycled.
50% reduction	• *Paper* Replace 50 per cent of newspaper and magazine subscriptions with online subscriptions. Students and staff need to reduce their personal paper and cardboard consumption by half. Currently the school uses six reams per person annually, or approximately 3,000 sheets of paper per person per year. • *Recyclables* Reduce plastic bottle consumption by 50 per cent through buying in bulk. Staff and students should be encouraged to drink from nearby pubs and cafés where drinks are served in reusable glasses. Minimizing the consumption of soft drinks and energy drinks is vital in reducing plastic bottle use. This can be discouraged by eliminating vending machines in the building. • *Organics* Produce no food waste that could have been eaten, or revert to using worm farm compost bins.
90% reduction	• *Paper* In conjunction with the 25 per cent and 50 per cent reduction schemes, SoAD can further reduce paper use by 50 per cent through double-sided printing and photocopying, and by another 25 per cent by printing two pages per sheet. This is already a 75 per cent reduction in paper use. A policy of having all submitted work, pdfs and ebooks, online journals and lecture notes available online for students, all writing and editing of staff and student work to be done on the computer and scanning digital copies instead of photocopying can save on paper use. Paper that cannot be double sided can be reused for note-taking in lectures. A 90 per cent reduction leaves 276 sheets (half a ream) per person per year. • *Recyclables* A 90 per cent reduction eliminates virtually all drinks bought from the store and relies on drinking water or home-made beverages in refillable glass bottles. Ensure that there are easily accessible water taps that can be used to refill glass water bottles. These schemes are healthier and cheaper options in the long run. • *Organics* After reducing as much food waste as possible, there is no other option but to rely on worm farm compost bins to take care of the rest of the organic matter, avoiding foods that the worms cannot process.

This brief overview of means of achieving reductions shows that it is very often more a matter of behaviour than one of technology. Some of the proposals mean quite large behavioural changes, such as giving up air travel and avoiding waste. However, even a 90 per cent reduction in overall emissions does not guarantee 'carbon neutrality', as there is still the remaining 10 per cent. It may be wise to attempt to deal with this by asking staff and students to reduce their personal emissions when they leave work. This ensures that behavioural changes are continued at home. These behavioural changes can be converted into measurable 'offsets' by calculating the savings achieved compared to the usual behaviour, such as savings gained from using more energy efficient appliances and turning them off when not in use. This is something typical carbon offset programmes do not offer, and represents the only cost-effective way of becoming truly 'carbon neutral'. Another option is to purchase a wind turbine off-site (solar energy is currently too expensive and Wellington is known for its abundant wind resource) to achieve carbon neutrality in energy use. This would come at a very high cost and would be out of SoAD's control. In addition, it might be beneficial to allocate specific personnel to monitoring, consistent review of performance and reporting of emissions reduction. This ensures savings can be guaranteed over time and gives the opportunity to analyse which reduction schemes are most effective for the school.

Conclusion

This research examines ways in which the School of Architecture and Design, as an example of a typical institution in the service sector, can become truly carbon neutral by dramatically reducing its most significant contributions to GHG emissions (and hence its ecological footprint) in order to meet three reduction goals, thereby minimizing its reliance on purchased carbon offsets. What this study also indicates is that, in order for SoAD to be truly carbon neutral, other important factors which are outside the institution's control need to be considered. It raises the issue that, in order for institutions to live within their ecological footprint, government policies need to change to address the bigger issues and perhaps to question the very nature of the way in which the current economic system operates.

A comparison and analysis of the different solutions available using today's technology has been provided. Considering the cost implications of each solution, the findings are that a 25 per cent reduction could be achieved through simple behavioural changes which cost very little, as they are mainly related to avoiding wastage; 50 per cent could be achieved through a combination of low- and high-cost measures, mainly to improve efficiency; and 90 per cent could be achieved through considerable investment in renewable technologies or drastic reduction in use and convenience. User behaviour is paramount in achieving and aiding the reduction goals, but in the absence of any incentives or rules it is difficult to change.

This research aims to provide a general guideline for institutions around the world by demonstrating a specific reduction programme for SoAD. At the same time, the challenge is for an institution to demonstrate that it can still remain part of a modern society by maintaining comfort and improving the quality of life even when its footprint has been reduced to be within the capacity of the planet. The most important thing to note is that, the longer the problem is ignored, the smaller the chance for a smooth transition to sustainability.

The focus also needs to shift from changing buildings to changing the very operation of the institutions which use those buildings. In the service sector it may be up to institutions to act as a catalyst for change and to set an example to be adapted by businesses and governments.

References

carboNZero (2009) *Certified Organisations*, http://carbonzero.co.nz/members/organisations_certified.asp#VUW, accessed 2 February 2009

Dos Santos Martins, Rui H. (2009) 'Sustainable Development Requires an Integrating Discipline to Address its Unique Problems – Design Thinking', *Proceedings of the 53rd Annual Meeting of the International Society for the Systems Sciences*, University of Queensland, Brisbane, Australia, 12–17 July 2009, http://journals.isss.org/index.php/proceedings53rd/article/viewFile/1156/459, accessed 9 July 2012

Faculty of Architecture and Design (2008) *Greenhouse Gas Reduction Plan*, Victoria University of Wellington, http://www.victoria.ac.nz/home/about/publications/2008_GHG_emission_reduction_plan_-_FoAD.pdf, accessed 13 February 2009

Fernandez, N. P. (2008) 'The Influence of Construction Materials on Life-Cycle Energy Use and Carbon Dioxide Emissions of Medium Size Commercial Buildings', PhD thesis, Victoria University of Wellington, New Zealand, http://researcharchive.vuw.ac.nz/bitstream/handle/10063/653/thesis.pdf?sequence=1, accessed 2 July 2009

Mardon, H. (2007) 'Schools in a Carbon Neutral World', in N. Harre and Q. Atkinson (eds) *Carbon Neutral by 2020: How New Zealanders Can Tackle Climate Change*, Craig Potton Publishing, Nelson, New Zealand, pp. 28–38

Ministry for the Environment (2008) *The Kyoto Protocol*, 20 March 2008, http://www.mfe.govt.nz/issues/climate/international/kyoto-protocol.html, accessed 5 July 2009

Monbiot, G. (2006) *Heat: How to Stop the Planet from Burning*, Allen Lane, London

Ryu, S. (2010) 'Guidelines to Make Victoria University School of Architecture Carbon Neutral through Minimising Its Reliance on Offsets', Master of Architecture thesis, Victoria University of Wellington, New Zealand, http://researcharchive.vuw.ac.nz//handle/10063/1367, accessed 18 June 2012

Smith, K. (2007) *The Carbon Neutral Myth: Offset Indulgences for your Climate Sins*, Transnational Institute, The Netherlands, February 2007, http://www.carbontradewatch.org/pubs/carbon_neutral_myth.pdf, accessed 16 March 2009

Wackernagel, M. and Rees, W. (1996) *Our Ecological Footprint: Reducing Human Impact on Earth*, New Society Publishers, Gabriola Island, BC, Canada

Part III
Footprints in the Past

10 A Study of Wellington in the 1950s

Carmeny Field (with Brenda Vale)

It is generally accepted that more modern lifestyles consume more resources, so there is a pressing need to discover how and why ecological footprints have changed over time. Analysing the almost certainly lower ecological footprint of a developing country will not always provide a relevant example for developed nations to follow, because for many people a low-impact lifestyle results from living with rural subsistence farming. For example, in India in the 1990s about 70 per cent of the population were involved in agriculture (EIU, 1997) when the country had an ecological footprint of only 0.38 global average hectares (gha) per person (Wackernagel and Rees, 1996).

One way to explore lower footprints is to go back in time. This chapter describes the ecological footprint (EF) of people living in Wellington, New Zealand in 1956. This date was chosen both because it coincided with census information and for the fact that in the 1950s life in Wellington was not so different from that of today – in both times people work to support the family and live in owner-occupied separate dwellings, the majority of which are detached houses; childhood education is free; health care is available; people have enough money to go on holidays; and so on. By talking to people about life in Wellington at that time through holding a series of focus groups linked to a survey questionnaire, it has also been possible to build a more detailed picture of life in the 1950s in Wellington and how acceptable people found the lifestyle. These results are compared with the 2005 ecological footprint, calculated using the same method, and information about quality of life in modern Wellington from another study (Nielsen, 2009). This comparison is significant because it may be possible to show whether the current EF can be reduced while still maintaining an acceptable lifestyle. It will also be possible to see whether this comes close to being a fair earth share footprint.

The scope of the study is the city of Wellington, as defined by the Wellington City Council boundaries, and includes the Wellington CBD (central business district) and surrounding suburbs, north to Tawa and Takapu Valley. This meant the whole of Wellington was included but not the Hutt Valley, which came under the jurisdiction of a different council.

The area also corresponded to the coverage of the tram routes in the 1950s.

The original intention was to use a 'bottom-up' or component-based analysis to calculate the ecological footprint for both years, using the components of housing, transport, food, consumer goods, and services (Close and Foran, 1998). However, as data were collected it became evident that lack of information specific to Wellington meant that national or regional data needed to be used and apportioned to the Wellington city population. For 1956, different sources were sometimes used to create a composite value for the EF calculation. One example of this is the 1956 food consumption data. These were initially extrapolated from FAO data first collected in 1961. The estimated weight of food consumed according to this source was checked against national Consumer Price Index data available for 1955, which recorded the weight of and money spent on food bought by the total population. A composite value was then produced that was used in the 1956 EF calculation (Field, 2011: 100–107). All data were averaged to a per capita level before the ecological footprint was calculated for each footprint category.

The EF was also apportioned into different land types. Two main types of land are used. The first is the actual land used, built on or converted to produce the resources needed to sustain the population being studied: consumed, crop, grazing and forest land. The second type is energy land, or the equivalent land area needed to produce the total energy required for the population's consumption of energy, including energy related to goods and services. Wackernagel and Rees (1996: 69) estimate an energy-to-land ratio of 100 gigajoules per hectare (GJ/ha) as a global value, based on a CO_2 absorption approach for fossil fuel use. This figure does not take into account the reduced impact of hydroelectricity generation, and about a third of New Zealand's primary energy generation is from renewable resources, such as hydroelectric, geothermal and wind energy. For this research, two land energy ratios were used for the Wellington footprint calculations. The first was the energy-to-land ratio for electricity, taking into account the renewable resources used to generate the primary energy. A productivity of 1,000 GJ/ha (Wackernagel and Rees, 1996: 69) was applied to the percentage of renewable primary energy, the majority of which in New Zealand is hydroelectricity. In 1956, hydroelectricity accounted for 96 per cent of all renewable energy in New Zealand (Palmer, 1974: 35). The energy-to-land ratio for the generation of electricity in New Zealand was 906.5 GJ/ha in 1956, when New Zealand's electricity was 89 per cent renewable (mostly hydro). The 2006 ratio was 711 GJ/ha when only 66 per cent of electricity was generated from renewable resources. The second energy-to-land ratio was for all other energy use and employed the higher value of 150 GJ/ha. New Zealand's forests are highly productive, with Bicknell and colleagues stating that the energy-to-land ratio for fossil fuels in New Zealand may be 150 GJ/ha (Bicknell *et al.*, 1998).

EF Calculation Results

The EF of those living in Wellington in 1956, as calculated, was 1.68 hectares per person (ha/person) and the land area of Wellington was 290,000 hectares (ha). This results in a biocapacity of 2.10 ha/person for the population of 138,297 in 1956, so Wellington was in ecological reserve. This is a significant finding, as it demonstrates that an urban population with a recognisably modern lifestyle, being based on wage earners, home ownership and travel to work and school, etc. could effectively live within the land area they occupied. Obviously this is not what happens, since even life in 1956 was based to a considerable extent on imports, with the money for these coming from New Zealand exports. However, the principle that urban life could be managed within ecological boundaries is very important.

However, the 1956 EF of 1.68 ha/person figure is not globally comparable because New Zealand's land has a higher productivity than the global average, by a factor of 2.5 (Ministry for the Environment, 2003: Executive Summary). Taking this into account and converting to global average hectares (gha), the 1956 EF becomes 4.19 gha/person, showing that even in 1956 life was not within the current fair earth share footprint of 1.8 gha/person (Wilson, 2001: 1). However, by 2006 the EF had risen by 43 per cent to 2.41 ha/person (6.0 gha/person) and, with the 30 per cent increase in population, this meant the urban population no longer lived within its ecological boundaries. The two footprints and their consumption categories are compared in Table 10.1.

Table 10.1 shows where the big increases have occurred. As might be expected the EF of food has changed very little, since there are limits to how much a person can consume. However, to look at where changes within even a small increase have occurred it is necessary to look at the land components of EFs, as shown in Table 10.2. A more detailed discussion of the five EF components and the changes in land follows.

Table 10.1 1956 and 2006 ecological footprint (EF) per person component comparison (values rounded)

Category	EF per person 1956	EF per person 2006	Percentage difference
Housing (ha)	0.07	0.12	+71%
Transport (ha)	0.23	0.47	+104%
Food (ha)	0.75	0.77	+3%
Consumer goods (ha)	0.40	0.72	+80%
Services (ha)	0.23	0.35	+52%
Total (ha)	1.68	2.41	+43%
Total, allowing for land productivity (gha)	4.19	6.03	+43%

Table 10.2 1956 and 2006 EF for land-use categories. Square metre values have been rounded to whole numbers

Land type	Year	Area (m²)						
		Food	Housing	Transport	Consumer goods	Services	Total
Consumed	1956	0	78	16	47	10	151
	2006	0	242	44	954	117	1,357
	Difference	0%	+209%	+169%	+1,934%	+1,101%	+798%
Garden	1956	—	182	—	87	—	270
	2006	—	186	—	167	—	353
	Difference	—	+2%	—	+91%	—	+31%
Crop	1956	544	—	—	4	—	549
	2006	638	—	—	1	—	639
	Difference	+17%	—	—	−85%	—	+16%
Grazing	1956	6,576	—	—	3,538	—	10,114
	2006	6,282	—	—	4,922	—	11,204
	Difference	−4%	—	—	+39%	—	+11%
Forest	1956	—	35	—	81	—	115
	2006	—	63	—	593	—	656
	Difference	—	+82%	—	+636%	—	+470%
Energy	1956	347	414	2,277	303	2,222	5,563
	2006	752	681	4,621	652	3,174	9,880
	Difference	+117%	+65%	+103%	+116%	+43%	+78%
Total	1956	7,467	709	2,293	3,973	2,319	16,761
	2006	7,672	1,172	4,665	7,123	3,458	24,116
	Difference	+3%	+65%	+103%	+79%	+49%	+44%

Food

Food is the largest component of both the 2006 (0.77 ha, 32 per cent of the total) and 1956 (0.75 ha, 45 per cent of the total) ecological footprints. Table 10.2 shows that there have been changes in the food EF that have balanced each other out: a drop in meat eating, reflected in the grazing land component, has been offset by a large increase in energy land. This energy is the embodied energy of agricultural activity and the chemicals used for producing food such as fertilisers. Generally, the modern diet consists of more processed food and more food eaten out than in the 1950s. There is also a greater variety of food, some varieties of fruit and vegetables only available during their season in the 1950s are now found in supermarkets all year round, imported from overseas. Produce that was often home-grown in the 1950s and supplemented with bought fresh produce is now available frozen and canned. Increasing amounts of processed food are being eaten today for 'convenience'.

The focus group survey of those living in Wellington in the 1950s found that 87 per cent of those asked grew vegetables at home and 60 per cent grew fruit. Owning poultry was more common in the 1950s (according to the 1956 population census, over 500,000 birds were owned by Wellington inhabitants, which is three hens per person; Department of Statistics, 1957), and 17 per cent of focus group respondents stated that they had chickens at home, either for meat or eggs. The majority of people (60 per cent) grew less than 25 per cent of their food consumed, 20 per cent grew approximately 25 per cent of their food, 12 per cent grew 50 per cent of their food consumed, and 8 per cent grew about 75 per cent of their food. A third of the respondents said they received food from relatives. Generally a case of fruit or vegetables was sent from relatives living on farms, often yearly or monthly depending on the produce.

Shopping habits have also changed. Now a regular weekly shop is possible because of having refrigerators and freezers at home, and often this shop is done at a large supermarket, accessed by car. In the 1950s, 54 per cent shopped at their local shops, 33 per cent would shop at the dairy (the New Zealand term for a local shop) and 13 per cent shopped at markets. Approximately half those asked shopped weekly and the rest daily.

In the 1950s eating out was a treat, whereas Wellington now has a higher number of restaurants per capita than New York and is known for its dining-out culture (Ministry for Culture and Heritage, 2012). Dinners in the 1950s were commonly cooked by the mother in the family, sometimes with help from the children. Only 17 per cent of the survey said they ate out, and this was usually fish and chips rather than eating at restaurants. Nevertheless, there were no adverse comments about the diet of the 1950s. Some participants stated that 'food was nothing fancy, was simple and was what is called slow cooking nowadays' and 'dinners similar to nowadays; meat and three vegetables was a typical dinner'. From the participants' comments it is evident that food was fresh and consisted of home-cooked meals or packed lunches. There was food wastage because of the lack of refrigerators; however, people shopped regularly, a behaviour which would have minimised wastage.

Thus, although the EF of food was not very different in 1956, more food was consumed at home than currently and people shopped more often.

Implications for food policy today

Food remains the biggest single component of every person's EF and there are limits to how much this can be reduced (see Chapters 1 and 2). What is obvious from the survey is that people in the 1950s had more appreciation of the effort taken to grow, prepare and cook food because more food was grown and eaten at home, which in turn leads to less wastage as well as an understanding of the seasonality of food production.

What can be seen from Table 10.2, however, is that food falls in significance as a part of the total EF between 1956 and 2006. Although conjecture, what can be drawn from this is that the food EF is going to feature as a very large component of a fair share EF. Although some reduction can be achieved by eating less meat and fish, the other components of the EF will need to undergo greater change.

Housing

The first thing to note with housing is the jump in consumed land, which is the land covered by the house and the landfill area needed to dispose of housing construction waste, as shown in Table 10.3. In 1956 the floor area of the average house was 115 square metres (BRANZ, 2010: 16) and this had risen to 145 square metres in 2006 (Quotable Value, 2012). At the same time, occupancy rates have fallen from an average 3.8 people per house in 1956 (Statistics New Zealand, 1957) to 2.6 in 2006 (Statistics New Zealand, 2007). Construction waste was found by allotting to this category a proportion of total waste to landfill. Compared to the 81 per cent increase in consumed land for housing the associated waste rose by almost 300 per cent. Although a social rather than a policy issue, this figure, which warrants more investigation, covers waste not just for new housing but also for house renovation activities, with the assumption being that these have increased significantly in the period between 1956 and 2006. This is part of the commodification of houses, which sees them not as a service but as an investment. In the past, beyond interior decoration, people were more ready to accept the house as it stood, whereas now a huge market has been created based on selling people materials and services to change their houses.

Table 10.3 EF of consumed land for housing

Consumed land for housing	1956 (ha/person)	2006 (ha/person)
House	0.0031	0.0056
Construction waste	0.0047	0.0186
Total	0.0078	0.0242

Of the other land categories for housing, forest land has risen the most, reflecting the increase in house size, as most houses in New Zealand are timber framed. Overall, the largest component of the housing footprint is energy land, which rose by 65 per cent. Household operational energy for 1956 was estimated using primary energy data from the Department of Scientific and Industrial Research (Palmer, 1974). This DSIR report discussed energy data from 1950 to 1974; data were available for primary energy sources for 1956 but energy consumption by sector was only available from 1962 onwards. However, a comparison between these data and that of the energy data file for 2006 shows that the percentage of total energy used by each sector (residential, commercial, industrial, agriculture and transport) has changed very little over the years, although there has been a significant increase in the amount of energy used per capita and by the country as a whole. For housing, this increase is the result of higher operational energy use and higher embodied energy (for construction and maintenance based on a 50-year life), both of which are related to having larger houses with more electrical fittings and appliances in them. This comparison is shown in the first two columns of Table 10.4. In the third column, the 2006 energy EF has been calculated assuming that New Zealand electricity does not have its large renewable (hydro) component, using a global average value of 100 GJ/ha (Wackernagel and Rees, 1996: 69). This more than doubles the energy land footprint, demonstrating the importance of generating electricity renewably at the national scale. Every wind farm turned down or even contested is another step away from lowering domestic energy footprint. In a country of 1.5 million dwellings in 2006 with less than 20,000 new dwellings – or 1.3 per cent of total – per year (Statistics New Zealand, 2007), reducing the impact of how electricity is supplied is arguably a quicker way to reducing housing footprint than building new zero energy houses. Zero energy houses make little difference.

Apart from size and occupancy rate, the survey found that houses in the 1950s were no so different from modern houses in Wellington. Of the people that answered the relevant question, 53 per cent owned (or their parents owned) the dwelling they lived in during the 1950s, while 33 per cent rented and 14 per cent did not know. For the whole sample, the average occupancy was 2 adults and 1.8 children in the household. The average dwelling had three bedrooms, a kitchen, living room and dining room, as well as bathroom and circulation space. Heating was most

Table 10.4 Housing EF

Category	Energy land for housing		
	1956 (ha/person)	2006 (ha/person)	2006 based on global average energy land of 100 GJ/ha (ha/person)
Construction and maintenance	0.0128	0.0232	0.0232
Operation energy	0.0287	0.0449	0.1379
Total	0.0415	0.0681	0.1611

commonly in living areas, with fewer stating that the kitchen was heated and even fewer the bedrooms. Common means of heating the living room were open fires (59 per cent), gas fire (11 per cent) and electric bar heater (15 per cent), although not everyone could remember what was used. Most people heated the living room in the evening between 5 pm and 10 pm. The kitchen was commonly heated by the oven (a coal range or gas stove) in the space and/or an electric bar heater, again generally in the evenings. However, this situation of limited heating is very similar to what happens now, as shown in the findings of the Household Energy End-use Project (HEEP), which set out to discover how people used energy in New Zealand houses (BRANZ, 2010). The HEEP verified that the rise in energy land comes from the increase in lights and appliances in New Zealand houses, not from an increase in heating.

Even in the 1950s it was not uncommon for some people in New Zealand to own a holiday home, known as a bach (pronounced 'batch', not like the composer). This building was usually small, simple and basic, often self-built, and situated on the coast or in a picturesque rural area. In the survey of people living in Wellington in the 1950s, 10 per cent owned a bach that was between 40 and 120 kilometres from Wellington. Having a second home will push up the housing EF.

Implications for housing policy today

The largest component of the housing footprint is domestic energy use. At present, beyond the ability to pay, there is no limit to how much energy a household can use. However, price is a very unsatisfactory mechanism for controlling use, since it advantages the wealthy and disadvantages the poor. A better policy for controlling demand is a sliding price scale whereby up to a certain level of use energy is cheap but beyond this it jumps in price, and for very high consumers the additional energy used is even more expensive. This is a use tax, which encourages frugal behaviour and does not disadvantage those who have difficulty in paying their bills; rather it holds up the behaviour of the poor as being the behaviour society values most. The problem with most use taxes is that they are based on a flat rate (GST,

VAT, etc.), and this taxes consumption rather than giving preference to lower consumption.

As Table 10.4 shows, construction and maintenance have increased since 1956. This is because houses are bigger than they used to be. There are also footprint savings to be made from building smaller houses and making the small house the desired norm. This will be difficult to achieve in a market economy which is geared to selling people stuff. Activating the desire for bigger houses ensures that more stuff can be sold. The danger is that the small house risks becoming only the socially provided house for those who cannot make their way in the traditional housing market.

That all new dwellings need to be built as zero energy dwellings goes without saying – though this is hardly yet policy, let alone widespread practice. Although Table 10.4 shows that this makes little difference over a year, over time it would certainly improve the situation. The UK had a policy of all new homes being zero carbon from 2016 but this has recently been changed (UK Green Building Council, 2011). In conjunction with policies for low-energy dwellings, there also needs to be a clear direction for increasing the renewable energy generation capacity of the national electricity supply. However, both approaches need comprehensive planning, not just of energy supply but also of national economies, to move them from the aim of increased growth to one of planned economies based on living within available resources.

Transport

Although there was a significant increase in the consumed land portion of the transport energy footprint, the major part of the impact of transport comes from the energy land. This is shown in Table 10.5, where transport is further broken down into private, public and goods transport (74 per cent, 5 per cent, and 21 per cent of the 2006 total footprint, respectively).

The consumed land is the land covered by transport infrastructure, such as roads, stations and the airport. The breakdown of this for 1956 and 2006 is shown in Table 10.6. Although the areas are very small in terms of the

Table 10.5 Transport EF by type of transport

	1956			2006		
Transport type	Consumed land (ha/person)	Energy land (ha/person)	Total (ha/person)	Consumed land (ha/person)	Energy land (ha/person)	Total (ha/person)
Private	0.0014	0.0432	0.0447	0.0036	0.3409	0.3446
Public	nil	0.1745	0.1745	nil	0.0216	0.0216
Goods	0.0002	0.0099	0.0101	0.0008	0.0995	0.1003
Total	0.0016	0.2277	0.2293	0.0044	0.4621	0.4665

Table 10.6 EF of consumed land for transport

Infrastructure type	1956 (ha)	1956 (ha/person)	2006 (ha)	2006 (ha/person)
Roads	75.8	0.0005	406.5	0.0023
Cycleways	—	—	1.9	0.0000
Footpaths	13.0	0.0001	244.2	0.0014
Wellington airport	110.0	0.0008	110.0	0.0006
Wellington railway station	28.0	0.0002	28.0	0.0002
Total	226.8	0.0016	790.6	0.0044

Sources: Data from Land Transport New Zealand (2006), Wellington City Council (2010), Infratil Assets (2006) and McCracken (2008)

total EF, the increase in land covered by roads, associated footpaths and car parks is the most important change in the 50-year period, giving rise to a 169 per cent increase overall (Table 10.2). This mirrors the change in energy land for private transport, shown in Table 10.5, which gave rise to a 103 per cent overall increase (Table 10.2): more private transport requires more public roads.

As a percentage of the overall personal EF, the drop in public transport is the most significant thing. This is surprising, as the 1956 Wellington tramway routes have been replaced with identical bus and trolley bus routes, so all places are equally accessible, with the addition of more bus routes by 2006, making public transport marginally more convenient but used less. This is apparent when the energy land is compared (Table 10.7).

The energy land depends on fuel used, embodied energy from the manufacturing, maintenance and disposal of vehicles, and the embodied energy of transport infrastructure. Of these, the fuel use is the largest component. In 1956, fuel use (0.1813 ha/person) was nearly five times the embodied energy of the vehicle fleet (0.0384 ha/person) and almost 80 times that of the transport infrastructure (0.0023 ha/person). By 2006, fuel use (0.2664 ha/person) is three times the embodied energy of the vehicle fleet (0.0809 ha/person) and 17 times that of the transport infrastructure (0.0153 ha/person). These figures again show the huge increase in transport infrastructure, mostly the building of roads, which has occurred in the 50-year period. Because this is paid for out of the public purse, it is hard for the individual to appreciate the consequences that this investment has, beyond seeing the impact of roads on the city when walking through it.

Public transport was commonly used in Wellington during the 1950s, even where people owned cars, as it was cheaper. Cars were generally used on longer trips to visit relatives or go on holidays. For daily travel, 59 per cent of people used buses and 48 per cent trains (some used both). Seventy per cent of survey respondents used trams frequently for daily travel to work or school. Other modes of transport listed by the participants were walking and bicycling. The majority of survey respondents (96 per cent) stated that

Table 10.7 Comparison of transport energy land

Transport type	1956 (ha/person)	2006 (ha/person)
Private	0.0432	0.3409
Public	0.1745	0.0216
Goods	0.0099	0.0995
Total	0.2277	0.4621

they thought public transport was affordable in the 1950s. Travel to school was also different, with 45 per cent walking, 28 per cent taking the tram, 10 per cent going by train, another 10 per cent cycling and 7 per cent going by bus. No one got dropped off in a private car – a very different situation from today. In the latest New Zealand travel survey, 56 per cent of school-aged children up to age 11 were driven to school, 25 per cent walked, 5 per cent went on the bus and 4 per cent cycled, with the remainder using combinations of these modes (Ministry of Transport, 2009: 16). For secondary school-aged children, the proportion being driven to school dropped to 35 per cent, with a further 4 per cent driving themselves, 26 per cent walking, 5 per cent using the bus and a further 5 per cent cycling, with the remaining journeys being combinations of modes, of which 16 per cent were bus and walking (Ministry of Transport, 2009: 17). The intensity of use of all the lower-footprint modes of travel to school dropped in the intervening 50 years.

Implications for transport policy

In 1956, the tram network in Wellington was electric, and at that time electricity had a high renewable component because of hydro generation. Although Wellington's trains and trolley buses are still electric in 2006, buses use diesel and private cars use mostly petrol, so not only has there been a change in mode but there has been a change in the fuels used as a result, with a consequent worsening of environmental impact. Quite simply, the 1956 EF shows that the way to reduce transport footprint is to use public transport and to power it with renewable energy. A well-used public transport system goes with less transport infrastructure. There are two policy issues to tackle here: making public transport attractive and public transport subsidies.

All public transport users are of necessity also pedestrians. The problem with Wellington in 2006 is that vehicular traffic is favoured over the pedestrian. Unless cities, and especially city centres, are a pedestrian realm, public transport will be disadvantaged. This means surfaces should be designed for pedestrian traffic and only occasional road use, and that traffic should be slowed. Access to public transport should always be given preference over other road traffic. Traffic lights within the city also need to be set up to favour pedestrians. These are simple measures that involve minimal cost and design changes, and would do no more than make Wellington much more like a European city. This is policy that local government can influence.

Without subsidies, public transport is not viable in a low-density city like Wellington. However, although the value of public transport subsidies has been debated (van Goeverden *et al.*, 2006), the money that goes into providing transport infrastructure for the majority private transport users is never questioned. In fact this is an expanding area. The New Zealand government policy statement for land transport planning 2009/10–2011/12 divides the budget into new infrastructure, maintenance, services, policing, training and management. Of the total budget, 57 per cent is allocated for new roads, 1.6 per cent for public transport infrastructure and 0.6 per cent to support walking and cycling (Ministry of Transport, 2011: 14). This is hardly showing support for the low-EF option. Unfortunately this is national policy that is hard to change, especially when political parties are tied to the idea of continuous growth, something that is impossible within the finite system of the earth.

Consumer Goods

Consumer goods formed the second largest component of both the 1956 EF (at 0.40 ha, 24 per cent of the total) and 2006 EF (at 0.72 ha, 30 per cent of the total), with an 80 per cent increase overall. Both consumed and energy land have increased significantly. The breakdown for all land types is shown in Table 10.8. The only categories to show reductions are the energy land for household contents and services, and the crop land for tobacco and cotton. The latter category probably reflects the rise in the consumption of artificial fibres. The calculation is based on weekly household expenditure and the energy intensity of each category, normalised for comparison.

The most striking increase is in the consumed land for waste. The throwaway society of today has probably had a large effect on the amount of waste entering landfills in Wellington. Many consumer goods are cheap and are often more convenient to replace than repair. In the survey of those living in Wellington in the 1950s, mention was made of repairing household items and clothes because they were expensive to buy, or often, in the latter case, homemade. In terms of household appliances, 76 per cent owned a fridge and 100 per cent an oven, which was mostly used daily. In addition, 79 per cent had a washing machine, usually used weekly, and 93 per cent a vacuum cleaner. Everyone had at least one radio, with some people owning up to three. Approximately half the people surveyed owned a gramophone and 86 per cent had a telephone. Listening to the radio or gramophone was a common evening activity, with 58 per cent of respondents stating they frequently listened to these as a family in the evening, while 39 per cent said they listened to them together occasionally. What has happened in 2006 is

Table 10.8 Comparison of consumer goods EFs by land type

Category	Land type (ha)										
	Consumed 1956	Consumed 2006	Crop 1956	Crop 2006	Grazing 1956	Grazing 2006	Forest 1956	Forest 2006	Energy 1956	Energy 2006
Tobacco and alcohol	—	—	—	—	—	—	—	—	0.0106	0.0159
Clothing and footwear	—	—	—	—	—	—	—	—	0.0086	0.0133
Household contents and services	—	—	—	—	—	—	—	—	0.0101	0.0063
Communication	—	—	—	—	—	—	—	—	—	0.0011
Recreation and culture	—	—	—	—	—	—	—	—	—	0.0101
Miscellaneous goods and services	—	—	—	—	—	—	—	—	0.0010	0.0080
Other expenditure	—	—	—	—	—	—	—	—	0.0046	0.0106
Consumer waste	0.0047	—	—	—	—	—	—	—	—	—
Tobacco and cotton	—	—	0.0004	0.0001	—	—	—	—	—	—
Wool	—	—	—	—	0.3538	0.4922	—	—	—	—
Paper	—	—	—	—	—	—	0.0081	0.0593	—	—
Total	0.0047	0.0954	0.0004	0.0001	0.3538	0.4922	0.0081	0.0593	0.0349	0.0653

a proliferation of household appliances and more expenditure on clothes, partly reflected in the increase in land associated with growing wool. There has also been an increase in the purchase of paper products such as books and magazines, and these products usually find their way to landfills, thus increasing the EF in two categories. There is talk of the paperless office because of the access to computers and associated technology, but the opposite seems to be the case. Energy land may also have risen because more consumer goods are now imported rather than made in New Zealand. As discussed above, the New Zealand energy mix, because of its hydro component, leads to a lower overall impact. Many imported goods are made in China, where energy comes from coal, leading to a higher overall energy intensity dollar figure, and hence a higher EF.

Implications for policy related to consumer goods

The simple way to reduce the EF of consumer goods is to spend less money and not consume. This means less need to earn money and more available time for people to do things for themselves. Obviously, this will also mean rethinking the overall economic strategy of a country like New Zealand, which at present is focused on full employment and economic growth – hardly a sustainable position. Doing things for yourself, whether making clothes or growing food, or even something as simple as cooking a meal, also reveals the time and effort that go into producing things. This in turn may be a spur to behavioural change, with more appreciation of what goes into making the consumer goods we purchase and hence less desire to throw them away, often before the end of their useful life.

Consumer goods also have knock-on effects. It is obvious that in the 1950s, before television was available in Wellington, the number of electricity-using appliances in the home was small – a radio, refrigerator, washing machine, and vacuum cleaner. Equipment like a lawn mower was more likely to be powered by hand than electricity. In the UK, the Energy Saving Trust (2007) has predicted that by 2020 consumer electronics will use 45 per cent of the electricity in the UK home. Not only do houses have more appliances, for example a TV in every bedroom, alongside improvements in efficiency these appliances are getting larger, neatly cancelling out efficiency gains. In addition, having more inevitably means that more finds its way into the waste stream. The leisure time that in the 1950s was occupied by reading, gardening, playing games or listening to the radio together, is now filled with consuming energy by watching TV, listening to an iPod or playing computer games. In economies tied to growth this is the inevitable outcome, as growth depends on selling people more and more stuff, even though they did not know they needed it. The only action the individual can take is not to consume.

Services

The services part of the EF is the land associated with being a citizen of Wellington, and hence New Zealand. As stated at the start of this chapter, the reason for choosing the 1950s as a point of comparison with the current EF is that the lifestyle was recognisably modern, with its health care, education and government welfare systems, all of which have an environmental impact. The increase in EF between 1956 and 2006 is not as big a jump for services as it is in some other categories. The services EF in 1956 was 0.23 ha (14 per cent of the total EF), which rose to 0.35 ha (15 per cent of the total) in 2006. Table 10.9 sets out the breakdown in land types (some categories vary between years). Many values are obtained from government expenditure figures converted to land using energy intensity figures for dollars, normalised to account for changes in the value of what money could buy over time.

Some categories in Table 10.9 show significant change. The footprint of waste emanating from commercial and public buildings has risen by 1,070 per cent. Tourism impact has risen by an even greater 2,167 per cent. The reason for this latter finding is simple: in the 1950s holidays were taken locally and New Zealand was not a holiday destination for those outside the country. The survey showed holidays were typically taken in the summer, the majority within New Zealand, and the common modes of holiday transport were car, bus and train. People stayed in camping grounds, baches or relatives' houses, with a few staying in motels or hotels. The rise in tourism in the last 50 years is as much to do with the need to grow the economy as it is with providing satisfying holidays. The impact of health services has risen by a more modest 467 per cent. From the survey comments, it was apparent that health-care services were accessible in the 1950s and available to most people. People typically went to the dentist on a regular basis, the doctor as needed and hospitals less frequently – generally for the birth of children. The majority of people had to pay for these services and several people perceived it to be expensive at the time. Several participants stated that health problems were commonly linked to living conditions: cold, damp houses often led to dust mites and asthma, something still found in modern Wellington houses (Howden-Chapman *et al.*, 2009) despite the fact that their larger footprint means more resources are going into them. One category that has seen only a very modest increase is the energy put into running government and other service sector buildings, so, despite the fact these have consumed more land because there are more of them, they are being operated more efficiently. Because this forms by far the largest part of the overall services EF this is also the reason that the latter has only increased by 52 per cent, despite some very large increases in some of the components.

Table 10.9 Comparison of services EFs by land type

Category	Land type (ha)					
	Consumed 1956	Consumed 2006	Garden 1956	Garden 2006	Energy 1956	Energy 2006
Defence	—	—	—	—	0.0013	0.0103
Administration	—	—	—	—	0.0026	0.0044
Social services*	—	—	—	—	0.0313	0.0096
Health and community services	—	—	—	—	0.0021	0.0119
Cultural and recreational services	—	—	—	—	0.0027	0.0048
Personal and other community services	—	—	—	—	0.0012	0.0032
Stabilisation	—	—	—	—	0.0006	—
Development of industry	—	—	—	—	0.0008	—
Tourism	—	—	—	—	0.0009	0.0204
Financial and insurance	—	—	—	—	0.0022	0.0033
Other expenditure	—	—	—	—	0.0013	—
Business communication and property services	—	—	—	—	—	0.0217
Service buildings – operation	—	—	—	—	0.1482	0.1536
Service buildings – construction, maintenance, disposal	—	—	—	—	0.0271	0.0770
Commercial and public buildings – waste	0.0010	0.0117	—	—	—	—
Parks and green spaces	—	—	0.0087	0.0167	—	—
Total	0.0010	0.0117	0.0087	0.0167	0.2222	0.3200

* Social services are counted differently in official statistics for the two years studied, so appear to have decreased

A Study of Wellington in the 1950s

Implications for policy related to service footprints

Simply, the 1956 EF shows that it is possible to live a modern lifestyle and have a lower footprint when it comes to the services component. Like the consumer goods EF, it is making do with less and causing less waste that are the critical steps that will affect all categories. The impact in certain categories such as tourism has risen dramatically because what is acceptable has changed. It is now the norm to take holidays abroad each year. This is something that is not necessary to human health and well-being but that has been 'sold' to people as being the normal thing to do, or to aim for. The consequences of this is that the much more egalitarian society of the 1950s, where the norm was to take a summer holiday, use public transport and go camping or stay with relatives, is displaced by a society made up of those who have the money to take foreign holidays and those who wish they did.

Footprints 1956 and 2006

Between 1956 and 2006, increases occurred in all the components of the ecological footprint of Wellington residents (Table 10.1). The largest percentage increase (104 per cent) occurred for transport. This is due to the energy land required between 1956 and 2006, during which time the ecological footprint associated with energy use (fuel) doubled, the embodied energy of vehicles (due to manufacturing, maintenance and disposal) doubled, and the embodied energy of buildings (due to construction, maintenance, demolition and disposal) was six times greater. These increases are due to increases in vehicle ownership and distances travelled, and also the development of transport infrastructure. The housing ecological footprint increase (71 per cent) emanates from several factors, including an increase in the size of house footprints of almost 30 per cent and lower occupancy. This results in a larger consumed land footprint per person. Larger houses also use more timber, so 82 per cent more forest land is required in 2006 to build new houses. Operational energy for houses is double that of 1956. This is due to the use of many more appliances, and the fact that larger houses require more electric lighting. The increase in EF of consumer goods of 80 per cent can mostly be attributed to the rise in consumed land, which relates to the amount of land associated with consumer waste. This indicates that people are not only spending and purchasing more, but throwing away more as a result. In 2006, Wellington residents live in a consumer society, and as technology and trends change there is pressure to keep up and constantly upgrade possessions. The ecological footprint for services also increased by approximately half, indicating that people expect more from their services, even if these are roughly of the same types as those available in 1956. Only the food footprint remained virtually the same, increasing by only 3 per cent. This relates to the embodied energy of food, calculated through expenditure. Energy intensity value or the amount of energy needed per dollar's worth of food (GJ/$) has increased significantly since the 1970s (the earliest data available). This is due to an increase in imported products and the energy needed for transportation, as well as more processed foods being available and consumed, and the energy needed to manufacture and process these.

Changes also occurred in the energy-to-land ratios for electricity, although the energy-to-land ratio for fossil fuel was kept at 150 GJ/ha for both calculations. However, the energy-to-land ratio decreased for domestic and commercial electricity, applied to the domestic and service buildings operational electricity calculations. This was the result of a decrease in the use of renewable resources for electricity generation since 1956. During the 1970s, gas production commenced. This affected the percentage of renewable resources used to generate electricity and since then it has fallen steadily. In 2006, the renewable component of electricity generation had decreased by almost 30 per cent to 66 per cent, and the remaining 34 per cent was from fossil fuel sources. The hydro component had also decreased to 55 per cent of overall electricity generation. Geothermal, wind and other sources have increased as a proportion of total renewable generation. The use of coal and oil has decreased, with gas now accounting for 64 per cent of fossil fuel sources.

A second comparison can be made in relation to the land-use categories. Increases occurred in all of the categories, the most significant over the 50 years being for consumed and forest land. The significant change for consumed land can be attributed to three components requiring this type of land. The first is housing, with an ecological footprint that almost doubled in 50 years. This is due to more houses in Wellington, with larger footprints and more construction waste. The second contributing component is consumer goods, with a significant increase in consumer waste requiring more landfill area. The third is services – as for housing, its contribution to the increase is due to more commercial and public buildings in Wellington and larger footprints for these buildings. Forest land use also increased significantly, and the majority of this came from the consumer goods component of the ecological footprint. Forest land is the area of forest needed to produce the paper resources used by the population. This includes paper for books, magazines, newspapers and office use. The area required was six times larger in 2006 than in 1956. This increase is despite the move to electronic means of communication, such as computers and mobile phones.

Another instance of increase, although not as large, was for energy land. Energy land increased by three quarters between 1956 and 2006, and significant increases occurred in the majority of the energy land consumption categories. Crop land increased due to higher consumption of foods associated with crop land, increasing the area needed for food by 17 per cent between 1956 and 2006. Grazing land for consumer goods increased 11 per cent over the 50 years despite a halving of the amount of wool being used, from 22 kilograms per person in 1956 to 12 kilograms per person in 2006.

Quality of Life Results

To investigate quality of life in 1950s Wellington, 30 people who had lived in the city during that time were contacted and asked to fill in a written questionnaire about life then and how it was perceived. This research was extended by holding four focus groups to discuss further the issues raised. Overall, 68 per cent of people rated their quality of life in the 1950s as good, 18 per cent extremely good, 11 per cent neutral, 3 per cent poor and no one selected extremely poor. This means that 86 per cent rated their quality of life in 1950s Wellington positively. People felt their family's quality of life was 'extremely good' during the 1950s because their family was not rich but was happy, they got on well with their parents who were intelligent and happy, and they had a large circle of family friends and good family relationships. Explanations for their family's quality of life being 'good' included that they felt loved and cherished, had a very pleasant family home and life, and in general they seemed to have everything they needed. Not all responses were positive with regard to why life was good. Some explained that life was hard going as there was not a lot of spare money; however, they made do with what they had. Reasons for the participants thinking their family's quality of life was 'neutral' included that they were not rich but had a well-rounded family unit, and as newcomers to New Zealand their life was about settling and adapting to a new way of life. Only one person stated that their family's quality of life in the 1950s was 'poor' due to lack of money; even so, they said they were happy, they had everything they needed and did not want for a lot of things, because the emphasis was on relationships rather than material wealth. These results can be compared with the 2008 Ministry of Social Development *Quality of Life* survey, in which 94.6 per cent rated their overall quality of life positively (Nielsen, 2009: 4), with 33.4 per cent rating it extremely good and 61.5 per cent good. So, for a doubling of EF, the perceived quality of life could be said to have risen by only 10 per cent.

These results suggest that there is no real connection between having a large EF, through being able to make use of more material resources, and having a satisfactory standard of living. This can be compared with other studies on perceived satisfaction with life. Easterlin and Angelescu (2009: 9–10) observed a correlation between growth in GDP and improvement in levels of happiness, but over the long term such a correlation was not visible. This argument has been further elaborated to the effect that it is relative difference between incomes that affects how happy a person feels; rather than growth it is relative income or income equality that should be considered if wider happiness is the goal (Clark *et al.*, 2008). The tragedy is that modern economics has widened the gap between rich and poor (Vale and Vale, 2009: 22). This is also true of New Zealand, as it is one of the countries in which the gap has widened most dramatically since the mid-1980s (New Zealand Institute, 2011). By contrast, in the 1950s, the myth was that there was full employment, because the Prime Minister knew the name of everyone drawing unemployment benefit, and that the New Zealand standard of living at the time was the envy of the world (Ministry for Culture and Heritage, 2011). Perhaps, therefore, although having less material wealth, Wellington in the 1950s did provide a better than good standard of living because the gap between the wealthy and the poor was much smaller. Quite simply, when everyone has a lower footprint it is possible to feel poor and be happy at the same time. The high average EF of many countries hides the fact that, as with income, it is the wealthy who are the big consumers of resources, not necessarily those struggling to make ends meet.

Another interesting result to emerge from the study of life in 1950s Wellington was that people then were much happier with their work-life balance. When asked how satisfied their family was with their work-life balance in the 1950s, the majority of people – 80 per cent – stated that they were satisfied, 16 per cent were very satisfied, 4 per cent were not satisfied and no one felt very unsatisfied. Of the people that gave reasons, some were satisfied because they never thought about it and just got on with life. Several stated that their mother was always home; however, their father tended to work long hours, and this was the reason stated for not being satisfied with their work-life balance. This finding can be compared with the 2008 survey, in which a slightly lower 72.7 per cent were satisfied with work-life balance (Nielsen, 2009: 7). Overall, life was not perceived as being significantly less satisfactory in the 1950s, even though the EF, and hence use of resources, was significantly lower.

One big difference in lifestyle between the 1950s and 2006 was the use of private transport, or rather the comparative lack of private transport in the 1950s. It is clear that public transport was more widely used by those taking part in the survey. This was influenced by cost and lack of car ownership. The participants' primary modes of transport in the 1950s were public transport and walking. While many still used public transport in 2006, they thought that there were many more private vehicles on the road, with public transport not as well utilised even though the services are better compared with the 1950s. However, it is clear that the lack of private cars did not stop people moving around Wellington in the 1950s, and the lack of cars made some modes, such as walking, safer. In the survey, 96 per cent of respondents felt public transport was affordable in the 1950s. This can be compared to 2008, when residents questioned agreed public transport was safe (86.3 per cent), easy to get to (81.6 per cent), frequent (62.9 per cent), affordable (62.1 per cent) and reliable (54.2 per cent), showing their perceptions of public transportation in Wellington were still mainly positive even though they were using it less (Nielsen, 2009: 7).

Conclusions

The ecological footprint results for those living in Wellington in 2006 suggest that a high-consumption lifestyle leads to a higher footprint, though with no significant increase in perception of quality of life. Several lifestyle changes need to occur if life in Wellington city is again going to be lived at a more sustainable level. Some of these are national changes and some relate to personal behaviour. Dealing with the latter first, since they are probably easier to put in place, an obvious way to lower EF is not to use the car for daily travel. Wellington has a good public transport system which will only get better if more people use it. Keeping the car out of the city would also lead to a much-improved environment for pedestrians and cyclists, since to be a public transport user is also to be a pedestrian. Fewer cars might also mean fewer roads. Linked to this is the whole business of economic activity and paid work. A lower-footprint lifestyle that uses fewer resources means less consumption (except in the case of food), so, for example, there will be less need to move goods around and less production of goods, all of which will impact on available jobs. The only way out of this is for everyone to work less in paid work, but at the same time do more things for themselves at home and in the community. These are the tasks, like gardening, that people claim not to have the time to do because they are so busy working. This, however, will mean a fundamental reorganisation of society.

A change to people's thinking and ideals will also mean a move away from putting value on material possessions, wealth and objects. If these are aspects of life on which people currently base their perception of quality of life, then they will perceive a reduction in their quality of life. However, if people have similar values to those of the 1950s, prizing family, relationships and community, these factors are going to remain unaltered when reducing the ecological footprint of those living in Wellington.

What emerges from the survey is that life as a whole in the 1950s was perceived as good because it reflected what people had, and everyone was in virtually the same situation. This method of measuring sustainability by using ecological footprint reinforces the point that everyone in the community contributes to and can make a change to sustainability and resource consumption in their area. Those now living in Wellington can change and move towards the lifestyle and consumption of those residents who lived in the city during the 1950s, and life can still be good.

References

Bicknell, K., Ball, R., Cullen, R. and Bigsby, H. (1998) 'An indicator of our pressure on the land: New Zealand's ecological footprint', *New Zealand Geographer* 54(2), pp. 4–11

BRANZ (2010) *Energy Use in New Zealand Homes: Final Report*, http://www.branz.co.nz/cms_show_download.php?id=a9f5f2812c5d7d3d53fdaba15f2c14d591749353, accessed 5 November 2010

Clark, A., Frijters, P. and Shileds, M. (2008) 'Relative income, happiness, and utility: An explanation of the Easterlin paradox and other puzzles', *Journal of Economic Literature* 46(1), pp. 95–144

Close, A. and Foran, B. (1998) *Canberra's Ecological Footprint*, http://www.cse.csiro.au/publications/1998/canberraecofoot-98-12-2.pdf, accessed 12 October 2010

Department of Statistics (1957) *New Zealand Census of 17 April 1956: Interim Returns of Population and Dwellings, Appendix A: Census of Poultry*, Department of Statistics, Wellington

Easterlin, R. and Angelescu, L. (2009) *Happiness and Growth the World Over: Time Series Evidence on the Happiness-Incomes Paradox*, IZA Discussion Paper 4060, ftp://ftp.iza.org/RePEc/Discussionpaper/dp4060.pdf, accessed 19 June 2012

EIU (1997) *India, Nepal: Country Profile*, The Economist Intelligence Unit, London

Energy Saving Trust (2007) *The Ampere Strikes Back: How Consumer Electronics Are Taking Over the World*, Energy Saving Trust, London

Field, C. (2011) 'The ecological footprint of Wellingtonians in the 1950s', Master's thesis, Victoria University of Wellington, New Zealand

Howden-Chapman, P., Viggers, H., Chapman, R., O'Dea, D., Free, S. and O'Sullivan, K. (2009) 'Warm homes: Drivers of demand for heating in the residential sector in New Zealand', *Energy Policy*, 37(9), pp. 3387–3399

Infratil Assets (2006) *Infratil Assets Wellington Airport: Frequently Asked Questions*, http://web.archive.org/web/20070928035456/http://www.infratil.com/wellington_international_airport_faqs.htm#q2, accessed 6 November 2010

Land Transport New Zealand (2006) *Network Statistics.*, http://www.nzta.govt.nz/resources/land-transport-statistics/docs/2005-2006.pdf, accessed 12 July 2010

McCracken, H. (2008) *Wellington Railway Station*, http://www.historic.org.nz/The Register/RegisterSearch/RegisterResults.aspx?RID=1452&m=advanced, accessed 12 November 2010

Ministry for Culture and Heritage (2011) 'The 1950s', http://www.nzhistory.net.nz/culture/the-1950s, accessed 26 June 2012

Ministry for Culture and Heritage (2012) 'Overview – Wellington cafe culture', http://www.nzhistory.net.nz/culture/the-daily-grind/overview-1920-1950, updated 30 August 2012, accessed 20 November 2012

Ministry for the Environment (2003) *Ecological Footprints of New Zealand and Its Regions*, http://www.mfe.govt.nz/publications/ser/eco-footprintsep03/html/index.html, accessed 14 March 2010

Ministry of Transport (2009) *How New Zealanders Travel*, http://www.transport.govt.nz/research/Documents/How%20New%20Zealanders%20travel%20web.pdf, accessed 11 June 2011

Ministry of Transport (2011) 'Government policy statement on land transport funding', http://www.transport.govt.nz/ourwork/KeyStrategiesandPlans/GPSonLandTransportFunding/, accessed 14 October 2011

New Zealand Institute (2011) *NZAhead*, http://www.nzinstitute.org/Images/uploads/nzahead-pdfs/NZahead_Full_Report_PDF.pdf, accessed 24 June 2012

Nielsen, (2009) *Quality of Life 2008 Wellington*, http://www.bigcities.govt.nz/pdfs/2008/Quality_of_Life_2008_Wellington.pdf, accessed 4 March 2010

Palmer, E. (1974) *Energy Consumption in New Zealand*, Department of Scientific and Industrial Research, Wellington, New Zealand

Quotable Value (2012), personal communication (QV collects information on New Zealand property but most of the data is not in the public realm)

Statistics New Zealand (1957) *Population Census 1956, Vol. IX: Dwellings and Households*, Statistics New Zealand, Wellington

Statistics New Zealand (2007) 'QuickStats about New Zealand's population and dwellings', http://www.stats.govt.nz/Census/2006CensusHomePage/QuickStats/quickstats-about-a-subject/nzs-population-and-dwellings/population-counts.aspx, accessed 14 September 2010

UK Green Building Council (2011) 'Government's U turn on zero carbon', http://www.ukgbc.org/site/news/show-news-details?id=398, accessed 7 October 2011

Vale, R. and Vale, B. (2009) *Time to Eat the Dog? The Real Guide to Sustainable Living*, Thames and Hudson, London

van Goeverden, C., Rietveldt, P., Koelmeijer, J. and Peeters, P. (2006) 'Subsidies in public transport', *European Transport*, 32, pp. 5–25

Wackernagel, M. and Rees, W. (1996) *Our Ecological Footprint: Reducing Human Impact on the Earth*, New Society Publishers, Gabriola Island, BC, Canada

Wellington City Council (2010) *Facts & Figures*, http://www.wellington.govt.nz/aboutwgtn/glance/index.html, accessed 6 July 2010

Wilson, J. (2001) *The Alberta GPI Accounts: Ecological Footprint*, http://pubs.pembina.org/reports/28_ecological_footprint.pdf, accessed 14 March 2010

Part IV

Footprints in the Present

11 A Study of China

Yuefeng Guo

Introduction

Any attempt to achieve global sustainability must take China into consideration. This is what Rees (2009) has called the 'China syndrome'. China has become the world's second-largest economy, with a 9.9 per cent average annual growth rate for GDP from 1979 to 2010. However, this recent success in economic growth is offset by the side-effects of resource depletion, environmental degradation and an increasing ecological footprint (EF). China is the world's largest consumer of coal, copper, iron and steel (Li, 2006: 1) and the biggest consumer of energy (International Energy Agency, 2010). This massive consumption of resources and energy does not necessarily improve the quality of life. In fact, while the material living standards are increasing, national happiness is actually declining (Brockmann *et al.*, 2009). According to the EF methodology, a means of measuring the impact of lifestyles on the natural environment, the current EF of China (2.4 global average hectares per person, or gha/person) more than twice overshoots its biocapacity (Global Footprint Network, 2011a), and is beyond the fair earth share of 1.9 gha/person (Paredis *et al.*, 2008: 62). Nevertheless, it is interesting to note that China's EF is still smaller than the global average of 2.6 gha/person (Global Footprint Network, 2011b). This may be the result of a large proportion of the population following a low-environmental-impact way of life in rural areas, as the EF of Chinese cities is 1.4 to 2.5 times greater than that of the countryside (WWF, 2010).

What are the differences in lifestyles between urban and rural areas that have produced different EFs? Why do urban dwellers feel less happy when they apparently enjoy more of the effects of the growing economy and rising income? Why is rural living more conducive to reducing EFs and improving quality of life? This chapter sets out to answer these questions. In addition, it provides a new approach to evaluating sustainability by means of the EF tool and quality of life indicators, suggests possible strategies for sustainable development of China and the world as a whole, and concludes that simple lifestyles or the Middle Way of life will not only have less impact on the earth but also bring greater happiness.

Comparison of EFs in Urban and Rural Areas of China

'Urbanisation' often refers to the transformation of land from rural uses to urban uses (Brown *et al.*, 2005). It is also defined as a process of population concentration in which rural people migrate to urban areas (Tisdale, 1942). However, Theobald (2004) argues that a merely human-demographic notion of urbanisation is inappropriate, especially when considering the EF of these areas. Cities appeared around five thousand years ago, but with very small populations living in them. Modern urbanisation occurred only in the nineteenth and twentieth centuries (Davis, 1955). Since then, it has transformed the world landscape and influenced almost every aspect of human living. By 2007, the majority of the world's population was living in cities (UNPD, 2005). Attractions and opportunities available to urban residents were once regarded as sufficient to overtake the disadvantages of urban life (Still, 1974: 1), but in today's cities this is no longer the case (Berger, 1978: 3). Urbanisation has been implicated in a wide range of problems in modern society, such as conditions like cancer (Greenberg, 1983) and obesity (Ewing *et al.*, 2003), crime (Shelley, 1981), unemployment (Zhang and Song, 2003), environmental degradation (Maiti and Agrawal, 2005) and resource depletion (Wheeler, 2004: 1).

Since the initiation of the opening up and reform policy in 1978, China has been experiencing rapid economic growth. This has accelerated the speed of its urbanisation. By the end of 2011, China's urbanisation rate had reached 51.3 per cent (China Association of Mayors, 2012). The changes in lifestyles that accompany urbanisation pose a great challenge in terms of managing the fast growth of EF for China. The per capita EF of urban areas is 0.9 to 1.8 gha higher or 1.4 to 2.5 times greater than rural areas (WWF, 2010: 25). This gap has increased sharply since 1985 and may continue to widen in the near future (WWF, 2010: 25). A study of the EF of Xianju County of Zhejiang Province reveals the per capita EF of urban areas to be 1.73 times bigger than rural areas (Min *et al.*, 2003). Increased urbanisation and associated changes in lifestyles may generate further growth in China's EF. To promote sustainable development, a study of the EFs of five Chinese cities demonstrates that adjusting urban residents' consumption style and structure and adapting environmentally friendly lifestyles are crucial (Min *et al.*, 2005).

Comparison of Lifestyles and Happiness in Urban and Rural Areas in China (from 1958 to Today)

Lifestyles serve as 'social conversation' through which people differentiate themselves from others, and signal their social position and psychological aspirations. Since many of the signals are mediated by goods, lifestyles are closely linked to material and resource flows in society (UNEP, 2010). Lifestyles include many aspects of living such as food, clothing, housing, travel, consumption and so on. As discussed above, changes in lifestyles are closely associated with changes in EFs. In fact, the per capita EF method is used to evaluate consumption and lifestyles and to measure this against the carrying capacity of planet earth (Wackernagel and Yount, 1998).

From 1958 to 1978, about 75 to 80 per cent of China's total population was rural (Bennett, 1978: 1), associating the then low EF of 1.0 gha/person (WWF, 2008: 4) with rural living. The Chinese model of controlling urbanisation was seen as a great alternative approach (Ma, 1977; Murphey, 1980: 1–4). The model of rural development was based on a belief in the necessity of collective benefits through equitable ownership of land, production means and other natural resources, and fair distribution of these resources and income (O'Leary and Watson, 1982). During this period, lifestyles were determined by the system of the people's commune and characterised by self-sufficiency (Zheng, 1981: 59), smallness, simple technology (Inkster, 1989; Durham, 1976: 262–266) and the Chinese traditional culture of living in harmony with nature (Guo et al., 2011).

Self-sufficiency reflected the belief that development must be pursued within a community with the full participation of all its members. Through self-sufficiency, the commune produced enough to feed and clothe an additional 300 million people (from 646.53 million in 1957 to 958.09 million in 1978; Zheng, 1981: 59). The focus on equity ensured that the commune structure was a form of community development in which all, rather than a minority, of the peasants in a collective could share in the benefits of economic development within the boundary of natural resources. Public accumulation could eventually provide social services and medical and welfare care for everyone. The absence of private ownership and hired labour guaranteed that social polarisation could not take place (O'Leary and Watson, 1982). Equity ensured a more equal standard of living for all members of the commune. Small industries within the commune were proven to be both workable and sustainable. They tended to use more renewable resources than urban-based industries, and alternative technology, emphasised in the Chinese tradition as a crucial element of China's economic history, was widely utilised in the commune (Inkster, 1989). For example, insect traps in the fields were commonly used to bring pests to the poison, rather than taking the poison to the plants (Durham, 1976: 262–266). The peasants did not need specialised training to utilise such simple technologies, so control of technology was actually in their hands.

During these twenty years, the main lifestyle of the Chinese people was conventional and rural. Western influences and almost everything related to capitalist links were prohibited. Many Chinese thought their lifestyles had dramatically improved compared to the standard prior to 1949, and they were not motivated for further change because others had the same (Guan and Hubacek, 2004) due to the equity mentioned previously. Table 11.1 shows that the lifestyle of peasants remained roughly the same over a twenty-year period.

188 Yuefeng Guo

Table 11.1 Peasants' expenditure categories 1958–1978

	Expenditure (%)						
Year	Food	Clothing	Fuel	Recreation	Housing	Other	Total
1957	67.75	13.44	10.03	1.74	2.10	6.94	100
1963	63.30	11.21	9.32	2.67	4.71	8.79	100
1965	68.46	10.51	8.31	2.71	2.83	7.18	100
1978	67.71	12.70	7.11	2.71	3.16	6.57	100

Source: Yang, 1986: 186

The developmental model of China from 1958 to 1978 has received criticism both in the West and from within China (O'Leary and Watson, 1982; Yu, 1981), mainly due to the occurrence of the Chinese famine, from 1958 to 1961 (Jowett, 1991). Besides the poor natural conditions (Li, 2000), the major cause of the famine was the attempt, during the Great Leap Forward (1958 to 1961), to follow the Soviet Union's developmental mode of socialism based on large industries in cities (Chan, 1992: 41–65; Fung, 1981).

It is interesting to note that from 1961 to 1973 China's EF was well below its biocapacity (Global Footprint Network, 2011a). During this period there was no big difference in consumption levels between rural and urban areas (Figure 11.1). The low EF of China was certainly associated with rural lifestyles.

Since the policy of opening up and reform took effect in 1978, China has undergone a transition from a planned economy to a mixed system with elements of central planning and market mechanisms, and has been

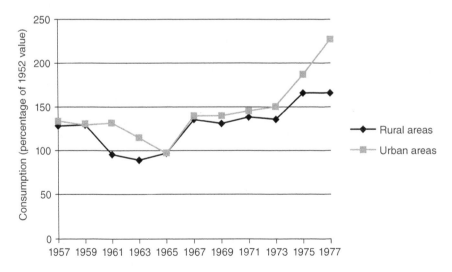

Figure 11.1 Consumption levels in China, 1958–1978

Source: China Statistics Bureau (1986)

striving to fit into the Western concept of social progress rooted in economic growth, scientific and technological advancement and exploitation of nature. The most significant result of China's rapid economic growth is the improvement in the standard of living for many Chinese. They are no longer satisfied with sufficient provision of food and clothing, but are eager to adopt Western lifestyles (Hubacek et al., 2009). This has led to China's EF growing from 1.2 gha/person in 1978 to 2.4 gha/person in 2010 (Global Footprint Network, 2011a). Modern China 'overshoots' its carrying capacity by a factor of two (Global Footprint Network, 2011a). Tables 11.2 and 11.3 show that consumption dramatically increased in both rural and urban areas between 1985 and 2010, and that consumption in cities was greater than that in the countryside, reaching over three times that of rural areas in 2010. The figures for the categories listed in the tables changed significantly over the period, and thus are used as indicators revealing lifestyle changes since 1978.

Although food and clothing still dominated Chinese expenditure both in rural and urban areas, from the perspective of food structure, the diet contained more meat and less cereal products (Guan and Hubacek, 2004). Urban per capita consumption of pork and red meat were much greater than rural consumption (Wu, 2003), and meat is the biggest contributor to the EF of people's daily diet (Vale and Vale, 2009: 40) The average per capita living space expanded from 8.1 to 32.8 square metres between 1978 and 2010 (China Statistics Bureau, 2002, 2011). In terms of sustainability of housing, size does matter: doubling the size of houses, given the same building materials, means doubling the EF (Vale and Vale, 2009: 146). If home is the place where people live and which reflects their creativity and aesthetics, rather than being thought of as an investment ('put it under the hammer') or a status symbol ('keeping up with the Joneses'), then 'small is beautiful' should be the rule. More spacious houses also allow consumers to buy and store more household appliances, among which air-conditioners, previously a sign of wealth, increased to about 30 systems per 100 households

Table 11.2 Rural consumption expenditure in selected categories

Year	Expenditure											
	Total		Food and clothing		Transport		Housing		Household appliances		Education and recreation	
	Yuan	%	Yuan	%	Yuan	%	Yuan	%	Yuan	%	Yuan	%
1985	317	100	214	78	5	2	58	18	16	5	12	4
1990	584	100	388	75	8	1	101	17	31	5	31	5
2001	1,741	100	930	53	110	6	279	16	192	11	248	14
2010	4,382	100	1,797	41	526	12	920	21	263	6	438	10

Source: Data from China Statistics Bureau, 1996, 2002, 2011

Table 11.3 Urban consumption expenditure in selected categories

	Expenditure											
	Total		Food and clothing		Transport		Housing		Household appliances		Education and recreation	
Year	Yuan	%	Yuan	%	Yuan	%	Yuan	%	Yuan	%	Yuan	%
1985	673	100	449	67	14	2	32	5	58	9	55	8
1990	1,279	100	865	67	40	1	61	7	108	10	112	9
2001	5,309	100	2,547	48	457	9	547	10	395	8	974	12
2010	13,471	100	6,248	47	1,983	15	1,332	10	908	7	1,637	12

Source: Data from China Statistics Bureau, 1996, 2002, 2011

by 2000, and almost trebled between 2000 and 2008 (Hubacek, 2009). Besides having large energy consumption and carbon emissions, air-conditioning systems also use refrigerants which contribute to global warming and climate change. Air-conditioning turns buildings into refrigerators which burn fossil fuels, emit greenhouse gases, raise global temperatures and create a craving for more air-conditioning. The sealed world created by air-conditioning requires constant readjustment, meaning that when people leave this man-made environment they have to adapt to a new one, physically and socially. This can lead to discomfort, stress, illness and isolation from nature and communities. Ownership of private cars is another status symbol in China and has increased phenomenally, from almost zero to 13 cars per 100 households by 2010 in urban areas (China Statistics Bureau, 2011). CO_2 emissions from cars have replaced coal as the major source of air pollution in major Chinese cities (Turner and Ellis, 2007). In Beijing, the total EF of driving private cars is over five times that of using existing public transportation (WWF, 2008). With the dramatic increase in income, the rich in urban areas also aspire to luxury leisure activities such as golf. The number of golfers in China is estimated to be between 300,000 and 3 million (about the total population of Lithuania). But golf is a land-hungry sport, which has huge impact on land and water (Vale and Vale, 2009: 256)

Over the last 34 years, China's astounding and enduring economic and income growth, while significantly improving the standard of living, has not increased the quality of life or happiness of the nation as a whole. According to Easterlin (1995), a society's average happiness is a constant that remains unchanged even if per capita income increases for most of the population. From 1990 to 2000, the percentage of Chinese who described themselves as very happy plummeted from 28 per cent to 12 per cent. Measured on a scale from 1 (low) to 10 (high), life satisfaction decreased in both urban and rural areas and in every income group, from an average of 7.3 to 6.5, although life in rural China appeared to be happier than life for urban residents (Brockmann *et al.*, 2009). Despite the relatively low living standards from 1958 to 1978, people tended to be much happier (Zhang, 2012). There is considerable consensus on a model or indicator for assessing quality of life (Bigelow *et al.*, 1991; Baker and Intagliata, 1982; Andrews and Withey, 1976), which is divided into five categories:

- physical well-being
- material well-being
- social well-being
- emotional well-being
- development and activity (Felce and Perry, 1995).

The current model of economic growth contributes only to material well-being, whereas real happiness is achieved by considering all five categories, within a holistic approach. A survey conducted by the NEF in Europe reveals that quality of life has little to do with consumption (Simms and Johnson, 2010: 21). Above a certain level of GNP per capita (approximately $US14,000), the relationship between wealth and well-being actually disappears (Inglehart, 1999: 88–120). City life often appears to be unhappy. In America, the most affluent nation, with 77 per cent of people living in cities, more than half of the population suffer from mental illness in any one year (Monbiot, 2000). While the living standard of urban citizens increases in China, they are actually less happy than those living in the countryside. Beijing and Shanghai now have the lowest happiness index (Global Times, 2010) and the highest number of insomniacs (Financial Express, 2006). It seems likely that happiness can only be achieved through sustainable living, or what Schumacher (1973) called the Middle Way of life, meaning knowing the right amount and practising moderation. It must be balanced in a way appropriate to the attainment of quality of life rather than the satisfaction of unlimited desires. The Middle Way of life is often found in rural living (Guo *et al.*, 2009).

Sustainable Living

The traditional Chinese way of life, which can be considered to be sustainable and which still exists in parts of the countryside of today's China, has been guided by the wisdom of harmony and balance as proposed by Confucianism, Taoism and Buddhism over many centuries (Guo *et al.*, 2009). These three philosophical approaches, although different, are interrelated and complementary, together forming a basis for sustainable living founded on a holistic and organic worldview (Jenkins, 2002). In terms of Confucianism, for achieving harmony with nature, the essence of the universe – or the balance between *yin* and *yang* – must be the foundation of human society. This concept was utilised to evaluate the moral and ethical standards of mankind. A person who obeyed the natural rule was regarded as a sage. Taoism holds the same point of view, stating that to follow the balance of *yin-yang* provides the meaning of life, and this can be achieved by means of *wuwei* (non-striving and non-attachment to the things of the world). Through *wuwei*, the physical development of human society may not get very far but at least it develops in the right direction. This is one of the reasons why Lao Tzu, the founder of Taoism, believed that the sage should avoid excess, extravagance, indulgence and attachment. For Buddhism, the idea of harmony is interpreted as the concept of the Middle Way, the way to fulfilling the idea of 'right-livelihood', through which quality of life, well-being and happiness can be achieved (Schumacher, 1973).

Sustainable living or the Middle Way of life is characterised by several key features:

- the holistic and organic worldview
- a harmonious attitude towards nature
- self-sufficiency
- the Middle Way to consumption
- accepting constraints for living
- fulfilling work and leisure time.

In contrast to the mechanistic worldview, the holistic and organic worldview maintains that the universe is made up of a multitude of objects, but has to be pictured as one indivisible, dynamic whole whose parts are essentially interrelated and can be understood only as patterns in a cosmic process. The focus of this ecological worldview is relationships rather than objects (Bateson, 1979: 94). Under its guidance people lived in small, cohesive communities and experienced nature in terms of an organic relationship, characterised by the interdependence of spiritual and material phenomena and the subordination of individual needs to those of the community (Capra, 1983: 53). This worldview will lead to a harmonious attitude towards nature, centred on two aspects: the frugal use of natural resources and recognition of the difference between renewable and non-renewable resources. This attitude will guide self-sufficient living and the Middle Way approach to consumption, which is simple and non-violent. Sustainable living also means accepting constraints for living. Constraints may not appear desirable for those who have been accustomed to modern, affluent lifestyles. They may even argue that these constraints are violations of individual freedom. However, Hardin (1968) believes that true freedom is the recognition of necessity, and that the most important aspect of necessity that modern people must now recognise is the necessity of accepting the need for constraint. Moreover, sustainable living involves the realisation that time spent doing fulfilling work is also leisure time, and enjoyable. For instance, growing food at home can be viewed as both work and leisure at the same time, thus reflecting the Buddhist view that work and leisure cannot be measured as separate 'things' but are just two sides of the same activity.

A concrete example of sustainable living would be the rural lifestyle in *tulou* (earth house or roundhouse) communities, located in the mountainous areas of south-western Fujian in China. Most of the Fujian *tulou* have been built and developed by the Hakka people, a Han Chinese ethnic subgroup that migrated into the region from central China during the Northern Song dynasty (690 to 1126) to protect themselves from bandits and wild animals, and retain communal living and coherence as proposed by Confucianism. The resilience and sustainability of the *tulou* community rests largely on the self-sufficient lifestyle of the residents. Most courtyards of Fujian *tulou* are cluttered with privies, pigsties and secondary kitchens. This is where children are permitted to play with chickens and ducks, and also forms a space for hanging laundry. Sometimes rice is also dried there. There is at least one well in the courtyard to supply water for the residents. Although chaotic and primitive to modern Western eyes, this way of living has sustained the villages for centuries, and through this sharing of space and resources it has had a low impact on the environment. Once the crops have been harvested and stored, the residents can live self-sufficiently within their *tulou* (Laude, 1992).

Urban or Rural?

Economic growth, urbanisation and lifestyle changes have contributed to the growth of China's EF. Table 11.4 shows that there is an immediate and relevant connection between urbanisation and high EFs. In the case of China, while it is increasingly urbanised, and 19 per cent more so than Vietnam, it seems to be achieving similar levels in terms of the other listed indicators. The low EF of China, lower than the global average (2.6 gha/person), is the result of a majority of the population still living in a traditional way in rural areas, while lifestyles in Chinese urban settings have become very similar to those of Western cities (Peng, 2010). If China increases its ranking on the United Nations Human Development Index to the very high levels of New Zealand and Australia, this will probably result in a large jump in EF. Given the high urbanisation rates of New Zealand and Australia, education would seem to be easier to achieve. However, although a majority of people live in rural areas in Vietnam and China, both countries have literacy rates over 90 per cent. This comparison shows that life in cities may not be much better or happier, in spite of its much larger environmental impact.

The issue as to whether China should give priority to the development of large cities or rural areas is a matter of allocation of its limited natural resources. If the national goal is to achieve a high level of economic growth in the short term, then efforts should certainly be made to concentrate the bulk of the nation's resources into a few large cities. City-based industrialisation means spatial concentration of production and is generally the fastest way to growth in the short term (Ma and Hanten, 1981: 7–8).

Table 11.4 Comparative footprints and statistics related to the Human Development Index for four countries

Country	Urbanisation (% of population)	EF (ha/ person)	Average life expectancy (years)	Adult literacy rate (% of population)	Undernourishment (% of population)	Human Development Index
Vietnam	27	1.3	73	90.3	13	0.733
China	46	2.4	72.5	93.3	12	0.772
New Zealand	86	8.35	80	99	< 5	0.95
Australia	89	6.6	81.4	99	< 5	0.97

Source: Guo *et al.* (2010)

However, such a strategy will only benefit urban residents – the elite – and further widen the gaps already existing between city and countryside in China. If the goal is to produce a more equitable pattern of economic development, then natural resources must be distributed more widely in space. Sustainable development of the countryside could thus have profound implications for China's long-term economic development. In the long run, an emphasis on the development of rural areas will better contribute to improvement of the quality of life of the vast rural masses.

It is no wonder that the visions of many green thinkers such as Schumacher (1973) and Shaw (2006, first published 1884) were often rural (Dearling and Meltzer, 2003: 23); what Marsh (1982) called 'the cult of the countryside', and what Todd (1977) called 'a revitalization of the countryside', as distinct from urbanisation. It has been argued that cities are closely linked to decline of ecological systems, and that they are not by themselves sustainable (Rees and Wackernagel, 1996), and that the purpose of modern cities is to prepare consumers and workers for capitalist exploitation (Brolin, 1976: 15–17). As a result of confronting the problems of urbanisation and the negative impact of 'affluent' lifestyles on the natural environment, the traditional ecological wisdom of Taoism and Confucianism has helped to give rise to the 'revitalisation of traditional Chinese philosophies' in today's China. Furthermore, the Chinese government has launched the 'Socialist New Village Movement' to develop rural areas in a sustainable way.

The Yanhe Village Project is an example of this. Yanhe is located in the north of Hubei province, and is representative of hilly rural communities in the centre of China, with an area of about 1.2 square kilometres, 222 families and 840 residents. The villagers' income comes mainly from agriculture, organic tea farming and eco-tourism. The village has become a role model for other eco-villages and has received a number of 'green' awards from central government. A study of the behaviour of the resident farmers has shown the sustainability of their actions, in relation to energy saving and pollution reduction (Yang and Wang, 2008). In the village, wastes are sorted and recycled; solar lights are used; organic tea farming is practised; and the ecosystem is under restoration. Rather than being educated for a raised awareness of sustainable issues, as happens in the West, the villagers are instructed in revitalisation of their traditional cultural habits. This is because concepts like global warming and climate change are considered too difficult for them to comprehend. Sunjun, the creator of the project, claims 'the villagers fell asleep, and they need to wake up to their own traditional culture'. When teaching sustainability, he simply tells them that sustainable development is about following your ancestors, and that it is good for your children. When instructing them to sort out and recycle wastes, he just demonstrates that there is a difference between 'wet' rubbish and 'dry' rubbish', as the farmers find it difficult to understand terms such as 'organic wastes'. The Yanhe Village Project shows that China's small EF per person derives from lifestyles rooted in traditional cultural habits in the countryside, even under the Chinese market economy.

Rural lifestyles have greater potential to be sustainable than urban. However, if development of cities is necessary, then China ought to follow the rule that small is beautiful. According to a recent report from the Urban China Initiative, small cities are more liveable than big cities such as Beijing and Shanghai in terms of air quality, waste treatment capacity, built environment, quality of life and sustainability (China Daily, 2012). Furthermore, 'rural' living, or sustainable living, in the urban setting could also have less impact on the natural environment. An example of this would be the Dongcheng district of Beijing city. Dongcheng has proposed an action plan for constructing a low-carbon city and green development by reducing energy consumption, counting the carbon footprint, organising low-carbon activities and cooperating with international communities to build green Dongcheng (Low-Carbon City China Alliance, 2011). In Dongcheng there is a low-carbon park showcasing low-carbon knowledge, sustainable housing and technologies for utilising renewable energy and energy efficiency, consumption of grey water and waste disposal. It is a park that not only provides a place to relax but also disseminates the idea of sustainable living. Residents are motivated to grow their own vegetables on their apartment balconies, use and exchange second-hand goods and organise low-carbon community days, when small, innovative ideas for sustainable living receive awards. This is remarkably similar to the 'Greenest Street' project on the Kapiti Coast in New Zealand (see Chapter 16). It is worth mentioning that one of the favourite leisure activities for residents, especially for the elderly, is dancing in the park. The case study of Dongcheng demonstrates that, although the urban environment provides fewer opportunities for sustainable living, it is still possible to reduce the per capita EF by 'living rurally' in big cities.

Conclusion

This chapter has investigated the changes in China's EF between 1958 and the present, and the reasons behind these changes. From 1958 to around 1973, the EF of China was smaller than its biocapacity. This resulted from the majority of the population living in the countryside. However, since 1978, the adoption of the Western developmental model and lifestyles in urban areas has led to the fast growth of China's EF. While the material standard of living has increased, the quality of life and happiness in the nation as a whole is actually declining, with urban residents appearing to be less happy than those in rural areas. The fact is that happiness has little to do with increased income and consumption and a growing economy, and that it can only be attained by means of the Middle Way of life, or sustainable living, characterised by the six key features and measured using the five categories of indicators proposed in this chapter. While the *tulou* community represents the long tradition of living in harmony with nature, the Yanhe village case study demonstrates the possibility of sustainable living in the countryside, even within the Chinese market economy. To prevent over-urbanisation and minimise the EF, it is sensible for China to direct more attention to preserving, encouraging and promoting the culture of 'rural living' in rural areas. For the development of urban areas, it is rational to promote small and medium-sized cities, and sustainable living in cities of all scales. The Dongcheng example demonstrates the feasibility of this approach. These strategies may be a prescription for treating the 'China syndrome'.

References

Andrews, F. M., and Withey, S. B. (1976) *Social Indicators of Wellbeing: Americans' Perceptions of Life Quality*. Plenum Press, New York

Baker, F. and Intagliata, J. (1982) 'Quality of Life in the Evaluation of Community Support Systems'. *Evaluation and Programme Planning*, 5, pp. 69–79

Bateson, G. (1979) *Mind and Nature*. Dutton, New York

Bennett, G. (1978) *Huadong: The Story of a Chinese People's Commune*. Westview Press, Boulder, CO

Berger, A. S. (1978) *The City: Urban Communities and their Problems*. Wm. C. Brown Company Publishers, Dubuque, IA

Bigelow, D. A., McFarland, B. H. and Olson, M. M. (1991) 'Quality of Life of Community Mental Health Program Clients: Validating a Measure'. *Community Mental Health Journal*, 27, pp. 43–55

Brockmann, H., Delhey, J., Welzel, C. and Yuan, H. (2009) 'The China Puzzle: Falling Happiness in a Rising Economy'. *Journal of Happiness Studies*, 10(4), pp. 387–405

Brolin, B. C. (1976) *The Failure of Modern Architecture*. VNR Company, New York

Brown, L., Gray, R., Hughes, R. and Meador, M. (2005) 'Introduction to Effects of Urbanisation on Stream Ecosystems'. *American Fisheries Society Symposium*, 47, pp. 1–8

Capra, F. (1983) *The Turning Point: Science, Society and the Rising Culture*. Wildwood House, London
Chan, K. W. (1992) 'Post-1949 Urbanization Trends and Policies: An Overview', in G. E. Guldin (ed.) *Urbanizing China*, Greenwood Press, Westport, CT, pp.41–64
China Association of Mayors (2012) 'China Urbanisation Rate Exceeds 50%'. http://www.chinadaily.com.cn/business/2012-05/30/content_15421183.htm, accessed 24 May 2012
China Daily (2012) 'Small is More Beautiful for Chinese Cities'. http://www.chinadaily.com.cn/china/2012-05/03/content_15203158.htm, accessed 12 May 2012
China Statistics Bureau (1986, 1996, 2002, 2011) *China's Statistical Yearbook*. China Statistics Press, Beijing (in Chinese)
Davis, K. (1955) 'The Origin and Growth of Urbanisation in the World'. *The American Journal of Sociology*, 60(5), pp. 429–437
Dearling, A. and Meltzer, G. (eds) (2003) *Another Kind of Space: Creating Ecological Dwellings and Environments*. Enabler Publications, Lyme Regis, England
Durham, T. (1976) 'Think Big, Think Little', in P. Harper and G. Boyle (eds) *Radical Technology*. Wildwood House, London
Easterlin, R. A. (1995) 'Will Raising the Incomes of All Increase the Happiness of All?' *Journal of Economic Behaviour and Organisation*, 27, pp. 35–47
Ewing, R., Schmid, T., Killingsworth, R., Zlot, A. and Raudenbush, S. (2003) 'Relationship between Urban Sprawl and Physical Activity, Obesity and Morbidity'. *American Journal of Health Promotion*, 18(1), pp. 47–57
Felce, D. and Perry, J. (1995) 'Quality of Life: Its Definition and Measurement'. *Research in Developmental Disabilities*, 16(1), pp. 51–74
Financial Express (2006) 'Lifestyle Changes Make More Urban Chinese Insomniacs'. http://www.financialexpress.com/news/lifestyle-changes-make-more-urban-chinese-insomniacs/148781/, accessed 14 February 2006
Fung, K. I. (1981) 'Urban Sprawl in China: Some Causative Factors', in L. J. C. Ma *et al.* (eds) *Urban Development in Modern China*. Westview Press, Boulder, CO, pp. 194–222
Global Footprint Network (2011a) 'Country Trends: China'. http://www.footprintnetwork.org/en/index.php/GFN/page/trends/china/, accessed 12 August 2012
Global Footprint Network (2011b) 'Ecological Footprint and Biocapacity in 2008'. http://www.footprintnetwork.org/en/index.php/GFN/page/footprint_data_and_results, accessed 15 May 2012
Global Times (2010) 'Survey: The More Developed Areas, the Lower Happiness Index'. http://life.globaltimes.cn, accessed 12 November 2010
Greenberg, M. R. (1983) 'Urbanisation and Cancer: Changing Mortality Patterns?' *International Regional Science Review*, 8(2), pp. 127–145
Guan, D. B. and Hubacek, K. (2004) *Lifestyle Change and Its Influences on Energy and Water Consumption in China*. http://www.unescap.org/esd/environment/mced/tggap/documents/2RPD/bgm/lifestyle%20chages%20and%20consumption%20in%20China.pdf, accessed 6 October 2004
Guo, Y. F., Vale, B. and Vale, R. (2009) 'Yin and Yang: The Battle for a Sustainable Built Environment'. AASA (Association of Architecture Schools in Australasia) Conference, 4–5 September, Wellington, New Zealand

Guo, Y. F., Vale, B. and Vale, R. (2010) 'A Sustainable Future for Housing Settlements: Urban or Rural?' 5th Australasian Housing Researchers' Conference, 17–19 November, Auckland, New Zealand

Guo, Y. F., Vale, R. and Vale, B. (2011) 'How Changing Economic Attitudes Have Affected China's Ecological Footprint: Implications for the Built Environment'. *Journal of Creative Sustainable Architecture and Built Environment*, 1, pp. 15–26

Hardin, G. (1968) 'The Tragedy of the Commons'. *Science*, 162(3859), pp. 1243–1248

Hubacek, K. (2009) 'Lifestyle, Technology and CO_2 Emissions in China'. http://www.eoearth.org/article/Lifestyle,_technology_and_CO2_emissions_in_China?topic=49473, accessed 18 January 2010

Hubacek, K., Guan, D., Barrett, J. and Wiedmann, T. (2009) 'Environmental Implications of Urbanisation and Lifestyle Change in China: Ecological and Water Footprints'. *Journal of Cleaner Production*, 17, pp. 1241–1248

Inglehart, R. (1999) 'Trust, Well-Being and Democracy', in M. E. Warren (ed.) *Democracy and Trust*. Cambridge University Press, Cambridge

Inkster, I. (1989) 'Appropriate Technology, Alternative Technology and the Chinese Model: Terminology and Analysis'. *Annals of Science*, 46, pp. 263–276

International Energy Agency (2010) 'China Overtakes the United States to Become the World's Largest Consumer'. http://www.iea.org/index_info.asp?id=1479, accessed 9 July 2010

Jenkins, T. N. (2002) 'Chinese Traditional Thought and Practice: Lessons for an Ecological Economics Worldview'. *Ecological Economics*, 40, pp. 39–52

Jowett, A. J. (1991) 'The Demographic Responses to Famine: The Case of China 1958–1961', *GeoJournal*, 23(2), pp. 135–146

Laude, O. (1992) 'Hekeng Village, Fujian: Unique Habitats', in R. G. Knapp (ed.) *Chinese Landscapes: The Village as Place*. University of Hawaii Press, Honolulu, pp. 163–172

Li, J. C. (2006) *China's Rising Demand for Minerals and Emerging Global Norms and Practices in the Mining Industry*. Foundation for Environmental Security and Sustainability, Falls Church, VA

Li, R. J. (2000) 'The Influence of Natural Disaster and Natural Conditions on the Creation of Famine in the Post-Great Leap Period'. *Contemporary China History Studies*, 7(5), pp. 17–23 (in Chinese)

Low-Carbon City China Alliance (2011) 'Dongcheng, Beijing Municipality'. http://www.low-carboncity.org/index.php?option=com_flexicontent&view=items&cid=16%3Adongcheng&id=84%3Astrategies-a-action-plans-&lang=en, accessed 18 June 2011

Ma, L. (1977) 'Counter-Urbanisation and Rural Development: The Strategy for Hsia-Hsiang'. *Current Scene*, 15(8–9), pp. 1–11

Ma, L. and Hanten, E. (eds) (1981) *Urban Development in Modern China*, Westview Press, Boulder, CO

Maiti, S. and Agrawal, P. K. (2005) 'Environmental Degradation in the Context of Growing Urbanisation: A Focus on the Metropolitan Cities of India'. *Journal of Human Ecology*, 17(4), pp. 177–187

Marsh, J. (1982) *Back to the Land: The Pastoral Impulse in England from 1880 to 1914*, Quarter Books, London

Min, Q. W. *et al.* (2003) 'Ecological Footprint-Based Comparison of Consumption Differences of Xianju's Urban-Rural Residents'. *Urban Environment and Urban Ecology*, 16(4), pp. 86–88 (in Chinese)

Min, Q. W. *et al.* (2005) 'Ecological Footprint-Based Comparison of Consumption of Living Consumption of Meso-Scale Cities Residents in China Taking Taizhou, Shangqiu, Tongchuan and Xilin Gol as Examples'. *Journal of Natural Resources*, 20(2), pp. 286–292 (in Chinese)

Monbiot, G. (2000) 'Dying of Consumption'. http://www.guardian.co.uk/Columnists/Column/0,5673,415777,00.html, accessed 10 May 2000

Murphey, R. (1980) *The Fading of the Maoist Vision: City and Country in China's Development*. Methuen, New York

O'Leary, G. and Watson, A. (1982) 'The Role of the People's Commune in Rural Development in China'. *Pacific Affairs*, 55(4), pp. 593–612

Paredis, E., Goeminne, G., Vanhove, W., Maes, F. and Lambrecht, J. (2008) *The Concept of Ecological Debt: Its Meaning and Applicability in International Policy*. Academia Press, Gent, Belgium

Peng, G. W. (2010) 'China's Ten Most Luxurious Cities'. http://cq.people.com.cn, March 2010 (in Chinese)

Rees, W. E. (2009) 'The Ecological Crisis and Self-Delusion: Implications for Buildings'. *Building Research and Information*, 37(1), pp. 300–311

Rees, W. E. and Wackernagel, M. (1996) 'Urban Ecological Footprint: Why Cities Cannot Be Sustainable – and Why they Are a Key to Sustainability'. *Environ. Impact Assess. Rev.* 16, pp. 223–248.

Schumacher, E. F. (1973) *Small is Beautiful: A Study of Economics as if People Mattered*. Blond & Briggs, London

Shaw, G. B. (2006, first published 1884) 'Economics', in *Fabian Essays in Socialism*. Cosimo, New York

Shelley, L. I. (1981) *Crime and Modernisation: The Impact of Industrialisation and Urbanisation on Crime*. Southern Illinois University Press, Carbondale, IL

Simms, A. and Johnson, V. (2010) *Growth Isn't Possible: Why We Need a New Economic Direction*. NEF, London

Still, B. (1974) *Urban America: A History with Documents*. Little Brown and Company, Boston, MA

Theobald, D. M. (2004) 'Placing Exurban Land-Use Change in a Human Modification Framework'. *Frontiers in Ecology and Environment*, 2, pp. 139–144

Tisdale, H. (1942) 'The Process of Urbanisation'. *Social Forces*, 20(3), pp. 311–316

Todd, N. J. (1977) 'Bioshelters and their Implications for Lifestyle', *Habitat*, 2(1/2), pp. 87–100

Turner, J. L. and Ellis, L. (2007) *China's Growing Ecological Footprint*. http://www.globalcitizen.net/Data/Pages/2813/papers/20100215141245290.pdf, accessed 14 March 2007

UNEP (United Nations Environment Programme) (2010) *Sustainable Lifestyles and Education for Sustainable Consumption*. http://esa.un.org/marrakechprocess/pdf/Issues_Sus_Lifestyles.pdf, accessed 11 May 2010

UNPD (United Nations Population Division) (2005) *Population Challenges and Development Goals*. UNPD, New York

Vale, R. and Vale, B. (2009) *Time to Eat the Dog: The Real Guide to Sustainable Living*. Thames and Hudson, London

Wackernagel, M. and Yount, D. (1998) 'The Ecological Footprint: An Indicator of Progress towards Regional Sustainability'. *Environment Monitoring and Assessment*, 51, pp. 511–529

Wheeler, S. (2004) *Planning for Sustainability: Creating Livable, Equitable, and Ecological Communities*. Routledge, New York

Wu, Y. (2003) *Demand for Feedgrain in China: Implications for Foodgrain Consumption and Trade*. University of Western Australia, Perth

WWF (2008) *Report on Ecological Footprint in China*. CCICED, Beijing

WWF (2010) *China Ecological Footprint Report 2010: Biocapacity, Cities and Development*. http://assets.panda.org/downloads/china_ecological_footprint_report_2010_en_low_res.pdf, accessed 18 November 2010

Yang, F. and Wang, Y. (2008) 'Study on Residential Behaviour During Construction of Energy-Saving and Pollution-Reduction Community in China'. World Sustainable Building Conference, 21–25 September, Melbourne

Yang, S. M. (1986) *Research on Chinese Consumption Structure*. China Social Science Press, Beijing (in Chinese)

Yu, Z. Y. (1981) 'Remarks'. *Economic Perspectives*, 1, pp. 25–9 (in Chinese)

Zhang, K. H. and Song, S. F. (2003) 'Rural-Urban Migration and Urbanisation in China: Evidence from Time-Series and Cross Section Analyses'. *China Economic Review*, 14, pp. 386–400

Zhang, X. G. (2012) 'Lifestyle Changes of New China's Peasants over the Past Sixty Years'. http://www.nongli.com/doc/0910/13231816-2.htm, accessed 12 May 2012 (in Chinese)

Zheng, L. (ed.) (1981) *China's Population: Problems and Prospects*. New World Press, Beijing

12 A Study of Surburban Thailand

Sirimas Hengrasmee

Introduction

Self-sufficiency and self-reliance are the key motives behind the national concept of sustainability in Thailand, as outlined in the philosophy of 'Sufficiency Economy' and 'New Theory Agriculture'. This chapter focuses on the implementation of these ideas in existing suburban households in Thailand, in order to illustrate the possibility of an alternative fair earth share lifestyle. Self-reliance in food, water and energy as a suburban basic need is investigated, and the ecological footprint is used as an indicator to see how implementation can contribute to lessening human impacts on the environment, as well as to illustrate a better use of available resources.

It is true to say that self-sufficiency may not be thought of as necessary for life in suburban areas, since, unlike in rural areas, suburbs are normally close enough to facilities that people can find all they need. But better practice in self-reliance in suburbs or elsewhere will support sustainable development. In order to validate this argument, it is necessary to point out that the earth is a finite resource, especially when it is still the only known planet that can support human life. The processes of urbanisation and rapid population growth increase the urban population, as well as transforming much of the countryside into new suburban areas. This also means that in the near future there will be less productive land and fewer primary resources. There could not be a better time to call for more self-reliant, self-sufficient and sustainable lifestyles among present and future populations, at all levels. Being more self-sufficient means that a person can rely more on himself or herself, and that they are removed from the competition for resources. Less competition then gives others who are in need more room.

Sufficiency Economy: The Sustainability Concept in Thailand

Sufficiency Economy is a philosophy which originated in Thailand, and which is about self-sufficency, self-reliance and personal contentment. The philosophy was initiated and named by the present King of Thailand,

Rama IX. Taking ideas from Buddhist teachings of the 'middle path', or moderation, Sufficiency Economy is about encouraging people to learn to be reasonable and to know what is enough. Moderation, reasonableness and the need for a self-immunising mechanism, to enable people to cope with internal and external changes, are the most important principles of the philosophy (NESDB, undated: 3).

Sufficiency Economy has been used as a Thai version of sustainable development and is meant to serve as an underlying principle for all activities, for a sustainable future. Though there is not enough evidence from comparative study of sustainable development approaches and Sufficiency Economy, it is possible to point out the potential sharing of key interests. Sustainable development lays stress on the limits of resources, while the main idea of Sufficiency Economy – to live within what one can have and not be extravagant – suggests that when one only has what one should, the sharing of resources is fairer and more reasonable. In this study, the ecological footprint is used as an indicator to see how the implementation of the national sustainability concept in Thailand, Sufficiency Economy, can contribute to lessening human impacts on the environment, and to provide a connection between Sufficiency Economy and the concept of sustainability.

The Implementation of Sufficiency Economy

In terms of implementation, the famous 'New Theory Agriculture' was initiated, especially for farmers, by King Rama IX. This theory was developed taking all principles of Sufficiency Economy into account. Fundamentally, it focuses on guiding farmers to be able to provide themselves with enough food and water for agriculture and everyday life, before aiming for trade. This idea responds to the principles of self-immunisation and reasonableness. The theory recognises land capital and uses it as the basis of resource availability. The key is to make the most of one's immediate environment at a manageable scale, while maintaining harmony with it at all levels. To practise and achieve the goal of the theory, acts of moderation and reasonableness are crucial, especially in achieving reasonable use of land, labour and consumption. It outlines the idea of what one can have.

The theory has three levels, from the first step, to be taken by a family at the smallest scale, to the third step, to be taken for the wider community. However, the essential part of the theory is the proper management of land use in the first step, based on producing enough for the family with some extra to sell. This step is about the agricultural system, and encourages farmers to grow a diversity of plants to balance the ecological system and to meet the needs of everyday life. The theory proposes farmers should manage their land by dividing it proportionally into four parts, a rice field (30 per cent), a pond (30 per cent), an area for farm plants or crops (30 per cent) and land for housing and other necessary buildings (10 per cent).

This land-use apportioning is suggested, rather than prescribed. The key is for farmers to divide their land reasonably, according to individual needs, ability and labour. Surplus land can be used for producing extra products for sale, if the family can manage to do this without overstretching themselves. The recognition of the middle path and self-sufficiency play an important role in a farmer's decision-making. So, to be successful in practising the New Theory Agriculture and to build a good economic foundation for a family, the combination of good judgment in terms of self-sufficiency and good practice in land management is the basic requirement. Knowing one's own capabilities and resources, then practising accordingly, is the basis of success.

After successful practice of the first step, in which basic needs of a family are fulfilled, the second step is to pool efforts and look for cooperation within a community. This will help to create a strong community relationship based on interdependence. The third step is to create relationships outside the community, for improving local economic status. This step can help a community to reduce its capital costs and progress to a stronger position in production and marketing. It will lead farmers to be more efficient and provide them with an economic marketing operation.

Since Sufficiency Economy is a philosophy, implementation beyond the rural boundary should be possible. This chapter aims to introduce an alternative adaptation of the philosophy, for the more complex suburban society. It should also be pointed out that the implementation could also be applied beyond suburban boundaries, as is the intention of the philosophy of Sufficiency Economy.

Implementation in Suburban Thailand

Sufficiency Economy can be implemented in suburban contexts by taking the successful New Theory Agriculture as a model. The principle of land management in the first step of New Theory Agriculture seems an appropriate beginning, given that it is applied to the individual level. The theory suggests, first, that farmers physically divide their land into several parts, each with a different purpose and producing different products, to fit their needs. This suggestion for land management is based on the amount of food each family needs to consume each year, as well as the need for water for food production and consumption, especially during the dry season. As the New Theory Agriculture suggests, farmers are less dependent on others for food and water, which represent basic rural needs; the fundamental way for a suburban family to increase their self-reliant lifestyle would be in the areas of food, water and also energy. Therefore, the implementation of Sufficiency Economy in suburban areas will take these three areas of suburban basic need into account, and will consider the extent to which self-sufficiency in existing suburban houses can be achieved.

After a careful consideration of the official definitions as well as general physical characteristics of suburban areas, existing suburban areas in the lower northern Phitsanulok province (Tesban Thambon Ban Mai and Tesban Thambon Wang Thong) were selected to represent the generality of suburban areas in Thailand. Information gathered from several government offices along with aerial maps and site visits confirmed that the most common land holding in the selected suburban areas ranges from 200 to 400 square metres, with an average 50- to 100-square-metre building footprint. It is necessary to point out that the size of land holding in the chosen areas seems quite large, compared to the big cities, especially Bangkok, where a single house plot can be as small as 120 square metres. However, this study is aimed at the majority of suburban single houses in Thailand, where the potential for self-sufficiency can be easily promoted.

Within the range of the housing footprints, a number of housing plans were selected to represent suburban houses in Thailand, based on standard plans provided free of charge to the public by the Department of Public Works and Town and Country Planning (undated). Only single houses which could fit into the typical land holding were selected from these standard plans, as single houses are assumed to be the preferred choice for families living in suburban areas. As a result, six standard house plans with footprints ranging in size from 77 to 146 square metres were chosen. These models also illustrate a general understanding of contemporary housing trends for Thai people and provide a foundation to work from, especially in terms of roof styles, when considering the capabilities for energy and water production. In order to cover the whole range of housing footprint, a house with a 50-square-metre footprint was added, based on the average footprint sizes. These houses are assumed to be on five plot sizes – of 200, 250, 300, 350 and 400 square metres – based on the typical land holding in the selected suburban areas.

To investigate the extent to which the existing houses can supply their inhabitants' basic needs, the average consumption of a family is used as a standard for demand. The physical characteristics of the house and land availability for food production can then be tested against this demand. According to a series of national surveys, average household size for the country as a whole is between 3 and 4 people per family. There has been little change in number of family members since 2004, when the survey showed that there were 3.4 people per family. To be on the side of overestimation rather than underestimation, 4 people per family will be used for consumption calculations throughout. The potential for self-reliance in the selected suburban houses based on the standard demand for a family is shown in Table 12.1. It can be seen from the table that the rate of self-reliance is quite high in all categories.

Table 12.1 Potential for self-reliance in food, energy and water for selected suburban houses

	Supply (% of yearly household consumption)							
							House with 50-m² footprint	
Category of demand	Urban House Type 2	Urban House Type 3	Urban House Type 4	Urban House Type 5	Urban House Type 6	Urban House Type 7	Hipped roof	Gable roof
Food (fruit and vegetables)								
200 m² plot	n/a	40	35	n/a	n/a	n/a	50	
250 m² plot	n/a	70	60	80	n/a	40	80	
300 m² plot	80	100	100	100	n/a	80	100	
350 m² plot	100	100	100	100	100	100	100	
400 m² plot	100	100	100	100	100	100	100	
Energy (electricity)								
PV: monocrystalline/polycrystalline silicon								
Low user	≥ 100	≥ 100	≥ 100	≥ 100	≥ 100	≥ 100	≥ 100	≥ 100
Intermediate/high user	≥ 100	82	100	80	88	≥ 100	70	100
PV: amorphous silicon								
Low user	≥ 100	≥ 100	≥ 100	≥ 100	≥ 100	≥ 100	≥ 100	≥ 100
Intermediate/high user	73	49	62	48	53	70	42	64
Water (rainwater)								
Conventional fixtures	73	48	63	54	63	71	42	
Water-saving fixtures	≥ 100	68	88	76	88	≥ 100	59	
Water-saving fixtures, grey water reuse	≥ 100	≥ 100	≥ 100	≥ 100	≥ 100	≥ 100	≥ 100	

Food

The calculation for the standard demand for a family in the food category considered only fruit and vegetable demand. Consumption of other food categories was not included here, because of limited land in suburban areas, as well as the level of skill involved for suburban inhabitants. Fruit and vegetable consumption was calculated based on suggested figures from the FAO and WHO (2004), of a minimum of around 400 grams per person per day. A portion higher in fruit and lower in vegetables is also mentioned. As a result, a family will need around 350 kilograms of fruit and 234 kilograms of vegetables each year.

The study selected three types of fruit (guava, papaya and banana), 13 vegetables (cabbage, cauliflower, Chinese kale, Chinese flowering cabbage, leaf lettuce, Chinese cabbage, water spinach, yard long bean, cucumber, pumpkin, tomato, angled gourd and okra) and 10 herbs (coriander, long coriander, kitchen mint, galangal, ginger, chili, lime, lemon grass, sweet basil and holy basil) to represent the types of fruit and vegetables to be produced in suburban areas. The processes of planting and harvesting were studied, together with the consideration of making food available for the whole year. As a result, an area between 133 and a maximum of 150 square metres is needed for all plants. The different areas needed at different times allows for plots to be shifted between short-life plants. Vegetables are assumed to be planted first, if there is not enough area for all plants, due to their higher productivity compared to fruit for the same area.

From the investigation, it becomes apparent that the current sizes of suburban properties in Thailand have great potential for providing food. It can be seen from Table 12.1 that the minimum that can be produced from available open areas of the selected housing models is 35 per cent of yearly household consumption. When the housing plot size reaches 350 square metres, all housing models reach the stage of complete self-reliance for fruit and vegetables. This level of production, however, was calculated based on only the unpaved open area clear from overhangs. In general circumstances, the unpaved shaded areas could also be used for planting, as each patch is likely to catch sunlight at some point in the day, if not as much as in the open areas. On the other hand, if the shaded unpaved areas are included to raise the minimum level of 35 per cent productivity for the 200-square-metre plot with Urban House Type 4, total production will increase to 50 per cent. This is made up of an increase in vegetable and herb production to 100 per cent of need and a 17.5 per cent increase in fruit production. The inclusion of shaded unpaved areas can thus increase fruit and vegetable production by 15 per cent. This example shows that even a house with a small plot, if carefully and efficiently planned to use every inch of land wisely, could be very productive. Potted edible plants on paved areas will also increase productivity levels.

Energy

The investigation of energy demand is based on estimation of types of appliances and periods of use in a household. Electricity is considered the only source of energy, to avoid complication. For the calculations, demands were set at two levels, 'low user' and 'intermediate/high user', according to the types of appliance owned. Types of appliance selected for this study are from a series of ownership surveys carried out by the National Statistics Office (1996, 1998, 2000, 2002 and 2004). Types owned by low users comprise 11 basic appliances: television, electric fans, electric pot, rice cooker, lighting, washing machine, electric cooker, refrigerator, water pump, radio and video player. For intermediate/high users, 15 appliances were chosen, including all of the appliances owned by low users and four additional appliances: air-conditioner, water heater/boiler, microwave oven and computer. The major difference in energy use between these two types of family lies in the use of air-conditioning as against electric fans. It was found that a family with air-conditioning is likely to use three times more electricity than one using only electric fans: the study estimated the use of 5.9 kilowatt hours per day (kWh/day) for a low user and 18 kWh/day for an intermediate/high user.

The capacity for energy production in suburban areas was estimated by testing the capacity for electricity generation from immediate natural resources. The location of Thailand, where the average solar radiation is about 18 megajoules per square metre per day (MJ/m^2/day; Janjai et al., 1999), provides a good opportunity for exploiting solar radiation by using photovoltaic cells (PV) for electricity generation. For the selected location, Phitsanulok, the data show that average solar radiation ranges between 15.8 and 22.3 MJ/m^2/day, or 4.4 and 6.2 kilowatt hours per square metre per day (kWh/m^2/day; Regional Office of Energy Development and Promotion, undated). The study investigated the opportunity for electricity generation via grid-connected systems. The use of standalone systems was not seen as necessary, as suburban areas, unlike rural areas, are normally close to facilities. The use of grid-connected systems was a more appropriate choice, considering both the avoidance of batteries and the availability of governmental support through the Very Small Power Producer (VSPP) programme, by which electricity surpluses can be supplied to the grid.

The approximate area requirement for installing a nominal power of 1 kilowatt peak (kW$_p$) was based on projects carried out by the Electric Generating Authority of Thailand (EGAT) and the Department of Alternative Energy Development and Efficiency (DEDE). A project carried out by EGAT and DEDE started in 1997 with 10 houses and continued later, in 2002, with 50 houses. The project introduced the use of several types of PV to the general public (EGAT, 1998, 2004). It showed that the area for PV with a nominal power output of 1 kW$_p$ using monocrystalline silicon or polycrystalline silicon is around 9 square metres, while amorphous silicon requires around 15 square metres. Only the southern side of a roof is used for PV installation in this investigation, since this is the most suitable in Thailand.

Since almost all the roof types of the standard house models provided by the government are hipped, the roof has only a small area for PV installation, compared to its solar-efficient area. The hipped roof seems to be increasingly popular among Thai people, according to the trend in standard houses, but a gable roof provides more area for PV installation because the rectangular roof allows for more panels. To confirm this, the investigation into the 50-square-metre-footprint house used both hipped and gable roofs to show differences in energy production.

It can be seen from Table 12.1 that all the housing models can produce enough electricity for a low-user household. The use of monocrystalline or polycrystalline panels allows the intermediate/high users to produce 70 per cent or more of need from a grid-connected system. The amorphous type requires greater area, so it is harder for the intermediate/high users to produce enough electricity with the current, hipped roofs. The use of a gable roof would provide more roof area and therefore more electricity.

Water

Data for water consumption per capita from 1997 to 2006 reveal that around 240 litres per person per day are used within the area covered by the Metropolitan Waterworks Authority (MWA, 2006), while 150 litres per person per day are used in areas served by the Provincial Waterworks Authority (PWA, 2005). Average water consumption for a Thai is therefore around 200 litres per day. This generalised figure is used here, to allow application throughout the country. Alternative water conservation practices of installing better fixtures and reusing grey water for out-of-house use are included in the study because they are common practices that can be easily achieved by all. For the investigation, average water consumption for different activities in Thai households, as shown in Table 12.2, is needed. The table shows patterns of water consumption in Thailand (Manager Online, 2003; Ngernmoon, 2003). In total, a family with conventional fixtures will require around 292,000 litres per year, a family with water conservation fixtures which is not reusing grey water will use 208,780 litres per year, and a family both with water conservation fixtures and reusing grey water will need only 122,640 litres per year.

Potential for self-reliance in water is based on the use of rainwater harvested on-site. There is much concern for human health regarding poor rainwater quality because of the increase in pollution, causing alarm to many Thai people living in urban or industrial areas. However, rainwater still remains one of the main sources of drinking water for many people (Department of Health, 2007). The main reason for impurities is understood to be from improper design of collection and storage systems, as well as lack

Table 12.2 Estimate of household water consumption

		Consumption per capita					
		Conventional fixtures		Water-saving fixtures			
				Without reusing grey water		Reusing grey water	
Use	Proportion (%)	litres/ day	litres/ year	litres/ day	litres/ year	litres/ day	litres/ year
Household cleaning	2	4	1,460	4	1,460	4	1,460
Kitchen	5	10	3,650	10	3,650	10	3,650
Washing clothes	9	18	6,570	18	6,570	18	6,570
Washbasin and shower	23	46	16,790	27.6	10,074	27.6	10,074
Flushing toilet	26	52	18,980	13.5	4,928	13.5	4,928
Out-of-house use	35	70	25,550	70	25,550	10.4	3,796
Total	100	200	73,000	143	52,232	84	30,478
Household consumption (four people)	—	800	292,000	572	208,780	336	122,640

of maintenance, rather than the quality of rainwater itself (Department of Health, 2007). The investigation into the capacity for water production is divided into two parts. The first part is testing the estimated amount of water for out-of-house use against the estimated productive area. Then the possibility of rainwater harvesting from the available roof area can be tested against the total water requirement.

The estimate of water needed for plants over and above natural rainfall, from the New Theory as described by Wallop Promthong (2001), is around 625 litres per square metre per year. A 150-square-metre garden will therefore need 93,750 litres per year. This quantity of water is well below the quantity of water needed for out-of-house use per household (102,200 litres per family per year), so having a productive garden need not be a burden on the existing water supply system.

Average annual rainfall in Phitsanulok is 1,300 millimetres. Only rainwater harvested from the roof of the housing model is calculated as alternative water supply. It can be seen from Table 12.1 that only a household using water conservation practices involving both water-saving fixtures and reuse of grey water can harvest enough rainwater to cover its annual consumption. Households in the other categories will have to combine rainwater collection with conventional water supply. Even if a family could harvest enough rainwater for their yearly consumption, which is not a common practice in Thailand, the big storage tank needed will be an obstacle to overcome. In general, Thai people normally store water sufficient for up to two days' consumption as a precaution in case of water cuts. It is true that the average annual rainfall is quite high at the selected location, but the dry period is also quite long. The rainwater tank will need to be as large as 38 to 48 cubic metres, depending on the catchment area. This size of tank will greatly affect the land area available for food production if it is installed above ground.

The Ecological Footprint of Suburban Lifestyles

In this section, the environmental impacts of different suburban lifestyles based on the level of self-reliance are examined. Here, a simple calculation for a component-based ecological footprint (EF) analysis using the EF conversion factors provided by Chambers *et al.* (2000) is used. This is because the appropriate EF for assessing the sustainability of different suburban lifestyles suitable for Thailand is not yet available. The footprint conversion factors suggested by Chambers *et al.* (2000) are based mainly on EU and UK data. So, the footprint calculated using this method and dataset will give only a rough picture of the situation in suburban Thailand. Nonetheless, it provides a method for evaluating possible changes in lifestyle. The results of the EF calculations are estimated from yearly consumption rates and are given in hectares (ha/year). Table 12.3 compares the EF of a conventional family with that of more productive families.

Ecological footprint of food

The analysis of the EF of food here accounts only for different levels of productive garden and food consumption based on the Dietary Reference Intakes (DRIs) suggested by the Nutrition Division, Department of Health (2003). The energy and life-cycle effects involved in household fruit and vegetable production are not included; however, these are understood to be small in comparison. Table 12.3 shows that differences in productive level lead to significant differences in EF. For example, as seen in Table 12.1, a family living in Urban House Type 3 built on a 200-square-metre land plot or Urban House Type 7 built on a 250-square-metre land plot both have the potential to produce only 40 per cent of their yearly fruit and vegetable consumption. However, even at this low level of production, these families contribute a reduction in environmental impacts as measured by EF of up to 1,000 square metres per year (Table 12.3). This level of reduction could be considered a crucial improvement.

Table 12.3 Ecological footprint for suburban basic needs according to level of family self-reliance

Ecological footprint component	Ecological footprint of a conventional family (ha/year)	Ecological footprint of a productive family (ha/year)						
		40%	50%	60%	70%	80%	90%	100%
Food								
Normal diet	2.7	2.6	2.6	2.5	2.5	2.5	2.5	2.4
Energy								
Low user (6 kWh/day)	0.24	0.17	0.15	0.13	0.11	0.09	0.07	0.05
Intermediate/high user (18.5 kWh/day)	0.75	0.51	0.46	0.40	0.34	0.28	0.22	0.16
Water								
Conventional fixtures	0.027	0.026	n/a	n/a	n/a	n/a	n/a	n/a
Water-saving fixtures	0.020	0.020	0.019	0.018	n/a	n/a	n/a	n/a
Water-saving fixtures and grey water reuse	0.013	0.012	0.012	0.012	0.012	0.012	0.011	0.011

Ecological footprint of energy

The investigation into reduction of environmental impacts from energy considered both sources of energy, central supply and PVs. Adjustment of EF conversion factors was carried out proportionally, to fit Thailand's situation. As a result, 110 hectare-years per gigawatt hour is used as the footprint conversion factor for central supply. For the footprint conversion factor of PVs, both built land (direct land use) and energy land (land required for absorbing both direct and indirect emissions of CO_2 during construction, operation and maintenance) are already accounted for in the conversion factor.

It can be seen from Table 12.3 that a low-user family that can produce about 40 per cent of its electricity need will reduce EF by as much as 700 square metres per year. However, complete self-reliance should not be seen as the priority at this stage, due to the potential of the alternative practice of demand reduction on the rate of energy consumption, which can be viewed as an efficient way to reduce human impacts on the environment. A shift from the energy consumption of an 'intermediate/high user' to that of a 'low user' for a conventional family shows a significant EF reduction, of as much as 5,100 square metres per year. The practice of moderation and reasonableness from the theory of Sufficiency Economy could contribute to the change.

Ecological footprint of water

The investigation into the water footprint looked at the water consumption rate (for water supply from authorities and rainwater) and estimated the EF of a concrete rainwater tank, including its embodied energy and the land area needed for tank installation. Operating energy to supply water to any area within a house using a water pump was calculated as part of electricity consumption. Treatment can be carried out, if necessary, by passing water through filters under gravity, so operating energy was excluded from this calculation to avoid double counting.

According to New Zealand-based studies by Mithraratne and Vale (2007a, 2007b), the life-cycle energy and CO_2 emissions of a concrete tank are much lower than for a plastic tank, and the concrete tank has a longer lifespan. It is assumed that water tank technology and water systems in Thailand and other countries, including New Zealand, are quite similar. Therefore a concrete tank is selected instead of a plastic tank for this study. The results of the EF investigation into the actual footprint as well as the embodied energy for different lifestyles are shown in Table 12.3. Similar to the result for the energy footprint, reduction of consumption by behaviour change, shifting from a conventional to an alternative consumption style, is more likely to reduce environmental impacts effectively than just being more productive. Changing to water-saving fixtures and grey water reuse reduces EF by as much as 140 square metres per year, which is far better than the 10 square metres per year achieved by retaining conventional practice and moving to 40 per cent self-reliance.

Conclusion

The three areas of environmental impacts investigated in this study respond to the application of increasing self-reliance, self-sufficiency and sustainable lifestyles in suburban areas, implemented using the uniquely Thai approaches of Sufficiency Economy and New Theory Agriculture. The success of the implementation and the benefits of a more self-reliant lifestyle in reducing human impact on the environment illustrate the possibility of an alternative fair earth share lifestyle. The ecological footprint analysis, which is widely used as an indicator for sustainability, shows that development based on the philosophy of Sufficiency Economy could help lower the rate of human impact on the environment. The national Sufficiency Economy and mainstream sustainability are therefore strongly linked through a common interest in the environment.

The results of the study show that the implementation of the philosophy of Sufficiency Economy, whether in terms of the reduction of consumption principle or the promotion of self-reliant lifestyles, can significantly reduce human impacts on the environment. However, the change in attitudes and behaviours of suburban people needed if they are to embrace more self-reliant practices are a vital part of the lower footprint lifestyle. Beyond the boundary of suburban areas, the findings also suggest what a household in Thailand with similar physical characteristics could achieve. According to this investigation, physical changes which encourage more self-reliance and the lifestyle of the household's members are as important as the change of attitude.

Although the findings suggest a self-reliant lifestyle in suburban areas is possible, complete self-reliance should not be seen as a requirement, given the location and lifestyles of suburban people. However, it is important to increase the level of household self-reliance, because this can be used to foster better understanding of resource consumption and teach families to manage their resource availability.

References

Chambers, N., Simmons, C. and Wackernagel, M. (2000) *Sharing Nature's Interest*, Earthscan, London

Department of Health (2007) *Household Drinking Water Quality Assessment Project* (in Thai), Department of Health, Ministry of Public Health, Nonthaburi, Thailand

Department of Public Works and Town and Country Planning (undated) *Standard Housing for Thai People* (in Thai), http://subweb.dpt.go.th/pip/house_model/framehome.html, accessed 3 August 2006

EGAT (1998) *Demonstration Project for Electricity Generation from Rooftop by Photovoltaic Cells, Phase 1 (10 houses)* (in Thai), Electric Generating Authority of Thailand, Bangkok

EGAT (2004) *Demonstration Project for Electricity Generation from Rooftop by Photovoltaic Cells, Phase 2 (50 houses)* (in Thai), Electric Generating Authority of Thailand, Bangkok

FAO and WHO (2004) *Fruit and Vegetables for Health: Report of a Joint FAO/WHO Workshop, 1–3 September, Kobe, Japan*, Food and Agriculture Organization of the United Nations, Rome

Janjai, S., Laksanaboonsong, J. and DEDE (1999) *Development of Daylight Potential Maps from Satellite Data for Thailand* (in Thai), Department of Alternative Energy Development and Efficiency and Silpakorn University, Bangkok

Manager Online (2003) 'Interesting Facts about In-House Water Consumption, 4 June' (in Thai), http://www.manager.co.th/asp-bin/mgrView.asp?NewsID=4698880440591, accessed 10 September 2005

Mithraratne, N. and Vale, R. (2007a) 'Conventional and Alternative Water Supply Systems: A Life Cycle Study', *International Journal of Environment and Sustainable Development*, vol. 6, no. 2, pp. 136–146

Mithraratne, N. and Vale, R. (2007b) 'Sustainable Choices for Residential Water Supply in Auckland', 2nd International Conference on Sustainability Engineering and Science: Talking and Walking Sustainability, http://www.nzsses.auckland.ac.nz/conference/2007/manuscripts.htm, accessed 6 December 2007

MWA (2006) *Annual Report 2006*, Metropolitan Waterworks Authority, Bangkok, Thailand

National Statistical Office (NSO) (1996) *Report of the 1996 Household Energy Consumption Survey*, National Statistical Office, Office of the Prime Minister, Bangkok

National Statistical Office (NSO) (1998) *Report of the 1998 Household Energy Consumption Survey*, National Statistical Office, Office of the Prime Minister, Bangkok

National Statistical Office (NSO) (2000) *Report of the 2000 Household Energy Consumption Survey*, National Statistical Office, Ministry of Information and Communication Technology, Bangkok

National Statistical Office (NSO) (2002) *Report of the 2002 Household Energy Consumption Survey*, National Statistical Office, Ministry of Information and Communication Technology, Bangkok

National Statistical Office (NSO) (2004) *Report of the 2004 Household Energy Consumption Survey*, National Statistical Office, Ministry of Information and Communication Technology, Bangkok

NESDB (undated) *An Introductory Note: Sufficiency Economy*, http://www.sufficiencyeconomy.org/en/files/4.pdf, accessed 10 May 2005

Ngernmoon, J. (2003) 'Water is Not Abundant', *Energy World Journal*, vol. 6, no. 21

Nutrition Division, Department of Health (2003) *Dietary Reference Intake for Thais 2003*, Nutrition Division, Department of Health, Bangkok

Promthong, W. (2001) *New Theory Agriculture*, Thai Wattana Panich, Bangkok

PWA (2005) *Annual Report 2005*, Provincial Waterworks Authority, Bangkok, Thailand

Regional Office of Energy Development and Promotion (undated) 'Solar Radiation' (in Thai), http://www.dede.go.th/dede/index.php?id=428, accessed 12 September 2007

13 Kampung Naga, Indonesia

Grace Pamungkas (with Fabricio Chicca and Brenda Vale)

The Ecological Footprint of Indonesia

Indonesia is not only a populous country (234.7 million in 2007; CIA, 2011) of many islands, it also has the rare characteristic of just living within its biocapacity. A 2007 report showed the average ecological footprint (EF) was 1.07 global average hectares per person (gha/person) and the available biocapacity 1.12 gha/person (Ministry of Public Works, 2010: 29). However, within these averages there are wide differences. Although the island of Java, which contains the capital Jakarta, has an average EF of 1.01 gha/person, its biocapacity is only 0.2 gha/person, and the tourist island of Bali is also in deficit, with an EF of 1.76 gha/person and a biocapacity of 0.24 gha/person. All other major islands in Indonesia are in ecological reserve (Ministry of Public Works, 2010: 31). The Ministry of Public Works (2010: 11) notes that the higher footprint of Bali is due to its extensive local and international tourism activities. This should send a real warning to those who believe that western-style development can be had without higher environmental impact.

Within the island of Java there are also big differences in lifestyle and footprint. The high footprint of the province of DKI Jakarta comes largely from the energy consumed for transportation and industrial purposes (Ministry of Public Works, 2010: 19). The same situation is true of the province of West Java: although it is still in ecological reserve when it comes to the use of forest land, all other footprint land categories are in deficit, giving it the second highest EF in Java (Ministry of Public Works, 2010: 21). Within this West Java average there are further differences. This province contains the capital Jakarta and other large cities such as Bogor and Bandung. It also contains some settlements where life is more traditional and footprints are lower, and one of these, Kampung Naga, is the subject of this chapter. The problem with averages is that they can hide some very low and very high values when it comes to environmental impact.

Kampung Naga

A visit to Kampung Naga in Java starts in a car park, proceeds down 300 steps that drop to the valley floor and effectively ends in a different world, one that is obviously living lightly on the earth. Kampung Naga is a traditional village of the Sundanese people of West Java, who are mainly Muslim (Newland, 2001). 'Sundanese hamlets tend to be close groupings of houses lying either in the oxbow of a river or between two streams' (Schefold et al., 2003: 430). Kampung Naga is one of these traditional Sundanese hamlets, and its geographical position is in a valley, in the oxbow of the River Ciwulan. The village is thus contained by the river, on the other side of which lies the sacred forest, which contains the graves of the ancestors and which the villagers believe is a place of all the spirits.

The traditional indigenous sacred forest of the community is an example of using the principle of forbidden access to exclude an area from exploitation (Wessing, 1999: 59), a rule of conduct that is obeyed by all the members of this community. This is only one example of the strict customs and traditions at Kampung Naga that stir curiosity about how and why this community chooses to keep honouring their ancestors by living in the traditional settlement when others in Indonesia are moving into the modern era.

The village is a community of 101 families occupying an area of 100 hectares (Jakarta Post, 2001), who have built their traditionally constructed houses on a stage-shaped rock foundation. The basic materials for the body of a house are bamboo and wood, with the roof made of palm leaves installed without nails, and instead mounted with a series of bamboo cross pieces. With the main pillars of the house only resting on the stone foundation, local wisdom has kept all of these houses intact; they were unaffected by the 7.3 Richter scale earthquake in September 2009 because their flexible construction was able to move with the quake vibrations (Sundalander, 2010).

It is not only the house construction that makes this community culturally unique, but also their daily activities, such as maintaining the communal well for drawing water, wearing traditional clothes (*sarong* and *kebaya*), choosing not to use electric power, and politely refusing to allow visitors to stay in the village. The fact that this village is located only 30 kilometres from the two growing developments of Garut and Tasikmalaya makes it valuable as an example of an indigenous community with a low-footprint way of life that is in touch with modern living.

Life in Kampung Naga

The first thing to note is that the population of the village is limited. If the population expands beyond these limits, people have to move out. When others die, people can move back into the village. This represents true understanding of living within the local biocapacity. It has led to a version of the village situated at the top of the steps, just off the car-parking area, where children who have moved out have made their homes. In style these modern houses are similar to those of the village, although some of the materials are more modern and electricity is available.

Within the village limits food is grown, water is collected for people, agriculture and fish culture, and the resources to maintain the houses are also generated. Because the village houses are clustered on the valley floor most rice is grown in terraces on the slopes of the surrounding hills. The hills provide the necessities of life – water, food and wood (including bamboo) – with the river on the valley floor taking away the waste (Mihalyi, 2007). Within the village vegetables are grown, fish are raised in ponds and chickens are kept to give a diet of rice, fish, chicken, eggs, fruit and vegetables. However, the systems for doing these things are much more interconnected than would initially appear. The fish ponds are situated between the river and the village. There is a fence between the village and these ponds, with places for washing just outside the fence. This means the waste water flows into the fish ponds. Rice is also threshed by hand in a small structure above another pond, so that any waste falls through for the growing fish. Goats are also kept in small thatched shelters by these ponds. The fence separates activities that seem 'dirty' from the 'clean' village. Vegetables are grown in pots on raised racks around the edges of the ponds.

Once through the fence, the houses are clustered together either side of narrow lanes running up the slope, which are shaded because of the overhanging thatched roofs of the houses either side. These lanes have stones in the centre as a raised central section for when it rains. Houses are 30 to 60 square metres, and are entered by taking a step up from the lane through a door into a visitors' room next to a kitchen, which gives onto a living space with sleeping space off it, all within a simple rectangular plan. The chickens range free under the houses, especially under the kitchen, where the bamboo floor finish allows any dropped food to fall through. To keep the eggs safe, the hens have to fly up to baskets hung near the top of the door frame to lay. Cut wood for cooking is stacked against the wall of the house, under the overhanging thatched eaves to keep it dry. The wood comes from the forests or bamboo gardens within the village (Mihalyi, 2007).

At the centre of the village is an open space with the mosque and meeting house, which are larger versions of the simple rectangular dwelling houses. The construction method for all buildings is the same, with the foundation stones taken from the river, the timber and bamboo for the posts, wall, floor and roof structure from village land, and the palm for thatching from another village (Mihalyi, 2007). The houses are consequently light in weight. The roof lasts about 15 years and the woven walls finished with lime wash 10 to 15 years, but all is easy to replace as necessary. Because of the light construction the houses cannot be filled with possessions, as in the west, so furnishings are used carefully and the heavy cooking stove stands on a stone foundation placed directly onto the earth. The houses are uniform in initial appearance but, like much vernacular building, vary in detail, so different patterns appear in the woven walls, which are made by women. Many houses have verandahs. Without electricity most tasks are done in daylight, with oil lamps lit at night.

Despite this life of seeming simplicity and low environmental impact the village has plenty of contact with the world at the top of the steps. Children attend local schools from the age of seven years and the village uses local health services, even if sometimes the infirm have to be carried up the 300 steps. Glass has also appeared in the village, as have portable radios (Mihalyi, 2007). One house visited had a television (black and white) powered by a car battery but, as the owner explained, watching was limited as the battery had to be carried up to the top every time it needed charging.

The village has therefore absorbed something of modern Indonesian life, even while keeping to the very simple way of life of their ancestors, believing the village to be inherited from them, with the need to pass it on to subsequent generations in a similar state – a true echo of the famous definition of sustainability in the Brundtland Report (World Commission on Environment and Development, 1987: 8). Some of the beliefs and values that keep the residents of Kampung Naga following their traditional way of life despite contact with the modern world are discussed below.

Beliefs in Kampung Naga

The nature and form of tribal-ancestor relations are central elements in this community. The spirituality that prevails between another world and this tribe both explains and gives rise to the features of their current settlement. How the particular hamlet is constructed, dwelt in and maintained reflects the way they treat the land as a sacred place. If Wessing (1990) is right that 'in West Java, land used to be (and sometimes still is) seen as belonging to the spirits, that is the embodiment of the fertility of the land', then ancestor worship might still be present in Kampung Naga as a means of securing the well-being of the whole community.

The humble great ancestor

The boundaries, layout, orientation, shape, materials and many aspects of dwelling tradition are defined with reference to *Sembah Dalem Singaparna*, the great ancestor whose grave is the most sacred place in the area, in the forbidden forest to the west of the village. As a result, the physical appearance of the dwellings and settlement expresses his spiritual concepts: *Teu Saba, Teu Soba, Teu Banda, Teu Boga, Teu Weduk, Teu Bedas, Teu Gagah, Teu Pinter* (Disbudpar, 2008: 78). These concepts translate into 'nobody should be better than the others with regards to materialism'. Historical research by the West Java Provincial Office of Culture and Tourism in 2008 revealed some information regarding the great ancestor of the Kampung Naga people. It is believed that Sembah Dalem Singaparna was one of seven brothers. Each of them had a 'powerful' quality, the first six being listed as physical strength, wealth, intelligence, knowledge of Islam, knowledge of agriculture and immortality, while the particular Kampung Naga great ancestor was blessed with no special power apart from the ability to live a humble life (Disbudpar, 2008: 78).

The way in which Naga people are restricted from building more houses, after reaching a population of 110 households, provides a rich basis for investigating the role of this myth in their understanding of the carrying capacity of the land and forming a sustainable lifestyle. Because of their great respect for their ancestor's philosophy of life, every Naga person believes they should not have anything more than their ancestor gave them at the beginning. In this sense, they are aware of the limits of the available land area and water resources that they have inherited from previous generations with the same beliefs. Once the leaders realised the need to build more houses for the growing population, they started to build these new houses without enlarging the area of the existing residential zone. They did not consider crossing the border of this zone and using any of the land used for rice cultivation, since they believed doing this would affect the water resources in their land (Disbudpar 2008: 38). This conflict between population growth and limited land resources was resolved by having a more dense housing arrangement. Furthermore, the leaders also decided to let the new community members build their houses outside Kampung Naga, hence they have never changed the land zoning that has been part of their sustainable life for a very long time.

'Rice and fish on a plate': the myth of Nyai Pohaci (Dewi Sri)

In his observations on myths of fertility and generation, Wessing (1990: 251) quoted O'Flaherty's explanation of the function of the Javanese goddess Dewi Sri's death and her transformation into useful crops, among which rice is prominent, as a symbolic expression of the cultural change from hunting (consuming living flesh) to agriculture. If O'Flaherty is correct that 'Rice ... functions as a ritual expression of the triumph of agriculture over hunting' (O'Flaherty, 1988: 116), then it is relevant to query the function of another story, Dewa Antasari's transformation into a fish in the Sundanese myth of Nyai Pohaci, the local name for Dewi Sri.

The Sundanese people have their own version of the myth of Dewi Sri, as retold by Suryani (2002). In this myth, Nyai Pohaci is sent to the earth by the Greatest God to visit a man called Dewa Antasari because he had been cursed and suffered a terrible skin disease after killing his brother, Dewa Antarasa. Despite looking very handsome, he had to be veiled to hide the leprosy on his body. Since Dewa Antasari had been good looking, Dewi Sri fell in love with him. Dewa Antasari also fell in love with her, because she was a most delightful companion. One day his veil fell away, and as soon as Dewi Sri realised that Dewa Antasari had leprosy, she felt embarrassed about falling in love with him and was very much saddened. Because of her deep grief, Dewi Sri killed herself, and her body transformed into a rice crop. Dewa Antasari felt ashamed, and killed himself. Before he died and transformed into a fish, he said that someday he and Dewi Sri would meet in Tegal Panjang. The Sundanese believe that Tegal Panjang is a plate where they place the rice and fish that forms their everyday meal. Having this belief means the act of consuming rice and fish together has a symbolic function as a ritual, despite its more mundane biological necessity. Thus the complex arrangements at Kampung Naga for recycling nutrients into the fish ponds and the water that feeds down to them through the rice crops are not just a technical solution to the problem of growing food for the village, but also part of the spiritual beliefs of the villagers.

Rituals of the great ancestor and the rice deity

How rice is planted, harvested, dried, threshed and kept in a granary all happens under the authority of Nyai Pohaci (the local name for Dewi Sri). There are rituals and customary laws that apply throughout the life cycle of a rice crop, in terms of treating this plant as a magical being. Given that the Kampung Naga villagers are Muslims, the syncretism between Islamic and traditional beliefs shows how their culture has been transformed. Almost all rituals for their ancestor and the rice deity are held according to the Islamic calendar.

Six times a year Kampung Naga people make a pilgrimage, called *Hajat Sasih*, to the great ancestor's grave in the sacred forest (Disbudpar 2008: 74). The purpose of these rituals is to pray and give an offering, followed by maintenance of the grave. Most of the dates for these customs coincide with Islamic celebrations, that is, the months of Muharam (Islamic New Year), Rayagung (Idul Fitri), Mulud (the birthday of Muhammad), Ramadhan (fasting month) and Jumadil Akhir (the Islamic middle of the year that is also the end of dry season), with the only one not based on an Islamic month being Rewah.

Three important rituals that show the villagers' respect for the rice crops are *Ngajuruan* (harvesting), *Nyimpen* (keeping rice in the granary) and *Ngaleuseuhan* (threshing and grinding the rice). The focal point of the harvesting ritual is choosing the best rice seed, as the best of the harvest to be kept for the next planting season. Kampung Naga people still only grow the particular variety of rice called *Pare Gede* (big rice), and they have kept this instead of replacing it with the new variety introduced by the Indonesian government. In the granary ritual, a rice shaman will ask Nyai Pohaci to protect and bless the rice in the granary, after which he will distribute the harvest to each family. In the grinding ritual, the focal point is the wooden mortars and pestles that have to be blessed by the rice shaman before each family can start grinding the first of the harvest. Grinding the rice together will be accompanied by chanting songs in praise of the rice deity.

Beside the rituals for rice crops, Kampung Naga people show great respect in their attitude towards rice. Only women are allowed to take rice from the house granary, called the *Goah*. This room is the most sacred part of every house, where no man should go. The women must kneel when they enter this room to get the rice they need for cooking. To step over a rice seed is forbidden, and they must take every seed that falls on the ground to avoid the deity becoming upset by being taken for granted. In another ritual, when they pray for blessing of the sacred weapons, powerful knives are placed in a bamboo bowl lined with banana leaves, with some rice placed beneath the leaves. This shows that rice has a symbolic function in protecting the whole community from harm.

The myths and legends that relate to these traditions give a full account of how, by whom, and in what manner the whole community is affected. Malinowski's classic work *Magic, Science, and Religion* is an anthropological milestone in studies of traditional religion. It recognises that magic and religion are a special mode of behaviour. They are a mode of action, built up of reason, feeling and will (Malinowski, 1954: 24). The link between the beliefs of the people of Kampung Naga and their lifestyle and consumption patterns shows how concepts with their origin in mythology and traditional religion have a clear regard for the limited resources of the sacred land. These same beliefs enable the villagers to live simply, so that they can pass the land on to their children in the same condition in which it was passed to them.

The Ecological Footprint of Kampung Naga

The footprint of Kampung Naga was assessed using a tool developed and validated by Fabricio Chicca (see Chapter 19) for the estimation of the EFs of settlements. The EF of a Kampung Naga resident is 0.75 gha/person, of which 0.7 gha is the food footprint. By all standards this is a low footprint, but it does raise a further issue. Most of the food EF for Kampung Naga comes from the cultivation of rice. Because this is a wet process methane is produced during the cultivation cycle, and methane is a greenhouse gas more than 20 times as effective at trapping heat in the earth's atmosphere than carbon dioxide. About 13 per cent of global methane emissions come from rice production (Graham, 2002). However, those who live at Kampung Naga have been responsible for the same levels of greenhouse gas production for centuries. In contrast, methane from landfills, which accounts for 6 per cent of global methane (NASA, undated) is an almost entirely avoidable set of emissions. First, landfills are largely a phenomenon of the developed world and its superfluity of stuff, and as such are avoidable.

There are no landfills in Kampung Naga. Second, the gas from landfills can be trapped and used. Sewage treatment is responsible for a further 5 per cent of global methane (NASA, undated). The fact that methane collection and use does not happen in all landfills and sewage treatment systems is simply one example of the developed world not being prepared to pay the cost of setting up these systems. This is because cheaper, and more polluting, alternatives exist and as yet the polluter does not pay. In fact the methane emissions from coal, oil and natural gas mining and extraction, at 19 per cent of total global emissions – the single largest category (NASA, undated) – put the whole of the world's methane emissions from rice production into a different perspective. Just because a footprint could be lowered by changing lifestyle does not mean that it should be. In the case of Kampung Naga, the low-footprint lifestyle has continued virtually unchanged for many centuries. It is those who have changed their lifestyle rapidly in the last 100 years without thought to the impact it is having who should be the ones to now think carefully about what they are doing for the future of all human life on earth.

Stewardship

Kampung Naga is not owned; it is held in trust by those living it in for their children and the succeeding generations. A house is 'my house' because I live in it, not because I own it. This is a very different situation from the western world where ownership models imply not just the right of tenure but 'the right to do what I like with my property'. This attitude to property has probably done much to place humanity in its present perilous position. Kampung Naga not only offers a physical model of how it is possible to live in harmony with the natural world and still lead a satisfying life, but also an alternative social model for how this might be achieved. Kampung Naga shows that sustainability is not an issue of technical innovation or development but rather one of understanding what the land can support perpetually and devising a way of living that can exist within these limits.

References

CIA (Central Intelligence Agency) (2011) *CIA World Fact Book*, http://www.indexmundi.com/g/g.aspx?c=id&v=21, accessed 30 July 2012

Disbudpar (2008) *Sejarah Kampung Naga (Suatu Kajian Antropologi Budaya)*, Dinas Kebudayaan dan Pariwisata Provinsi Jawa Barat, Balai Pengembangan Kemitraan dan Pelatihan Tenaga Kepariwisataan, Bandung, Indonesia

Graham, S. (2002) 'Rice Paddy Methane Emissions Depend on Crops' Success', *Scientific American*, 20 August 2002

Jakarta Post (2001) 'Kampung Naga: A Traditional Village in West Java', *Jakarta Post*, 11 February

Malinowski, B. (1954) *Magic, Science, and Religion and Other Essays*, Doubleday Anchor Books, New York

Mihayli, G. (2007) 'The Sundanese House', http://www.architectureweek. com/2007/0307/culture_1-2.html, accessed 30 July 2012
Ministry of Public Works (2010) *Ecological Footprint of Indonesia*, Ministry of Public Works, Jakarta
NASA (National Aeronautics and Space Administration) (undated) 'Education: Global Methane Inventory', http://icp.giss.nasa.gov/education/methane/intro/cycle.html, accessed 30 July 2012
Newland, A. (2001) 'Syncretism and the Politics of *Tingkeban* in West Java', *The Australian Journal of Anthropology*, vol. 12, no. 3, pp. 312–326
O'Flaherty, W. D. (1988) *Other People's Myths*, McMillan, New York
Schefold, R., Nas, P. and Domenig, G. (2003) *Indonesian House: Tradition and Transformation in Vernacular Architecture, Vol. 1*, KITLV Press, Leiden, The Netherlands
Sundalander (2010) *Traditional Building Construction of Kampung Naga*, http://www.sundalander.com/2010/11/traditional-house-building-construction-of-kampung-naga/, accessed 30 October 2011
Suryani, E. (2002) *Calakan Aksara, Basa, Sastra, Katut Budaya Sunda*, Ghalia, Boger, Indonesia
Wessing, R. (1990) 'Sri and Sedana and Sita and Rama: Myths of Fertility and Generation', *Asian Folklore Studies*, vol. 49, no. 2, pp. 235–257
Wessing, R. (1999) 'The Sacred Grove: Founders and the Owners of the Forest in West Java, Indonesia', *L'homme et la foret tropicale*, pp. 59–73
World Commission on Environment and Development (WCED) (1987) *Our Common Future*, Oxford University Press, Oxford, UK

14 A Study of Hanoi, Vietnam

Han Thuc Tran

Introduction

This chapter presents a case study of Hanoi, exploring how ways of using space in an urban setting, in particular sharing of space, can contribute to lower footprint living. Furthermore, it examines how changes in ways of living would significantly impact the environment (ecological footprint).

Hanoi, the capital of Vietnam, is in a transition time of booming economy associated with urbanization and globalization. As the result, there has been much change in terms of lifestyle, towards modernization and western standardization. Many people have become wealthier and are able to live in bigger houses, and consume more food and goods than before.

Nevertheless, the ecological footprint (EF) of Vietnam is still low: about 1.3 global average hectares per person (gha/person; Global Footprint Network, 2009), even lower than the 'fair share footprint' of 1.8 gha/person. Many traces of indigenous living still remain in Hanoi and can be seen in everyday life and the community forms of living, in which people use fewer resources and have high tolerance for this situation through acceptance and adaptability. Having a productive landscape and overlapping land-use patterns, every piece of land of Hanoi is intensively used for housing and business, and also for cultivation. For example, public space such as sidewalks can be used for sports, buying and selling goods, eating, and even growing vegetables in pots.

In this case study of Hanoi and some examples of developed countries, such as New Zealand and Finland, the three following issues are presented. First, ways of using space in Hanoi, in particular sharing of living space and sharing between public and private space, are described. Second, the current EF of Hanoi (its housing and transport footprint) in relation to ways of using space is presented. Third, how EF may change if lifestyles change is examined.

Sharing of Space in Hanoi

Mixed use of public and private space

There are many differences between the West and the East in ways of using space. Western modern cities tend to be larger in scale with mass transportation. The boundary between public and private realms is clear. The public sector is responsible for circulation space and public services; the private for housing. In contrast, Asian cities' development is often smaller in scale with mixed-use functions and high intensity. Using the example of the tube house in Hanoi, this chapter looks at the way in which public space (the street and pedestrian sidewalk) and private space (housing) combine efficiently and provide beneficial income as well as accessible services for local residents.

In Hanoi, due to officially low income, most people have more than one business and people use their houses as ideal places from which to run their businesses. Most dwellings in Hanoi use the ground floor for commercial purposes. Housing therefore plays a very important role in community livelihood, as the home is not only for private living. Dwellings have multiple uses which mix private and public functions through inclusion of productive activities (trading, manufacturing products, office work, recreation).

The tube house is the most common type of urban housing and has created the main urban fabric of Hanoi. Originally, the tube house, also called a shop house, dominated business districts in many Asian cities. In Hanoi, the tube house was born in the fifteenth century with the establishment of market towns by traders and craftspeople. The tube house is a mixed-use model, shared between living and working (trading and workshop) uses. Because shops in Vietnam were taxed according to their width of frontage or width of marketing area, they developed long, narrow shapes, and storage and living spaces were moved to the rear. The average size of traditional tube houses is about 3 metres in width and 30 to 40 metres in depth, with only one or two storeys. This model of mixed-use housing, combining living and trading, has been carried on into modern times and most dwellings now use the ground floor for commercial purposes. The average size of the 'neo-' or modern tube house is about 4 metres in width and 20 metres in depth, with more storeys (from 3 to even 10 floors).

It is important to emphasize the relationship between the building's ground floor and street activities. In Vietnam, putting aside the western assumption that the road is solely for traffic, streets become special places where vibrant public activities happen. Here, there is no strict separation between the street and the house: the functions are mixed (walking, bicycling, trading, working and living). The private and public factors are not in conflict, and even seem to promote each other because they are integrated in the ground floors of the houses.

The ground floor is the intermediate space between the street and the house. It can be seen as part of the street, or, conversely, the street can be seen as part of the house frontage. The function of the ground floor changes according to the different activities that take place in the course of the day: it is an open space in the daytime for trading purposes and a closed space in the evening for family life. The ground floor can even be switched to different commercial activities – it can be hired and run by different retailers or family members – within a day. For example, in the early part of the day, people set out implements for serving breakfast or coffee, or lunch at noon. Then during the afternoon the ground floor can become a 'mini-market' selling vegetables or fruit, or it may be open for miscellaneous services. 'Flexible' and 'resilient' are words to describe the retailers, and they know very well how to balance their business with the needs of people in the area. These productive activities are even extended to the sidewalks, for example for customers' and owners' motorcycle parking. Furniture is used as the tool for changing the function of space. Sun blinds, small chairs and tables, etc. are used to extend the space of the threshold to the sidewalk. At the end of the day, when all family members come back home, they tidy up the implements used for business and rearrange their furniture. Together, they then enjoy family life. During the night, the empty sidewalks, the absence of trading activities, and the closed entrances make the houses and the street totally separate.

This way of sharing and combining public and private space not only brings jobs and beneficial income for households but also provides accessible services to many residential areas, reduces the need for travelling and reduces EF; this will be discussed further below. More importantly, the space is used efficiently and most of the time is full of productive activities. The types of activities carried out and occupations of residents are shown in Tables 14.1 and 14.2.

Living space

This section discusses the interaction between house size, household size and environmental impact. The importance of shared living space in reducing EF is shown via comparative examples of Hanoi, New Zealand and Finland.

Changes in ways of living have driven change in settlement patterns. There has been an increasing difference in housing between the past and modern times. In the past, the home was smaller, with more family members, and people spent more time at home and also more time together for cooking, gardening, making clothes and other productive activities. The medieval house represents an autonomous model of living in which home and workshop were complementary, in which inhabitants lived sustainably in the same houses and worked in the same workshops as their ancestors (Benevolo, 1980).

Table 14.1 Functions of ground floors in a survey of 230 houses in Hanoi

Function	Number of houses	Percentage
Living room	30	14.1
Cafeteria	3	1.4
Restaurant	5	2.4
Hair salon	4	1.9
Grocery shop	12	5.6
Office	3	1.4
Dental clinic	1	0.5
Acupuncture clinic	1	0.5
Workshop	8	3.8
Tailor	4	1.9
Embroidery service	3	1.4
Childcare	1	0.5
Meeting room	1	0.5
Tea shop	3	1.2
Clothing shop and miscellaneous services	134	62.9
Total	213	100.0

Source: Hoang and Nishimura (1990: 50)

Table 14.2 Jobs of people interviewed in a survey of 70 households in Hanoi

Occupation	Number of people	Percentage
Craftsman	6	8.5
Entrepreneur	28	40.0
Retired	22	31.5
Government officer	12	17.2
Other	2	2.8
Total	70	100.0

Source: Hoang and Nishimura (1990: 24)

A larger house costs more to build and also carries a higher cost for its maintenance and operation. The modern home is not only larger with fewer people occupying it, but its occupants spend less time at home. Home now is the place people come back to after working hours for eating, bathing, watching TV, searching the Internet, sleeping and so on. In other words, home now tends to be the place solely of consumption of goods and energy rather than of production.

Declining household size and increasing numbers of one-person households are the typical trends underlying change in settlement patterns in many western countries. Increasing numbers of young people living alone, growth in an older population living alone, rising living standards (for example new constructions are usually larger in size and depend largely on external energy to operate) are all factors that have made individual footprints larger.

For instance, the average floor area of new houses in New Zealand in 1991 was 139 square metres; in 10 years this rose to 178 square metres (Statistics New Zealand, undated). In 1886, the number of people per dwelling in New Zealand was 5.2, and 100 years later, in 1996, this number was 2.8 (Cook, 1998). In a recent New Zealand government report (Bascand, 2009), it is predicted that household size will continue to fall, from 2.6 people in 2006 to 2.4 people in 2031. It is also predicted that the number of one-person households will increase, from 363,000 in 2006 to 619,000 in 2013. This means that the proportion of one-person households is projected to increase from 23.4 per cent in 2006 to nearly 30 per cent in 2013. Finland's housing statistics show that the percentage of one-person households in 2010 was over 40 per cent, compared to only 27 per cent in 1980 (Statistics Finland, 2009). The average dwelling size in Finland in 1960 was 51 square metres, and this increased to about 80 square metres in 2008 (Statistics Finland, 2008). Compared to these western countries, the average area of housing per person in Vietnam is only about 15 square metres, and the number of persons per house is over four (Statistics Vietnam, 2006b). The percentage of people living alone in 2009 was only 7.24 per cent (Nguyen, 2011: 49).

Having a smaller house with more people means more sharing and lower housing cost per person. More importantly, it reduces the use of resources and environmental impact. Tables 14.3 and 14.4 show that the individual housing footprint for Vietnam (0.01 gha/person/year) is much smaller than that of New Zealand (0.24 gha/person/year – 24 times bigger) and Finland (0.09 gha/person/year – 9 times bigger).

Table 14.3 Comparison of housing footprint in New Zealand and Vietnam. Details of the calculations used to generate Tables 14.3–14.8 can be obtained from the author

	New Zealand	Vietnam
Average house size (m^2)	197	57
Average household size (number of people)	2.6	4.2
Embodied energy at 4.5 GJ/m^2 (GJ)	900	260
Maintenance (GJ)	450	130
Household annual energy use (MJ)*	50,000	1,770
Total energy use over 100 years (GJ)	6,350	567
EF of household (gha/year)	0.64	0.06
EF per capita (gha/person/year)	0.24	0.01

* Most houses in New Zealand are heated in the winter and are not cooled in summer, as the climate is mild. The opposite is true in the case of Vietnam, where the winter is mild. Currently, most families in Vietnam use electric fans for cooling.

Table 14.4 Housing footprint per capita for different years in Finland: energy efficiency improvement vs. change in lifestyles

	1960	1970	1990	2008
Dwelling size (m^2)	51	60	74	79.1
Household size (number of people)	3.34	2.99	2.42	2.09
Heating energy use (MJ/m^2)	907	677	504	216
EF per capita (gha/person/year)	0.11	0.11	0.13	0.09
Increase in dwelling size since 1960 (%)	reference point	18	45	55
Reduction in household size since 1960 (%)	reference point	10.5	27.5	37.4
Heating energy efficiency improvement since 1960 (%)	reference point	25.4	44.4	76.2
Change in EF per capita since 1960 (%)	reference point	0	+18.2	−18.2

Table 14.4, showing the housing footprint in Finland in different years, illustrates how heating energy use has been reduced by 76 per cent compared with 1960, but a rise in house size combined with a reduction in occupancy means that the per capita housing footprint has been reduced by only 18 per cent. It seems that, in terms of environmental impact, technology alone can hardly compensate for changes in lifestyles, such as the move to much larger houses, unless there are guidelines for sustainable lifestyles and values.

The Ecological Footprint and Ways of Living

Using Hanoi as a case study, what would be the outcome if the ways of using space of Hanoi were applied to New Zealand's cities? This section explores how changes in patterns of living could affect the use of resources and hence the ecological footprint.

The ecological footprint of Hanoi

Housing energy footprint

Different ways of cooling the home or different attitudes to what building comfort means can make a huge difference to the environmental impact. Currently, most urban households in Vietnam use electric fans for cooling. However, the proportion of people who can afford air conditioners for cooling has increased sharply. According to the latest statistics (Statistics Vietnam, 2006a: 293, Table 7.4), 9.9 per cent of households had an air conditioner in 2006, whereas the percentage of households with an air conditioner in 2002 was barely 4.5 per cent. Table 14.5 shows that, keeping

Table 14.5 Impact of patterns of living – ways of cooling the home and vehicle use – on the EF of an individual in Hanoi, assuming a dwelling size of 57 m² and a distance driven of 6,600 km/person/year

Pattern of living	EF (gha/person/year)
Current	
Transport: motorbike/scooter	0.08
Home cooling: electric fan	0.01
Total	0.09
Trend	
Transport: car	0.26
Home cooling: air conditioner	0.05
Total	0.31

the average dwelling size (57 square metres) unchanged, the individual housing footprint would increase fivefold, from 0.01 gha/person to 0.05 gha/person, if users switched from electric fans to air conditioners for cooling.

Transport footprint

As for cooling in the home, use of transportation in Hanoi has changed significantly in recent years, moving away from walking and riding bicycles towards riding motorcycles and driving cars, which have become more and more popular in high-income groups. According to Schipper *et al.* (2008: 9, 32), in 1995 more than 70 per cent of all trips in Hanoi were made on bicycles. Today, 84 per cent of all households own a motorcycle, and 40 per cent have more than two. The rate of car ownership is only 1.6 per cent of households, but the figure has increased rapidly as economic growth has reached 11 per cent per year. Furthermore, due to the lack of public transport, it is predicted that the use of private cars will increase significantly in the next 10 years, from about 800 to around 6,000 kilometres per vehicle per year.

The average distance driven per person per year by motorbike or scooter in Hanoi is 6,600 kilometres (Schipper *et al.*, 2008: 32). Typical fuel consumptions range from 2 to 4 litres per 100 kilometres (scootermoped.net, undated). 'A well cared-for scooter should be able to go about 25,000 miles before it needs a major rebuild. The 50cc two-strokes work harder, so I'd think they should be overhauled by 15,000 miles. I have heard of some Helixes running over 100,000 miles' (Stanley, 2009). Assuming the weight of a scooter is 100 kilograms and it lasts for 40,000 kilometres, its embodied energy will be (100 megajoules per kilogram × 100 kilograms) ÷ 40,000 kilometres = 0.25 megajoules per kilometre (MJ/km). If it achieves a fuel consumption of 3 litres per 100 kilometres, it will use a further 1 MJ/km of fuel, giving a total of 1.25 MJ/

km. So the transport footprint of a scooter rider per year in Hanoi is 0.08 gha/person/year, and this would be reduced if more than one person is on the scooter.

Fuel consumption could be increased significantly if the use of cars became more popular, as in the western world. If the overall energy demand of a car is 4.0 MJ/km (midway between a VW Golf and a Holden Commodore [Vale and Vale, 2009: 77]), keeping the distance driven per person per year (6,600 kilometres) unchanged, the transport footprint of a person travelling by car could be 0.26 gha/person/year (assuming a conversion of 100 GJ/gha). This is more than triple the transport footprint of someone travelling by scooter.

Discussion

Table 14.5 reveals that changes in vehicles and ways of cooling houses could increase the EF of an individual in Hanoi more than threefold, from 0.09 to 0.31 gha/person/year. The housing footprint of an individual in Hanoi is much smaller than the transport footprint, so changing vehicle type (from scooters to cars) has greater impact on the environment than changing the way of cooling houses (from electric fan to air conditioner). It seems zero-energy building is meaningless if people still travel by car, use aircraft for long-distance holidays, and so on. In terms of architecture and design, a broader concept is needed for sustainable design, looking beyond the building.

The possibility of reducing the EF of the West: the example of New Zealand

Housing footprint

Houses in New Zealand have increased in size significantly since the 1970s, while their occupancy has simultaneously decreased, and this has had an effect on their EF. Table 14.6 below reveals that, since the 1970s, the New Zealand housing footprint has increased by nearly 50 per cent, from about 0.13 to 0.24 gha/person/year.

Travelling to work, shops and services

The tendency of New Zealand cities is to low density and sprawl, so people usually have to travel a long distance to work, study and shop. For example, Auckland is 'a city built around the car' and Auckland's rate of car ownership is one of the highest in the world (Harre and Atkinson, 2007: 114–115). Private transport use accounts for almost 90 per cent of New Zealand's total passenger transport energy use (Marth and Sean, 2007); and 45.2 per cent of total carbon dioxide emissions is contributed by domestic transport (Mithraratne *et al.*, 2007: 13). The Household Travel

Table 14.6 Ecological footprint of housing in New Zealand over an assumed life of 100 years, 1970–1973 and 2006–2008

	1970–1973	2006–2008
Dwelling size (m^2)	110	197
Household size (number of people)	3.6	2.6
Embodied energy at 4.5 GJ/m^2 (GJ)	500	900
Maintenance energy (GJ)	250	450
Operating energy over 100 years (GJ)	4,000	5,000
Total energy (GJ)	4,750	6,350
EF of house (gha/year)	0.48	0.64
EF per capita (gha/person/year)	0.13	0.24

Source: Tran et al. (2010)

Survey for 2006 indicates that travel to work is the largest travel category and also the most dependent on driving, and that motor cars were the main means of travel to work for approximately two thirds of the employed population (Statistics New Zealand, 2009: 1–2). Driving for shopping and services has increased significantly during the period 1989–1990 to 2006–2009; whereas driving for work and social or recreational purposes has stayed nearly the same (Ministry of Transport, 2010: 8).

If New Zealand also promoted mixed-use living/working/trading models, as used in Hanoi (and as described in earlier sections), people might not need to drive so frequently to the supermarket or shopping mall (Figures 14.1 and 14.2). Instead, they would be able to buy things in local shops run by their neighbours, hence reducing their transport footprint for shopping. Most of the time the distance between shops and houses would be within 2 kilometres, a distance suitable for car-free travel. For daily items and miscellaneous services such as food, shoes or key repair, health clubs, etc. people would not have to leave their residential areas. Currently 70 per cent of all car travel is for 'other' purposes, with 30 per cent for 'work' and 'commuting' (NZTA, 2007). The annual distance travelled by a driver is around 11,000 kilometres (Ministry of Transport, 2009: 10). This suggests that travel for 'other' purposes must be at least 7,700 kilometres per year. There are 0.74 vehicles per person in New Zealand (Ministry for the Environment, 2009), making the total distance per person for 'other' travel around 6,000 kilometres per year. If the total distance driven for shopping and similar trips in the 'other' category could be reduced by at least 50 per cent, there would be a reduction of 3,000 kilometres per person per year. For the remaining 3,000 kilometres, where the distance is probably less than 5 kilometres, switching to a different mode of transport (bus or scooter) could reduce further the footprint for travelling. This is discussed further below.

A Study of Hanoi, Vietnam 233

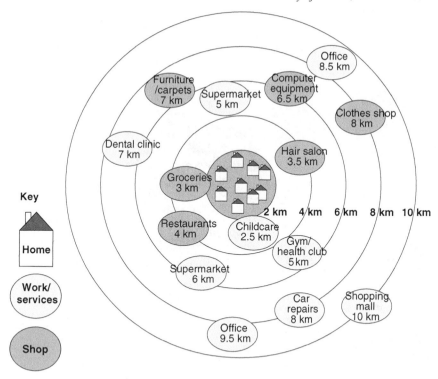

Figure 14.1 The current, separate-function and large-scale model, in which the average distance travelled to work, shops or services ranges from 3 to 10 kilometres

Source: Tran *et al.* (2010)

The question that arises is: how many New Zealanders would be able to run their own business and work in their own house, so that commuting travel to work could be reduced? Like most developed economies, the service sector in New Zealand is significant, accounting for over two thirds of GDP (72.2 per cent in 2003) and three quarters of all jobs (Ministry of Foreign Affairs and Trade, 2004). There is great potential for promoting local small-scale businesses as well as self-employment in order to enable many more New Zealanders to work at home. The level of self-employment in New Zealand is already high, accounting for 19.8 per cent of all employment compared to the OECD average of 14.2 per cent (Goodchild *et al.*, 2003: 8).

Assuming a mixed-use model similar to that of Hanoi was applied in New Zealand's cities, many people would be able to run their business at home and work at home for at least part of the time. It can be assumed that the total distance driven for work and commuting for those working at home would be reduced by 50 per cent. With average car travel of around 11,000

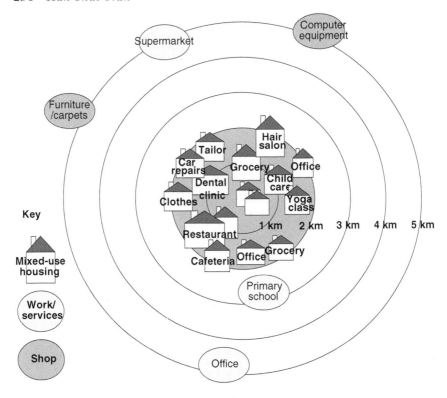

Figure 14.2 A mixed-use housing model that promotes small businesses and reduces travel distance. Life is more local and the average distance travelled to work, shops or services is reduced by 50 per cent. Most of the shops and services are within 2 kilometres of the home

Source: Tran *et al.* (2010)

kilometres per year and 30 per cent of this for work and commuting, an average person's work-related travel is around 2,400 kilometres a year, allowing for the fact that there are 0.74 cars per person. This could therefore be reduced to 1,200 kilometres a year among those working at home.

Transport footprint

The average distance travelled per person per year by car to work, shops and services is 8,400 kilometres. If this travelling is done in a car, the ecological footprint will be 0.34 gha/person/year. This could be halved by the travel reductions that would result from the application of mixed-use housing as found in Hanoi.

Fuel consumption could be reduced further if the use of scooters or small motorcycles became more popular in New Zealand, as it is in Vietnam. Table 14.7 looks at New Zealand and compares types of settlement (separate

Table 14.7 Current footprint of personal travel for a New Zealander, and reduction in footprint based on transport type, fuel type and settlement model (separate function or mixed use)

Pattern of living	Distance driven (km/person/year)	Total EF – vehicle using fossil fuel (gha/person/year)	Total EF – battery electric vehicle (gha/person/year)
Car driving		(4.0 MJ/km for a petrol car)	(2.0 MJ/km* for a battery electric car)
Current separate-function model	8,400	0.34	0.17
Mixed-use model	4,200	0.17	0.09
Scooter driving		(1.25 MJ/km for a petrol scooter)	(0.63 MJ/km for an electric scooter**)
Current separate-function model	8,400	0.10	0.05
Mixed-use model	4,200	0.05	0.03

* Based on BYD E6 which uses 21.5 kWh or 77.4 MJ per 100 km (Green Car Congress, 2010)
** Electric scooter energy use is given as half that of a petrol scooter, mirroring the figures for cars
Source: Tran *et al.* (2010)

function and mixed use), types of vehicle (car and motorbike or scooter) and fuel (petrol and battery) based on the assumptions and calculations given previously, to show the reductions in EF that can be achieved through different choices.

Reduced footprint of housing and transport

The various factors discussed – patterns of living, choice of transport mode, and house size and occupancy – are brought together in Table 14.8 in an attempt to discover which factors might make the greatest contribution to reducing the ecological footprint.

Discussion

It has been proposed that for a sustainable society the ecological footprint can be no larger than around 1.8 gha/person/year. This means a reduction of around 65 per cent in the current footprints of developed countries, including New Zealand. The results shown in Table 14.8 reveal that the current footprint of housing for a New Zealander (0.24 gha/person/year) is less than the footprint of driving a car. Oil is running out, so in a future of non-fossil fuel living, the results in Table 14.8 show that it would not simply be a case of replacing fossil fuel cars with electric cars and keeping the current patterns of living. A change to electric cars, keeping everything

Table 14.8 The effect of patterns of living, transport and house size/occupancy on ecological footprint

Pattern of living	Total EF – vehicle using fossil fuel (gha/person/year)	Total EF – battery electric vehicle (gha/person/year)
Car driving	*(4.0 MJ/km for a petrol car)*	*(2.0 MJ/km for a battery electric car)*
Current separate-function model (8,400 km/person/year driving)	0.34	0.17
Mixed-use model (4,200 km/person/year driving)	0.17	0.09
Current house	0.24	0.24
1970s house	0.13	0.13
TOTAL C1: separate function, current house, car driving	0.58	0.41
(Percentage reduction)	(reference point)	(−29%)
TOTAL C2: separate function, 1970s house, car driving	0.47	0.30
(Percentage reduction)	(−19%)	(−48%)
TOTAL C3: mixed use, current house, car driving	0.41	0.33
(Percentage reduction)	(−29%)	(−43%)
TOTAL C4: mixed use, 1970s house, car driving	0.30	0.22
(Percentage reduction)	(−48%)	(−62%)
Scooter driving	*(1.25 MJ/km for a petrol scooter)*	*(0.63 MJ/km for an electric scooter)*
Current separate-function model (8,400 km/person/year driving)	0.10	0.05
Mixed-use model (4,200 km/person/year driving)	0.05	0.03
Current house	0.24	0.24
1970s house	0.13	0.13
TOTAL S1: separate function, current house, scooter driving	0.34	0.29
(Percentage reduction)	(−41%)	(−50%)
TOTAL S2: separate function, 1970s house, scooter driving	0.23	0.18
(Percentage reduction)	(−60%)	(−69%)
TOTAL S3: mixed use, current house, scooter driving	0.29	0.27
(Percentage reduction)	(−50%)	(−53%)
TOTAL S4: mixed use, 1970s house, scooter driving	0.18	0.16
(Percentage reduction)	(−69%)	(−72%)

Source: Tran *et al.* (2010)

else unchanged, would reduce the EF by only 29 per cent (see Total C1 for battery electric vehicles). Since the 1970s, the New Zealand housing footprint has increased by nearly 50 per cent (from about 0.13 to 0.24 gha/person/year; see Table 14.6) due to the drop in number of people per household and the increase in the size of houses.

As Table 14.8 makes clear, the model of mixed-use housing, involving a more compact lifestyle, combined with more compact vehicles like scooters, could reduce the annual footprint of housing and transport sufficiently to meet the 65 per cent reduction target. The current separate-function model of residential development is possible only with a return to the housing standards regarding size and occupancy levels of the 1970s, combined with electric scooters such as are found widely in China (see Total S2 for battery electric vehicles). The mixed-use model, as found in Vietnam, would allow for a greater reduction in footprint when combined with electric transport, but housing size and occupancy still need to be reduced compared to those current in New Zealand.

Conclusion

This case study demonstrates that a lower ecological footprint is created when people use spaces productively as well as having flexible lifestyles. In particular, more sharing of both living and public spaces is key to the smaller footprint per capita in Hanoi, when compared to western countries. The other key, in terms of living space, is a smaller house area per person. In addition, transport and other aspects should be considered in order to achieve the reduction target for the typical western footprint of 65 per cent.

Technology and design cannot on their own make up for the environmental impact of lifestyle change, such as larger houses or the increase in the number of people living alone, unless guidelines for sustainable lifestyles and values are provided. The future of sustainable development greatly depends on people's behaviour and lifestyle. It requires a high level of acceptance and a culture of sharing.

References

Bascand, G. (2009) 'Subnational family and household projections: 2006 (base) – 2031', http://www.stats.govt.nz/browse_for_stats/people_and_communities/Families/SubnationalFamilyandHouseholdProjections_HOTP06-31.aspx, accessed 16 March 2010

Benevolo, L. (1980) *The History of the City*, Scolar Press, London, England

Cook, L. (1998) 'Household size continues to fall', http://www2.stats.govt.nz/domino/external/PASFull/pasfull.nsf/173371ce38d7627b4c25680900046f25/20e154cc0ad8efd74c2566dc00181813?OpenDocument, accessed 14 February 2010

Global Footprint Network (2009) 'Footprint factbook Vietnam 2009', http://www.conservation-development.net/Projekte/Nachhaltigkeit/DVD_10_Footprint/Material/pdf/Footprint_Factbook_Vietnam_2009.pdf, accessed 18 January 2012

Goodchild, M., Sanderson, K. and Leung-Wei, J. (2003) *Self-employment and Small Business Succession in New Zealand*, Business and Economic Research Ltd, Wellington

Green Car Congress (2010) '40 BYD e6-based electric taxis enter service in Shenzhen City', http://www.greencarcongress.com/2010/05/byd-20100518.html#tp, accessed 19 May 2010

Harre, N. and Atkinson, Q. D. (eds) (2007) *Carbon Neutral by 2020: How New Zealanders Can Tackle Climate Change*, Craig Potton Publishing, Nelson, New Zealand

Hoang, H. P. and Nishimura, Y. (1990) 'The historical environment and housing conditions in the "36 Old Streets" quarter of Hanoi', *Noi O va Cuoc Song cua Cu Dan Ha Noi – Settlement and Life of the Hanoians*, Culture and Information Publisher, Hanoi

Marth, T. and Sean, W. (2007) 'Reducing road transport carbon emissions: options for government policy', Research Report 23, School of Geography, Environment and Earth Sciences, Victoria University of Wellington, New Zealand

Ministry for the Environment (2009) 'Total distance travelled (total vehicle kilometres travelled)', http://www.mfe.govt.nz/environmental-reporting/transport/vehicle-km-travelled/total-vkt/, accessed 6 April 2011

Ministry of Foreign Affairs and Trade (2004) 'Overview of the Thailand and New Zealand economies', http://www.mfat.govt.nz/Trade-and-Economic-Relations/0--Trade-archive/0--Trade-agreements/Thailand/0-study-thainz-economies.php, accessed 14 November 2009

Ministry of Transport (2009) *How New Zealanders Travel: Trends in New Zealand Household Travel 1989–2009*, Ministry of Transport, Wellington

Ministry of Transport (2010) *Household Travel Survey – Driver Travel: In Cars, Vans and SUVs*, Ministry of Transport, Wellington, New Zealand

Mithraratne, N., Vale, B. and Vale R. (2007) *Sustainable Living: The Role of Whole Life Costs and Values*, Elsevier, Amsterdam

Nguyen, T. B. (2011) 'The trend of Vietnamese household size in recent years' in *2011 International Conference on Humanities, Society and Culture IPEDR*, Vol. 20, IACSIT Press, Singapore, pp. 47–52

NZTA (2007) 'Table A2.4 – Vehicle occupancy and travel purpose' in *Economic Evaluation Manual 2007, Volume 1, Amendment 1 (Road Infrastructure)*, New Zealand Transport Agency, Wellington

Schipper, L., Anh, T. L., Orn, H., Cordeiro, M., Ng, W.-S. and Liska, R. (2008) 'Measuring the invisible: quantifying emissions reductions from transport solutions', http://pdf.wri.org/measuringtheinvisible_hanoi-508c_eng.pdf, accessed 12 December 2011

scootermoped.net. (undated) 'Scooter fuel consumption', http://www.scootermoped.net/scooter-fuel-consumption.html, accessed 24 April 2011

Stanley, J. (2009) 'Scooter FAQ', www.jacksscootershop.com/FAQ.html, accessed 16 April 2011

Statistics Finland (2008) *Dwellings and Housing Conditions 2008: Overview*, Statistics Finland, Helsinki

Statistics Finland (2009) 'Buildings and dwellings', http://www.stat.fi/tup/suoluk/suoluk_asuminen_en.html, accessed 17 March 2010

Statistics New Zealand (undated) 'Table 12.1: Average floor area (m^2) of new residential dwellings', http://m.stats.govt.nz/browse_for_stats/economic_

indicators/NationalAccounts/long-term-data-series/social-indicators.aspx, accessed 31 July 2012

Statistics New Zealand (2009) 'Car, bus, bike or train: what were the main means of travel to work?' in *Commuting Patterns in New Zealand: 1996–2006*, http://www.stats.govt.nz/publications/populationstatistics/commuting-patterns-in-nz-1996-2006.aspx, accessed 12 February 2010

Statistics Vietnam (2006a) 'Fixed assets and durable goods', http://www.gso.gov.vn/default.aspx?tabid=512&idmid=5&ItemID=8182, accessed 17 October 2009

Statistics Vietnam (2006b) 'Housing, electricity, access to safe drinking water, sanitary and Internet', http://www.gso.gov.vn/default.aspx?tabid=512&idmid=5&ItemID=8182, accessed 4 October 2009

Tran, T. H., Vale, B., and Vale, R. (2010) 'Sustainable housing: a footprint comparison', *The 5th Australasian Housing Researchers' Conference*, The University of Auckland, New Zealand

Vale, R. and Vale, B. (2009) *Time to Eat the Dog? The Real Guide to Sustainable Living*, Thames and Hudson, London

15 A Study of Surburban New Zealand

Sumita Ghosh

Introduction

Sustainability of a human settlement is significantly influenced by a very wide range of factors, including spatial distributions of transport networks and land-use patterns, resource demand and associated environmental emissions, access to facilities and services, residential location choices, behavioural practices, community participation and knowledge, social qualities, economic factors and impacts of urban planning policies (Anderson *et al.*, 1996; Bertolini *et al.*, 2005). Residential land is 'one of the major determinants of urban structure', covers approximately 40 per cent of the total developed land in a city and is 'the generator of most types of urban traffic' (Romanos, 1976: 4). The two main patterns of urban development identified in sustainability literature are 'compact' and 'sprawl'. A significant debate continues around the relative sustainability performance of compact or high density developments versus sprawl or low density patterns (Williams *et al.*, 2000; Jenks and Dempsey, 2005; Newman and Kenworthy, 1989; Gordon and Richardson, 1997; Troy, 1996; Troy *et al.*, 2005). The proponents of urban intensification argue that high density developments are more sustainable compared to low and medium densities due to their potential to reduce transport use (Newman and Kenworthy, 1999), as they could be located in close proximity to work, protect ecologically sensitive land areas and water quality (United States Environment Protection Authority [EPA], 2006) and facilitate better social interactions by creating well-designed places where people can 'live, work and play' (English Partnerships and Housing Corporation, 2000). Most current urban planning policies are targeted towards further intensification of urban environments as compact developments are considered the sustainable solution (Guy and Marvin, 1999).

Low and medium density suburban residential developments, although they may use a greater share of private transport as they are located further away, could have some useful capacity to trade off this disadvantage through growing local food, being solar oriented, using efficient roof rainwater harvesting systems and other meaningful practices. Following many research studies, the sustainability potential of low density developments in producing on-site local food, promoting efficient water consumption and enhancing biodiversity and ecological functions is well established (Moriarty, 2002; Pauleit and Duhume, 2000; Daniels and Kirkpatrick, 2006; Loram *et al.*, 2008). It is argued that 'the concept of higher density is entirely relative' (Jenks and Dempsey, 2005: 304). In the absence of a standardised methodology for measuring densities, significant difficulties and controversies exist (Jenks and Dempsey, 2005: 293). Research in the quest for a sustainable urban form indicates the possibility of more than one alternative sustainable urban form (Williams *et al.*, 2000; Jenks *et al.*, 1996).

Urban and suburban forms are continually evolving, through dynamic processes of change. Research has identified that a significant percentage of households prefer to live in 'a single-family home with a private yard' (Gordon and Richardson, 2001: 140; Kaplan and Austin, 2004: 235), both important characteristics of suburban living. New Zealand suburbs are influenced by people's lifestyle choices and represent their unique values and cultural identities. Private garden areas have been mapped as covering 46 per cent of all residential areas in Dunedin city in New Zealand, with a ratio of gardens to houses of 1:089 (Mathieu *et al.*, 2007). Residential gardens as important elements of New Zealand residential suburbs are mainly associated with low and medium densities. Gardens form the immediate environs around the dwelling and are significant from a sustainability perspective. They could accommodate a number of sustainability practices and activities such as growing local food.

The suburban patterns already embedded in the urban spatial fabric are likely to continue for a considerable time, as the intensification of existing suburbs is not possible within a short timeframe. This raises several important questions in relation to the sustainability performance of suburbs. What could be the timeline for retrofitting existing suburbs with improved sustainability practices? Will it be possible for communities to adopt lifestyles and behaviour changes that utilise the sustainability potential of suburban developments? To address these questions, it is immensely important to understand the morphologies of different development patterns and their form-specific capacities for improving sustainability performance.

New Zealand is one of the most highly urbanised countries in the world, with 87 per cent of the total population living in what are defined as urban settlements (Ministry for the Environment [MfE], 2005: 2).The Auckland Region, New Zealand's largest urban area housing one third of the total population, is rich in cultural diversity, and is the economic hub of activities (Ministry of Economic Development and Auckland Regional Council, 2006: 1). The main aim of this chapter is to investigate whether the residential forms in suburbs could be effectively redesigned to emerge as alternative, sustainable urban forms. The prime objective of this chapter is to analyse and compare the potential for accommodating environmentally sustainable practices on-site in existing low, medium and high density residential developments in suburban Auckland, based on completed and ongoing research. It will examine potential applications of three main sustainability practices – local food production, localised solar energy generation potential and roof rainwater harvesting – for five selected residential case studies in Auckland. This study focuses on understanding how these practices could be used effectively for a better sustainability performance.

Sustainable Residential Developments: A Brief Review

National and international best practice examples of sustainable neighbourhoods have commonly applied the practices of roof rainwater collection, solar hot water and solar electricity from photovoltaic (PV) modules.

For example, the Earthsong Eco-Neighbourhood, a medium density development with 32 homes in suburban New Zealand, has on-site provision for individual home gardens, a common vegetable garden and edible landscaping throughout the development. Solar energy for hot water is generated from the roofs and rainwater is collected on-site in rain tanks (Earthsong Eco-Neighbourhood, 2012).

The five zero-emission earth sheltered houses in the Hockerton Housing Project, a low density housing development in the UK, have an organic garden and a polytunnel for growing vegetables all year round. In this development the solar PV modules are mounted on the roofs and passive solar design principles have been applied to retain heat generated by sun rays within the houses by constructing a conservatory towards the south. These houses have no heating but stay at a temperature of 18°C in the English winter (Hockerton Housing Project [HHP], 2012; for further discussion see Chapter 16).

Village Homes is a sustainable community development in California in the USA, built in 1970, with orchards for growing fruits, vineyards, and community and home gardens for growing vegetables. The orientations of the units (located in the northern hemisphere) are towards the south and the dwellings have solar water heaters for both space and water heating and PV panels for generating electricity on the roofs (Wack, 2005).

These examples demonstrate that sustainable neighbourhood design models acknowledge the significance of practical applications of on-site practices.

Selection of Case Studies and Data Collection

The New Zealand urban form classification system developed by Ghosh and Vale (2009) for residential developments comprises five urban scales identified in the New Zealand context:

- metropolitan/regional
- sub-metropolitan/city
- community/neighbourhood
- residential block/local scale
- houses/micro-scale.

The neighbourhood case studies selected for this study are at a residential block or local scale, with populations ranging from 150 to 650 people or 50 to 200 households (Ghosh, 2004: 226). This research links the case studies to specific urban forms in low, medium and high densities with defined built forms using this classification system. All five case studies are located within the boundary of Auckland City. Site selection criteria include:

- distance from the city centre (CBD)
- proximity to the main transport corridors and shopping facilities
- zoning patterns according to the district plans (Auckland City Council, 1999; Waitakere City Council, 2003)
- site configurations
- total number of households (ranging between 50 and 125)
- density patterns.

The selection criteria connect these five residential case studies to different typologies of residential forms across the urban and suburban continuum, transport accessibility, built forms and their linked density and site layout patterns, household sizes and other morphological characteristics useful for conducting and comparing their sustainability performance assessment.

As this chapter is focused on New Zealand suburbs, the five case studies are named according to their density pattern and prevalent built form characteristics, and thus link closely with the urban classification system developed for New Zealand. The five case studies are shown in Figures 15.1 to 15.5.

The distance gradually decreases from the CBD from Case Study 1 (10 kilometres from the CBD) to Case Study 5 (1 kilometre from the CBD). As the names signify, the 'low density – detached housing' has single- and two-storey large detached family houses with generous open spaces around the houses. The 'low density – mixed housing' contains mainly single-, two- and some three-storey large detached family houses with ample open spaces around the dwelling and also some housing units constructed as infill developments. 'Medium density – mixed housing' includes a combination of two-storey row housing units, single- and two-storey detached, semi-detached and attached houses, while 'medium density – heritage cottages' contains residential cottages from the Edwardian and Victorian eras aligning with the streetscape. The 'high density – row housing' is an inner city suburb containing row houses, two-storey and single-storey,

244 *Sumita Ghosh*

Figure 15.1 Low density – detached housing (Case Study 1)

Source: Drawing by Sumita Ghosh based on Auckland City Council Operative District Plan 1999 (City of Auckland, 2005)

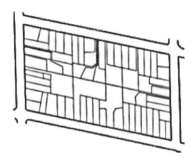

Figure 15.2 Low density – mixed housing (Case Study 2)

Source: Drawing by Sumita Ghosh based on Auckland City Council Operative District Plan 1999 (City of Auckland, 2005)

Figure 15.3 Medium density – mixed housing (Case Study 3)

Source: Drawing by Sumita Ghosh based on Auckland City Council Operative District Plan 1999 (City of Auckland, 2005)

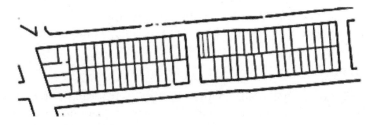

Figure 15.4 Medium density – heritage cottages (Case Study 4)

Source: Drawing by Sumita Ghosh based on Auckland City Council Operative District Plan 1999 (City of Auckland, 2005)

Figure 15.5 High density – row housing (Case Study 5)

Source: Drawing by Sumita Ghosh based on Auckland City Council Operative District Plan 1999 (City of Auckland, 2005)

semi-detached and attached houses. Thus the case studies differ in their morphological characteristics and urban design qualities.

Data were collected mainly from aerial photographs, Statistics New Zealand census data for different years (Statistics New Zealand, 1991, 1996), rapid visual survey, and national and international research reports and databases. The basis of the calculation was the year when the aerial photograph was taken. All the data for calculations were aligned with the same base year to achieve appropriate results. The residential block or local scale in the classification system corresponds to the 'mesh block' scale, as defined by Statistics New Zealand. A mesh block is the smallest geographic unit for which census data is collected (Statistics New Zealand, 2012). Each residential case study comprises one or more mesh blocks and therefore the socio-economic and other relevant census data could be collected from Statistics New Zealand. The existing land-use patterns of the residential blocks are calculated from aerial photographs using Geographic Information Systems (GIS) and new primary data generated.

Methodology

The methodology systematically examines the retrofitting potential (with the aim of becoming more sustainable) of the five residential case studies in Auckland. Using GIS, morphologies and spatial distributions of land-cover patterns for these residential case studies are calculated at a neighbourhood scale. The main land-cover categories assessed were:

- building roof areas
- road areas (including the half width of the perimeter road and full width of roads within the site)
- paved and unpaved paths and driveways
- tree canopy cover, including existing tree canopy cover (trees and shrubs)
- productive land cover, including remaining open spaces except all those in non-productive use.

This study analyses and objectively compares potential applications of three main sustainability practices – local food production (mainly vegetables), localised solar energy generation (solar PV panels and solar water heating) and roof rainwater harvesting – for these five residential case studies. The assessment includes development of a separate model for measuring each sustainability factor using mathematical models, GIS and applications of ecological footprint conversion techniques (Ghosh, 2004). New Zealand forests are highly productive and can produce 120 GJ to 150 GJ annually per hectare (Wackernagel and Rees, 1996). The New Zealand average land-energy ratio is calculated to be 135 GJ per hectare per year (Ghosh, 2004). The ecological footprint approach calculates the land area that would be needed to grow the consumption demand of a particular population by cultivating biomass or sequestering carbon in forests (Wackernagel and Rees, 1996). 'Land productivity is an excellent indicator of the locally, regionally or globally available natural "interest"' (Simmons and Chambers, 1998: 355).

For this study, the 'land area equivalent' or LEQ, a land based measure (Ghosh, 2004; Ghosh, Vale and Vale, 2008), is developed using the land-energy ratio based on the ecological footprint concept. The LEQ of the annual total demand of the resident population includes calculations for two different sustainability factors: domestic electricity and hot water use, and growing vegetables as a share of dietary energy. The total LEQ of 'available' energy that could be produced on site annually from photovoltaic modules, solar water heaters, growing vegetables and collecting roof rainwater is estimated. Then the LEQ of total 'deficit' energy required to be supplied from off-site locations or the LEQ of total 'surplus' energy that could be supplied after meeting the total annual demand of the resident population can be calculated. The measure of deficit or surplus LEQ per household per annum is calculated for each of the five case studies.

A study by North Carolina State University (1996) in the USA indicates that a home garden with an area of around 100 square metres (m^2), or 7.6 metres by 12.8 metres, could produce annually almost all the vegetables required for two people. The typical New Zealand suburban residential plot, with an area ranging between 500 m^2 and 1,000 m^2, could produce most of the vegetable food demand and surplus fruits for a household (Francis, 2010). In this study, the energy gain from on-site vegetable production is calculated by considering the embodied energy content of food. Embodied energy is different from life-cycle energy as it includes only the 'upstream' or 'front-end' component and does not include the energy required for operation and disposal (Commonwealth of Australia, 2012). The total energy demand for vegetables is 10 per cent of the total dietary energy demand in New Zealand (Ministry for Agriculture and Fisheries [MAF], 1995). The annual average dietary energy demand for vegetables is assumed to be 3.024 GJ per capita, including the embodied energy content. The local productivity of a domestic garden in Auckland is calculated to be 0.007 GJ/m^2 (Ghosh, 2004) based on Allen's (1999) research on the productivity of a vegetable garden of an area of 35 m^2 in Auckland and equivalent energy values of vegetable productivity by Chhima (2001). Using a specific embodied energy multiplier of 7.2 developed by Pritchard and Vale (1999), 1 m^2 or 1 hectare (ha) of on-site vegetable garden is equivalent to 7.2 m^2 or 7.2 ha of off-site productive land area allowing for the embodied energy content of food. The off-site land area equivalent deficit (or surplus) for local food production is calculated at the overall neighbourhood level and on a per household per year basis.

Solar water heaters (SWH) and photovoltaic (PV) modules can be used to provide hot water demand and to generate electricity required for household use. Any roof area oriented 45° either side of north was considered solar efficient (Breuer, 1994). Two solar water heating panels, 1 m by 2 m, equivalent to 4 m^2 per household, mounted on a solar efficient roof can supply at least 2200 kilowatt hours (kWh) or 7.92 GJ/year (EECA, 2001; Centre for Advanced Engineering [CAE] and Energy Efficiency and Conservation Authority [EECA], 1996: 186). However, SWH and PV modules can be mounted on frames in less solar efficient parts of building roofs and can be tilted as required to provide appropriate solar orientation. For this study, using GIS, a geometric method and considering the individual roof slope, roof form and orientation, available solar efficient roof areas are estimated. Based on the availability of different solar panels on the market, it is assumed that average PV panels of size 0.6 m by 1.2 m could be fitted on

248 *Sumita Ghosh*

the rest of the solar efficient roof areas after placing 4 m² of solar hot water panels. The annual yield for 1 kilowatt peak (kWp) of PV panels in Auckland is 1,300 kWh (IT Power [Australia] Pty Ltd for Ministry of Economic Development, 2009: 53). The area of 1 kWp PV panel is 10 m². If an area of 10 m² of PV can produce 1,300 kWh or 4.7 GJ in Auckland, 1 m² can produce 0.47 GJ. The corresponding energy generated from solar water heaters and PV modules were calculated in GJ per annum and converted to equivalent land areas, in hectares.

The Auckland Water Quantity Statement (Auckland Regional Council, 2006) confirms that the domestic water demand in the Auckland Region is 173 litres of water per capita per day for all household uses, such as drinking, cooking, washing and cleaning, and flushing toilets (Bannister *et al.*, 2004). Based on this, the per capita water demand per annum is 63,145 litres or 63 m³ in Auckland (Ghosh and Vale, 2006). Areas of building roofs for the case study sites were calculated using ArcGIS from aerial photographs. Using a formula devised by Ghosh and Head (2009), on-site roof rainwater collection potential was calculated and is expressed in cubic metres. The average annual rainfall for Auckland is calculated to be equal to 1,244 mm, based on the fact that in 2004, one of the wettest years, the Auckland Region received 1,331 mm of rainfall, which was 107 per cent of the average annual rainfall (National Institute of Water and Atmospheric Research [NIWA], 2004). It was assumed that 10 per cent of the collected rainwater would be lost by evaporation and 90 per cent could be collected for use (Waitakere City Council, 2008), and that all the houses would require rainwater for first flush diverters at the rate of 0.2 litres per square metre of roof area (Rain Harvesting, 2008) to prevent contaminants entering the rainwater tank. This study estimates the percentage of total water demand that could be supplied from on-site roof rainwater collection and the deficit that would need to be supplied by the city water supply. The average household size for Auckland adopted for this study is 2.9.

Results and Discussion

In this section, the sustainability performances of the neighbourhoods in local food production, localised solar energy generation and roof rainwater harvesting are calculated.

Using GIS, the land cover patterns of five residential case studies are determined first. Table 15.1 represents the broad land cover distributions in these case studies, the solar efficient roof areas of building roofs, total site areas, total numbers of households living in the neighbourhood and household density. New Zealand concepts of high, medium and low residential densities are different from international perceptions of these densities. New Zealand is characterised by mainly low and medium density developments, although the current urban forms are transforming at a much faster rate in the main cities, such as Auckland. The 'low density

Table 15.1 Land cover patterns, density and solar efficient roof areas for the case studies

Parameter	Low density – detached housing	Low density – mixed housing	Medium density – mixed housing	Medium density – heritage cottages	High density – row housing
Total site area (ha)	5.3	8.2	6.7	3.1	3.6
Total land cover* (ha) (Percentage of total site area)	2.61 (49.2%)	4.11 (50.1%)	3.81 (56.9%)	2.01 (64.8%)	1.57 (43.6%)
Total tree canopy cover (ha) (Percentage of total site area)	1.1 (20.8%)	2.15 (26.2%)	1.21 (18.1%)	0.52 (16.8%)	1.2 (33.3%)
Productive land area (ha) (Percentage of total site area)	1.59 (30.0%)	1.94 (23.7%)	1.68 (25.1%)	0.57 (18.4%)	0.83 (23.1%)
Total solar efficient roof area (ha) (Percentage of total building roof area)	0.21 (29.6%)	0.37 (23.1%)	0.48 (29.6%)	0.19 (21.1%)	0.29 (46.8%)
Total number of households	53	119	122	65	104
Household density (households/ha)	10	15	18	21	29

* Including building roofs, roads, paths and driveways

Sources: Ghosh and Vale (2004); Ghosh (2006) and Ghosh et al. (2007)

– detached housing', with 10 households per hectare, 'low density – mixed housing', with 14 households per hectare, and 'medium density – mixed housing', with 18 households per hectare, represent typical New Zealand suburban developments. The 'medium density – heritage cottages', with 21 households per hectare, and 'high density – row housing', with 29 households per hectare, represent the suburbs which are located close to urban centres.

Figure 15.6 presents the percentage distribution of different land covers in the five case studies as a comparative chart. Figure 15.7 shows the spatial mapping of land cover distribution and solar efficient building roof areas for the 'low density – mixed housing' case study.

The local food production, solar energy generation and total rainwater collection potential at a neighbourhood scale are presented in Table 15.2. The table indicates that the 'low density – detached housing', 'low density – mixed housing' and 'medium density – mixed housing' case studies are able to generate surplus vegetables on site. The local food production

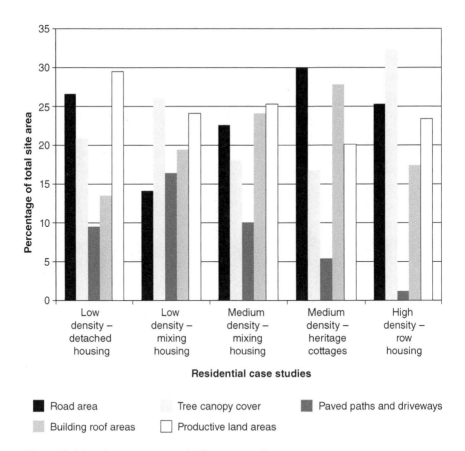

Figure 15.6 Land cover patterns in five case studies

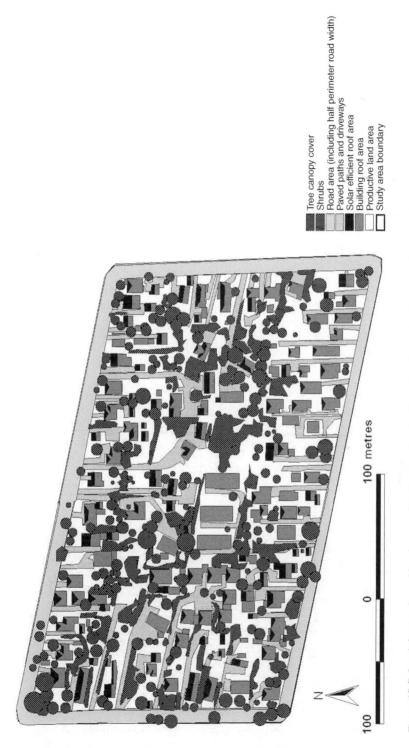

Figure 15.7 Spatial mapping of land cover distribution for the 'low density – mixed housing' case study

Table 15.2 Local food, solar generation and roof rainwater collection potential of the case studies

Sustainability factors	Low density – detached housing	Low density – mixed housing	Medium density – mixed housing	Medium density – heritage cottages	High density – row housing
Local food production including embodied energy content of food					
Total dietary energy demand from vegetables (GJ)	472	1,052	1,086	575	922
Total LEQ of demand (ha)	3.5	7.8	8	4.3	6.8
Total available LEQ of on-site production (ha)	7.2	11.1	10.2	3.9	5
Total LEQ of deficit or surplus (ha)	3.7	3.3	2.1	-0.3	-1.8
Local food available on-site as percentage of total demand	205.7%	142.3%	127.5%	90.7%	73.5%
Deficit or surplus local food as percentage of total demand	105.7%	42.3%	27.5%	-9.3%	-26.5%
Solar energy generation					
Total energy demand* (GJ)	2,207	4,924	5,080	2,689	4,316
Total LEQ of demand* (ha)	16.4	36.5	37.6	19.9	32.0
Total available LEQ of solar generation on-site (ha)	9.7	18.2	22.2	9.5	14.7
Total LEQ of deficit or surplus (ha)	-6.7	-18.3	-15.4	-10.4	-17.3
Solar energy available on-site as percentage of total demand*	59.3%	49.9%	59.0%	47.7%	46.0%
Deficit or surplus solar energy as percentage of total demand*	-40.7%	-50.1%	-41.0%	-52.3%	-54.0%

Sustainability factors	Low density – detached housing	Low density – mixed housing	Medium density – mixed housing	Medium density – heritage cottages	High density – row housing
Roof rainwater collection					
Total water demand (m³)	9,828	21,924	22,617	11,970	19,215
Total roof rainwater collection on-site (m³)	7,962	18,059	18,060	9,753	7,015
Total water deficit (m³)	1,866	3,865	4,557	2,217	12,200
Percentage of total water demand supplied from rainwater	81%	82%	80%	81%	37%
Deficit as percentage of total water demand	19%	18%	20%	19%	63%
Annual deficit in supplying water demand per household (m³)	35	32	37	34	117

* Including space heating

Sources: Calculations based on Ghosh (2004); Ghosh (2006) and Ghosh *et al.* (2007)

values obtained for these first three case studies indicate that New Zealand suburbs could very easily accommodate growing local food as an important sustainability practice. The other two case studies, 'medium density – heritage cottages' and 'high density – row housing', could generate a total of 91 per cent and 74 per cent of their total dietary vegetable demand on-site, respectively. These two case studies establish that a significant part of vegetable demand could be grown on-site in similar locations, which could improve sustainability to a greater extent.

The 'low density – detached housing' and 'low density – mixed housing' case studies could generate 59 and 50 per cent of their total demand for home energy uses – such as water heating, cooking, etc. – from solar energy resources, respectively. The 'medium density – mixed housing' case study is also able to generate 59 per cent of the total energy demand. The other two case studies, 'medium density – heritage cottages' and 'high density – row housing', could generate 48 and 46 per cent of their total energy demand, respectively. All the case studies have a significant energy deficit that could not be supplied by solar-oriented roof areas in the existing neighbourhoods, when space heating demand is included. Space heating demand could be reduced through appropriate solar orientation of the houses, passive solar design, such as placing a conservatory towards the north, and use of relevant materials to store solar heat during the day in the thermal mass of the buildings and to release the heat during the night to keep the building envelope at a comfortable temperature. Space heating could also be provided by wood, a bio-fuel which already supplies 50 per cent of residential space heating energy in New Zealand (Camilleri et al., 2007). Solar energy from the roof could then supply nearly all other energy demands in all the case study neighbourhoods, as shown in Table 15.3.

The values in Table 15.3 indicate that, if space heating is taken out of the calculation, the 'low density – detached housing' and 'medium density – mixed housing' can supply over 90 per cent of their water heating, cooking, refrigeration, lighting and appliances energy demand from solar water heating and PV generation. The figures for 'low density – mixed housing' (78 per cent), 'medium density – heritage cottages' (75 per cent) and 'high density – row housing' (72 per cent) show that it is possible to supply roughly three quarters of energy demand from solar in a wide range of existing housing forms, which would be a significant step towards achieving neighbourhood sustainability. The use of more efficient lights and appliances could serve to reduce demand to the level that could be supplied.

The potential for roof rainwater collection (see Table 15.2) is higher in 'low density – detached housing' (81 per cent), 'low density – mixed housing' (82 per cent), 'medium density – mixed housing' (80 per cent) and 'medium density – heritage cottages' (81 per cent) compared to 'high density – row housing' (37 per cent). The rainwater collection potential values are determined by the total number of residents in the neighbourhood

Table 15.3 On-site solar energy generation potential of the case studies, excluding space heating demand

Solar energy generation sustainability factors	Low density – detached housing	Low density – mixed housing	Medium density – mixed housing	Medium density – heritage cottages	High density – row housing
Total energy demand excluding space heating (GJ)	1,412	3,149	3,249	1,720	2,760
Total LEQ of demand excluding space heating (ha)	10.5	23.3	24.1	12.7	20.4
Total available LEQ of solar generation on-site (ha)	9.7	18.2	22.2	9.5	14.7
Total LEQ of deficit or surplus (ha)	−0.8	−5.1	−1.9	−3.2	−5.7
Solar energy available on-site as a percentage of total demand excluding space heating	92.8%	78.0%	92.2%	74.6%	71.9%
Deficit or surplus solar energy as a percentage of total demand excluding space heating	−7.2%	−22.0%	−7.8%	−25.4%	−28.1%

influencing the total demand and the building roof area available for rainwater collection. It depends also on the space available on the site for installing rain tanks to capture all the rainwater available from the roofs.

Overall, existing suburbs with mainly low and medium density residential developments may be able to supply significant shares of their vegetable, home electricity and domestic water demands. Assuming an average household size of 2.9 for the Auckland Region, performance in two key sustainability factors – local food production and localised solar energy generation – is presented as land area equivalents in hectares per household per year in Figure 15.8 and Table 15.4. The deficit or surplus LEQ serves as the indicator of the degree of neighbourhood self-sufficiency and sustainability performance. LEQ, a land based indicator, provides an easy, objective comparison between different case studies with varying physical densities.

A study by Ghosh and Head (2009) compared the sustainability performance of home gardens in low density traditional and contemporary suburban residential developments in Australia. Results indicated that roof rainwater harvesting could supply 63.4 per cent of the total demand in the traditional development but the contemporary development could produce 35 per cent surplus after supplying the total demand, as the roof sizes were larger. This brings up the issue of having enough space in the gardens of contemporary developments to install rainwater tanks, as the garden spaces tend to be smaller than for traditional housing and also include other garden features such as swimming pools and paved surfaces for barbeques,

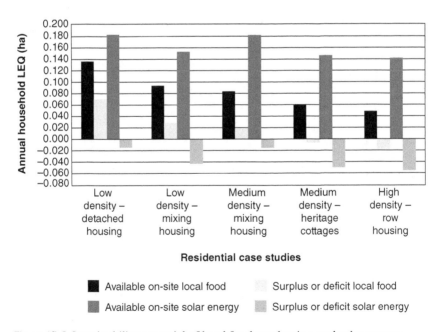

Figure 15.8 Sustainability potential of local food production and solar energy generation

Table 15.4

LEQ (ha)	Low density – detached housing	Low density – mixed housing	Medium density – mixed housing	Medium density – heritage cottages	High density – row housing
Available on-site local food	0.136	0.093	0.084	0.060	0.048
Surplus or deficit local food	0.070	0.028	0.018	−0.006	−0.017
Available on-site solar energy	0.183	0.153	0.182	0.146	0.141
Surplus or deficit solar energy	−0.014	−0.043	−0.015	−0.050	−0.055

etc. The traditional development was able to produce 288 kilograms of vegetables per capita, while the contemporary could produce only 93 kilograms (Ghosh and Head, 2009).

A study in Auckland on an existing low density suburb demonstrated that 92 per cent of building roof areas of separate dwellings could accommodate 4 m^2 of solar water heater panels and only 8 per cent of roof areas could not, due to their inappropriate solar orientations (Ghosh and Vale, 2006). Realistically, at a neighbourhood scale only 58 per cent of the total available solar efficient roof areas could be useful, and 42 per cent would be lost due to inappropriate roof designs, such as hipped roofs. Replacing hipped roofs with gable roofs could provide 26 per cent more solar energy generation, as a rectangular roof makes it much easier to fit rectangular solar panels (Ghosh and Vale, 2006).

These two studies establish the importance for sustainability of not only the dwelling but its immediate surroundings, such as garden spaces, as well as the impact of specific morphological characteristics of suburban forms, such as building roof orientations and roof forms (gabled or hipped).

The research study presented in this chapter addresses the importance of practices and technologies that could improve the sustainability of future residential neighbourhoods, if applied, and their comparative performances at low, medium and high densities. The New Zealand residential suburban developments studied here exhibit significant potential for accommodating meaningful practices for improved sustainability and consequently a lower ecological footprint.

Focusing only on environmental sustainability aspects could make social practices and people invisible in the sustainability assessment process. Behaviour change plays a significant role in the uptake of these sustainable practices and the application of renewable technologies which could impact significantly on overall settlement sustainability. For example, growing local food is directly dependent on personal and household motivation to grow food, while solar energy generation is mainly technological and includes indirect components of behavioural change shaped by sustainability knowledge and awareness. The community is currently unaware of the significant potential of suburban developments to contribute to household shares of food, energy and water demands. Therefore actions need to be focused also on implementation gaps and formulation of more targeted policy, appropriate directives and regulatory legislation essential for implementing efficient practices. Achieving meaningful collaborative partnerships between local authorities, power companies, communities and other stakeholders will be necessary. Future research should explore the form-specific comprehensive sustainability potential of both urban and suburban forms and should develop useful indicators for measuring and monitoring performance, such as the ecological footprint.

Conclusions

A disconnection between research, policy and practice makes it difficult to achieve a future vision of socially, economically, ecologically and environmentally responsive built environments. Applications of useful sustainable technologies, community involvement and participation and positive behavioural changes towards sustainability could drive much greater neighbourhood self-sufficiency in current and emerging suburbs, in New Zealand and elsewhere.

References

Allen, L. (1999) 'Gardening for the Kitchen', elective study (B. Arch.), School of Architecture, University of Auckland, New Zealand

Anderson, W. P., Kanaroglou, P. S. and Miller, E. J. (1996) 'Urban form, energy, and the environment: a review of issues, evidence and policy', *Urban Studies*, vol. 33, no. 1, pp. 7–35

Auckland City Council (1999) 'Part 7: Residential Activity', in *City of Auckland, District Plan, Isthmus Section, Operative 1999*, Auckland, pp. A11–A18

Auckland Regional Council (2006) *Auckland Water Quantity Statement June 2004–May 2005*, 2005 TP300, Auckland Regional Council, New Zealand

Bannister, R., Crowcroft, G. and Johnston, A. (2004) 'Auckland Water Resources Quantity Statement', Technical Publication No. TP 249, Auckland Regional Council, Auckland

Bertolini, L., Clercq, F. and Kapoen, L. (2005) 'Sustainable accessibility: a conceptual framework to integrate transport and land use plan-making – two test-applications in the Netherlands and a reflection on the way forward', *Transport Policy*, vol. 12, no. 3, pp. 207–220

Breuer, D. R. (1994) *Design for the Sun, Vol. 1: Working Manual*, Energy Efficiency Conservation Authority (EECA), Auckland

Camilleri M., French L. and Isaacs N. (2007) 'Wood and solid fuel heating in New Zealand', paper presented at the XXXVth International Association of Housing Science (IAHS) World Congress on Housing Science, Melbourne, 4–6 September

2007, http://www.branz.co.nz/cms_show_download.php?id=507d73bd7cf49233 4f345121f1dffc3089795747, accessed 17 July 2012

Centre for Advanced Engineering (CAE) and Energy Efficiency and Conservation Authority (EECA) (1996) *New and Emerging Renewable Energy Opportunities in New Zealand*, University of Canterbury, Christchurch, New Zealand

Chhima, D. (2001) 'Sustainable Development and Compact Urban Form', master's thesis, University of Auckland, Auckland

City of Auckland (2005) *City of Auckland – District Plan Isthmus Section – Operative 1999*, City of Auckland, New Zealand

Commonwealth of Australia (2012) 'Chapter 5.2: Embodied Energy', in *Your Home Technical Manual*, http://www.yourhome.gov.au/technical/pubs/fs52.pdf, accessed 12 February 2012

Daniels, G. D. and Kirkpatrick, J. B. (2006) 'Comparing characteristics of front and back domestic gardens in Hobart, Tasmania, Australia', *Landscape and Urban Planning*, vol. 78, pp. 344–352

Earthsong Eco-Neighbourhood (2012) 'Design and development section', http://www.earthsong.org.nz/design-development.html, accessed 12 February 2012

EECA (2001) *National Energy Efficiency and Conservation Strategy: Towards a Sustainable Energy Future*, Energy Efficiency and Conservation Authority, Wellington, New Zealand

English Partnerships and Housing Corporation (2000) *Urban Design Compendium 1: Urban Design Principles*, English Partnerships, London

Francis, R. (2010) 'Harvesting the Suburbs and Small-Space Gardens: Micro-Eden Series #2', Permaculture College of Australia, Nimbin, NSW, Australia

Ghosh, S. (2004) 'Simple Sustainability Indicators for Residential Areas of Auckland, New Zealand', PhD thesis, School of Architecture, University of Auckland, New Zealand

Ghosh, S. and Head, L. (2009) 'Retrofitting the suburban garden: morphologies and some elements of sustainability potential of two Australian residential suburbs compared', *Australian Geographer*, vol. 40, no. 3, pp. 319–346

Ghosh, S. and Vale, R. J. D. (2006) 'The potential for solar energy use in a New Zealand residential neighbourhood: a case study considering the effect on CO_2 emissions and the possible benefits of changing roof form', *Australasian Journal of Environmental Management (AJEM)*, vol. 13, no. 4, pp. 216–225

Ghosh, S. and Vale, R. J. D. (2009) 'Typologies and basic descriptors of New Zealand residential urban forms', *Journal of Urban Design*, vol. 14, no. 4, pp. 507–536

Ghosh, S., Vale, R. J. D. and Vale, B. A. (2007) 'Metrics of local environmental sustainability: a case study of Auckland, New Zealand', *Local Environment*, vol. 12, no. 4, pp. 1–23

Ghosh, S., Vale, R. J. D. and Vale, B. A. (2008) 'Local food production in home gardens: measuring onsite sustainability potential of residential development', *International Journal of Environment and Sustainable Development (IJESD)*, vol. 7, no. 4, pp. 430–451

Gordon, P. and Richardson H. W. (1997) 'Are compact cities a desirable planning goal?' *Journal of the American Planning Association*, vol. 63, no. 1, pp. 95–105

Gordon, P. and Richardson, H. W. (2001) 'The sprawl debate: let markets plan', *Publius: The Journal of Federalism*, vol. 31, no. 3

Guy, S. and Marvin, S. (1999) 'Understanding sustainable cities: competing urban futures', *European Urban and Regional Studies*, vol. 6, no. 3, pp. 268–275

Hockerton Housing Project (HHP) (2012) 'Hockerton Housing Project', http://www.hockertonhousingproject.org.uk/, accessed 12 February 2012

IT Power (Australia) Pty Ltd for Ministry of Economic Development (2009) *Assessment of the Future Costs and Performance of Solar Photovoltaic Technologies in New Zealand*, New Zealand Government, Wellington

Jenks, M. and Dempsey, N. (2005) 'The language and meaning of density', in Jenks, M. and Dempsey, N. (eds) *Future Forms and Design for Sustainable Cities*, Architectural Press, Oxford, England, pp. 287–309,

Jenks, M., Williams, K. and Burton, E. (1996) 'A question of sustainable urban form', in Jenks, M., Burton, E. and Williams, K. (eds) *The Compact City: A Sustainable Urban Form?* E. & F. N. Spon, London

Kaplan, R. and Austin, M. E. (2004) 'Out in the country: sprawl and quest of nature nearby', *Landscape and Urban Planning*, vol. 69, pp. 235–243

Loram, A., Warren, P. H. and Gaston, K. J. (2008) 'Urban domestic gardens (XIV): the characteristics of gardens in five cities', *Environmental Management*, vol. 42, no. 3, pp. 361–376

Mathieu, R., Freeman, C. and Aryal, J. (2007) 'Mapping private gardens in urban areas using object-oriented techniques and very high resolution satellite imagery', *Landscape and Urban Planning*, vol. 81, pp. 179–192

Ministry for Agriculture and Fisheries (MAF) (1995) *The National Food Survey 1993*, Wellington, New Zealand

Ministry of Economic Development and Auckland Regional Council (2006) *Urban Centres and Economic Performance Auckland Stocktake*, Auckland Sustainable Cities Programme, Auckland

Ministry for the Environment (MfE) (2005) *New Zealand Urban Design Protocol*, Wellington, New Zealand

Moriarty, P. (2002) 'Environmental sustainability of large Australian cities', *Urban Policy and Research*, vol. 20, no. 3, pp. 233–244

National Institute of Water and Atmospheric Research (NIWA) (2004) 'New Zealand National Climate Summary', January 2004, http://www.niwascience.co.nz/ncc/cs/aclimsum_04.pdf, accessed 6 September 2005

Newman, P. and Kenworthy, J. R. (1989) 'Gasoline consumption and cities: comparison of US cities with a global survey', *Journal of American Planning Association*, vol. 55, pp. 24–37

Newman, P. and Kenworthy, J. R. (1999*) Sustainability and Cities: Overcoming Automobile Dependence*, Island Press, Washington, DC

North Carolina State University (1996) 'Home Vegetable Gardening' http://www.ces.ncsu.edu/depts/hort/hil/ag-06.html, accessed 12 February 2012

Pauleit, S. and Duhume, F. (2000) 'Assessing the environmental performance of land cover types for urban planning', *Landscape and Urban Planning*, vol. 52, no. 1, pp. 1–20

Pritchard, M. and Vale, R. (1999) 'How to save yourself (and possibly the world) on 20 minutes a day', in *Proceedings of the 6th Conference of the Sustainable Energy Forum: Threshold 2000 – Can Our Cities Become Sustainable?* Sustainable Energy Forum, University of Auckland, Auckland

Rain Harvesting (2008) 'Rainwater harvesting: complete rainwater solutions: first flush water diverters', http://www.rainharvesting.com.au/determining_the_amount_of_water_to_diver.asp, accessed 13 November 2008

Romanos, M. C. (1976) *Residential Spatial Structure*, Lexington Books, Lexington, MA
Simmons, C. and Chambers, N. (1998) 'Footprinting UK households: how big is your ecological garden?' *Local Environment*, vol. 3, no. 3, pp. 355–362
Statistics New Zealand (1991) *New Zealand Census of Population and Dwelling (Super Map)*, Department of Statistics, Wellington
Statistics New Zealand (1996) *New Zealand Census of Population and Dwelling (Super Map)*, Department of Statistics, Wellington
Statistics New Zealand (2012) 'Meshblock Dataset', http://www.stats.govt.nz/Census/2006CensusHomePage/MeshblockDataset.aspx?tab=Download, accessed 14 January 2012
Troy, P. N. (1996) *The Perils of Urban Consolidation: A Discussion of Australian Housing and Urban Development Policies*, Federation Press, Sydney
Troy, P., Holloway, D. and Randolph, B. (2005) 'Water Use and the Built Environment: Patterns of Water Consumption in Sydney', Research Paper No. 1, City Futures Research Centre, University of New South Wales, Sydney
United States Environment Protection Authority (EPA) (2006) *Protecting Water Resources with High Density Developments*, EPA, Washington, DC
Wack, P. (2005) 'Village homes, Davis, California: a learning lab for future planners', *Focus*, vol. 2, pp. 36–39
Wackernagel, M. and Rees, W. (1996) *Our Ecological Footprint: Reducing Human Impact on Earth*, New Society Publishers, Gabriola Island, BC, Canada
Waitakere City Council (2003) 'District plan: human environments guidelines', Waitakere City Council Operative District Plan 2003, http://www.waitakere.govt.nz/AbtCnl/pp/districtplan/infosheets.asp#1, accessed 12 September 2012
Waitakere City Council (2008) 'Using rainwater', http://www.waitakere.govt.nz/Abtcit/ec/bldsus/pdf/water/usingrainwtr.pdf, accessed 28 December 2008
Williams, K., Burton, E. and Jenks, M. (2000) 'Defining sustainable urban form', in Williams, K., Burton, E. and Jenks, M. (eds) *Achieving Sustainable Urban Form*, E. & F. N. Spon, London

16 The Hockerton Housing Project, England

Brenda and Robert Vale

The Hockerton Housing Project (HHP), completed in 1998, was the first attempt in the UK to build a zero-emission community. This case study will look at how much difference its design has made to the overall footprints of its inhabitants. Is 'green' design the way forward, or are lifestyle changes more effective?

In 1993, a group of people set out a proposal to build houses on a 10-hectare site outside the village of Hockerton in Nottinghamshire, in the United Kingdom. The houses were to be the centre of a sustainable development using renewable energy which would attempt to have a low environmental impact. The houses, while 'green' in their own right, were also seen as the focus of a whole way of living that would be more sustainable, with the planting of trees, the production of organic food, and the creation of a lake which would encourage wildlife and where fish could be farmed. At the same time employment would be generated in ecologically sound businesses.

The houses themselves were eventually designed as zero-energy earth sheltered buildings arranged as a terrace of five. Two further houses were built some years later, independent of HHP. The houses are long and thin and south-facing to maximise the exposure to solar energy, as set out in Vale (1973), with the north side buried in the ground, and all the habitable rooms open into a conservatory to provide a climate buffer between the indoors and outside. The houses are connected to the electricity grid and their energy comes from photovoltaic (PV) panels along the edge of the roof, and from two small wind turbines sited in a field on the property. Recently a further PV array has been installed on the roof of the office. When there is surplus power it is exported to the grid, and when there is insufficient it is imported. Over a year, the houses put back into the grid as much as they take out, and the grid connection avoids the use of inefficient batteries. The first of the wind turbines caused some problems when initially proposed, being fiercely resisted not only by local people from Hockerton village, but even by objectors from 50 miles away, who would not have been able to see it because of the curvature of the earth. Given that the Trent Valley, where the Hockerton Housing Project is situated, is fairly full of power stations, pylons and other necessary attributes of electricity distribution, two small wind generators on masts 10 metres above ground were hardly an intrusion in the landscape. However, permission was eventually given and the wind turbine established. A second one, of similar size, followed without objection.

Since then Hockerton village residents have come together, purchased and installed a grid-linked second-hand 225-kilowatt turbine. This has been installed by HHP on behalf of Sustainable Hockerton Ltd. (2012). The 10-year-old turbine, which was no longer wanted at its original location because it had been replaced by a much larger machine, is owned by a number of shareholders, and the money generated from the feed-in tariffs both benefits the shareholders and can be used for other projects aimed at lowering carbon in the local community. The wind turbine now makes more electricity than the village consumes (BBC, 2011).

This shows the way that self-reinforcement of values held in a community can lead to action. The Hockerton Housing Project involved a self-selected group of people who wanted to invest their own money in a zero-energy, low-environmental-impact project. Because they were a group they were able to weather resistance to the project. However, once the project was built, monitored and shown to be a success (from the start the Hockerton Housing Project had a website where information was shared), others with similar values were emboldened to instigate further projects in the same village, to the benefit of all. This suggests two things: first, demonstration projects, if carefully monitored and reported, can be a catalyst for change; and second, it is easier to do things against the current paradigm in society if you do them in a group, because of the self-reinforcement this offers. So how successful was the Hockerton Housing Project (HHP)?

The Ecological Footprint of the HHP

The seven households at Hockerton were sent an ecological footprint questionnaire, with four returned complete. This was never going to be a statistically robust exercise, but the detailed answers to the four questionnaires still reveal useful information. All data were entered into the Centre for Sustainable Economy's ecological footprint quiz (Centre for Sustainable Economy, undated). This was chosen as being more thorough than some other available footprint calculators – not least because it probes energy supply in some detail – but results were also checked against other calculators, from which similar results were obtained. The energy issue is important, as since the electricity in Hockerton village comes from the wind turbine (which produces roughly as much as the village consumes in a year) and the HHP generates energy from wind and sun equal to its own annual consumption, this will lower the ecological footprint for energy consumed in the home and on-site. Some people at the HHP also generate their income from working at home, and this income will be based on renewable energy, which is not necessarily the case for work outside the site. This shows the problem of using conventional footprint calculators for projects like the HHP and Findhorn in Scotland (discussed later), which fall outside the current ways in which society is organised. However, as long as results are compared, rather than being used as absolute, they can be useful. The aim here is to see how close the HHP comes to being a fair earth share settlement.

Table 16.1 sets out the results for the four HHP households compared with the UK average (figures are in global hectares, gha, per person). The footprint of citizenship is not separated out in this calculator but is part of all figures.

There are savings in all categories against the UK average, the smallest being food. The savings for all categories based on the HHP average footprint are shown in the last column of Table 16.1.

The largest reduction (71 per cent) came in the goods and services category, with consistently low scoring across the four households. At the HHP almost everything that can be composted or recycled is, and all households described themselves as cautious consumers, only replacing household effects when absolutely necessary. Obviously living in a rural location has an effect, as it is more of an effort to consume since it involves travel. However, the results also suggest that the HHP can supply an environment where people feel sufficiently contented not to want to consume. Life at the HHP perhaps offers intrinsic rewards in terms of satisfaction received from participating in a joint activity (Ebreo *et al.*, 1999: 110). Being part of a project which is known to be an attempt to live sustainably may also be guiding individual consumer decisions, again showing the power of the group to behave differently, rather than just the individual or household. Some of the footprint reduction came from not creating as much waste as normal and recycling all or most unwanted materials. At the HHP each household contributes around 300 hours a year to the on-site business, which is a mix of consultancy services, educational tours and courses, all helping to promote the idea of sustainable living.

The second biggest reduction, of 67 per cent, came in the energy category, which is hardly surprising given that the settlement uses renewable energy. However, the largest differences in this small sample came within this category, because it also includes energy used for travel. Household 4 did no flying but drove a combi van for business and domestic use, though less than 20,000 kilometres per year. In fact all households had a car and there was no bus or train travel, which is hardly surprising given the rural location and the fact that the local branch line closed for passengers in 1929 and for goods in 1964 (another casualty of Dr Beeching's axing of rural railways in the UK to save money; Anderson and Cupitt, undated). The nearest bus service is a two-mile walk into Southwell. This lack of alternative transport modes emphasises the problem of not designing housing and transport in any planned manner, which has to be one goal for

Table 16.1 Hockerton Housing Project survey and UK average household ecological footprints

EF category	Household 1	Household 2	Household 3	Household 4	HHP average	UK average	Percentage difference
Energy – carbon (gha/person)	0.84	0.62	0.79	0.43	0.67	2.07	−67%
Food (gha/person)	1.39	1.34	1.29	1.32	1.34	1.62	−17%
Housing (gha/person)	0.45	0.31	0.44	0.35	0.39	0.63	−38%
Goods and services (gha/person)	0.34	0.33	0.38	0.30	0.34	1.18	−71%
Total (gha/person)*	3.0	2.6	2.9	2.4	2.73	5.50	−50%
Number of earths	1.64	1.46	1.64	1.32	1.5	3.07	—

* Component totals differ due to rounding

a fair earth future. This is not to say that everyone has to live in compact urban developments that are transit oriented, but rather that even rural developments need some alternatives to the car. In fact, Bradshaw's Railway Map from 1907 shows that almost everyone in the UK would have been within an hour's walk of a railway: 1907 marks the peak of railway building in the UK (Bradshaw, 2011). This suggests that investing in the right transport system could well allow travel for all. Once on the railway in the UK it is possible to go to Vladivostok (should you want to), and the Chinese have plans for a high-speed rail link to London (Phillips and West, 2010). This is travel that is much less damaging to the environment. The HHP tried to obtain permission to build another group of affordable zero-energy houses on a site near a local railway station but outside the designated building zone, but this was refused, with one councillor noting there was no way to force people to use the train. The evidence from the HHP, however, is that if there is a common collective motivation to live in a low footprint way then this would probably have been the result (HHP, undated). There was little international flying among the HHP residents surveyed, and no domestic flying at all. However, the prize at the HHP should probably go to the household that took an overseas holiday to France by bike.

The zero-energy approach to the houses has some effect, with a footprint 67 per cent lower than UK average. However, the houses are still relatively generous in size, having three bedrooms, with two out of the four households comprising two people, another two adults and two small children, and the last two adults and three teenagers. The category with the least saving compared to the UK average was that of diet, with an average reduction of 17 per cent. All households eat meat or fish, both high footprint parts of diet, although not many meals are eaten out, possibly for the reason that travel is also somewhat limited, meaning that to eat out driving (or cycling) would be necessary. The reductions in the food footprint came from vegetables and other food grown on-site and the efforts made to eat local and organic produce, though these were not as extreme as they could have been.

Overall, although the sample is very small, a 50 per cent average reduction in footprint was achieved against the UK average. Given that life in the HHP is recognisably western, this seems a commendable achievement, although it is still just over 50 per cent more than a fair earth share EF.

Could the HHP footprint be reduced sufficiently to be a fair earth share? Taking Household 4, with the smallest EF of 2.4 gha per person, the EF was recalculated assuming a vegan diet, in which no meat or dairy products are eaten. This reduced the EF to just under 2.1 gha per person, or 16 per cent above the fair earth share. This is probably the single change which, coupled with buying (or growing) more organic food, would bring the HHP to fair earth share living; it may also be the most contentious. As long ago as 1971 it was suggested that eating plants was perfectly alright for people and very good for the planet (Lappé, 1971). Low EF societies such as Indonesia have diets based on plants, the staple protein in Indonesian diets being tofu, which is eaten in many forms. The fact that soya beans, the source of tofu, do not grow in the UK does not mean that a diet of mostly plants could not give satisfactory nutrition to adults. Most peasant diets will contain meat and fish occasionally but not as a staple, and this is the real difference. Diet is a contentious issue but avoiding meat, at least some diary and fish is a quick way to reduce EF. Perhaps a useful rule of thumb would be to eat the dairy and fish you can grow at home, where at least you would be aware of the land and effort it takes to produce the foodstuff.

Comparison with Findhorn and BedZED

Two other published footprint studies of alternative sustainable settlements – those of Findhorn and BedZED – offer a useful comparison with the HHP.

Findhorn

Findhorn in Scotland is a much older community than the HHP, having been established in 1962. By 2006 the community had some 40 ecological buildings, wind turbines and an on-site sewage treatment system, and food is grown on site (Findhorn Foundation, undated). The Findhorn footprint study used 58 resident questionnaires and 276 visitor questionnaires (Tinsley and George, 2006: 10, 21), so has a very much wider base than the survey at the HHP. Table 16.2 shows the breakdown of the HHP and Findhorn resident EFs, adjusted in the latter case so the footprint of citizenship has been assigned between the three categories as a weighted percentage.

In fact visitors to Findhorn had a lower EF than residents, at 2.1 gha per person, giving a combined EF for Findhorn of 2.56 gha per person, which is slightly better than the HHP but still not a fair earth share. It seems that the diet at the HHP is raising the EF while at Findhorn the single biggest contributor to the EF is travel, at 0.5 gha per person, because of the number

Table 16.2 EF comparisons between the HHP, Findhorn and the UK average

EF category	HHP	Findhorn residents	UK average
Housing and energy – carbon* (gha/person)	1.06	1.30	2.70
Food (gha/person)	1.34	0.82	1.62
Goods and services (gha/person)	0.34	0.59	1.18
Total (gha/person)	2.73	2.71	5.50

* Housing and energy footprints combined for comparison and all travel included

of international flights taken by residents and guests, which is above the Scottish average (Tinsley and George, 2006: 29; see also Chapter 3). Energy is also not 100 per cent renewable at Findhorn, with natural gas use making the largest contribution to the energy EF (Tinsley and George, 2006: 14).

These differences aside, the encouraging result is that both communities, even with the EF of citizenship, are making a significant attempt at lowering the EF. Both are rural communities with a sense of collective endeavour, both offer educational opportunities, and both see the use of zero-energy or ecological buildings and systems for dealing with energy, water and waste as important to community identity. Taken together these results challenge not just the UK planning system with respect to where communities should be placed, but also much conventional talk about sustainability and the need to increase density, in order to reduce the impact of travel and benefit from housing forms in which more walls are shared, thereby lowering energy use. Findhorn is a collection of buildings, some joined together in terraces as at the HHP, but some detached, for example the famous recycled whisky vat houses (Hanfman, 2006). It seems that low EF comes not so much from what you live in but from a shared ideal, that because you live in an ecological house you live in an ecological way. This is a useful starting point for comparison with the much denser and more urban eco-village of BedZED in south London.

BedZED

BedZED is a larger community of 100 households. It was developed by Peabody, a housing association, in partnership with BioRegional, a charity involved in delivering sustainability solutions, and designed by Bill Dunster Architects (BioRegional, 2009: 2). This makes it different from the HHP and Findhorn in being set up by a third party, rather than by the residents (even though, as happens, the residents change over the years). The development is a mix of social housing (rented), and shared-ownership and owner-occupied dwellings. The original scheme contained workspaces so that some of the residents could be close to their place of work. Like the HHP and Findhorn the buildings were designed to be ecological, with significantly reduced energy demand, water collection, waste recycling and renewable energy generated on-site. The site of BedZED is close to bus services and a local station for trains to the centre of London. A detailed footprint study of BedZED residents was done seven years after the opening of the project in 2002. Table 16.3 shows the breakdown of the HHP EF and that of various categories of BedZED resident (BioRegional, 2009: 39). Again the figures in the latter case have been adjusted so the footprint of citizenship is assigned between the three categories as a weighted percentage.

The BedZED average certainly achieves a reduction on the UK average EF, of 15 per cent. Social tenants ('BedZED social'), with an assumed lower income, have an ecological footprint 10 per cent lower than that of

Table 16.3 EF comparisons between the HHP, BedZED and the UK average

EF category	HHP	BedZED average	BedZED keen	BedZED social	BedZED private	UK average
Housing and energy – carbon* (gha/person)	1.06	1.99	1.26	1.72	2.29	2.70
Food (gha/person)	1.34	1.62	1.06	1.64	1.59	1.62
Goods and services (gha/person)	0.34	1.04	0.69	1.07	1.03	1.18
Total EF (gha/person)	2.73	4.68**	3.01	4.43	4.91	5.50

* Housing and energy footprints combined for comparison and all travel included
** Totals from BioRegional (2009: 39) components; differences due to rounding

owner-occupiers ('BedZED private'), pointing to some link between available income and EF (BioRegional, 2009: 39). Perhaps of more interest are the criteria on which the assessment of the 'BedZED keen' households was based, since their EF is close to that of the HHP households. As at the HHP, all energy is renewable. The 'keen' household does no driving or flying, but travels 4,995 kilometres per year by train and 676 kilometres per year by bus. They are also frugal in their consumption, at 42 per cent of that typical of the UK, vegetarian and reduce waste by 30 per cent compared to the UK average (BioRegional, 2009: 38). This means the keen BedZED household is behaving more like the HHP households, with their no (or limited) flying and not a lot of driving, the latter matched by public transport use for the BedZED keen household. In fact there is more flying undertaken by BedZED residents in general than by the local community around it (BioRegional, 2009: 26), an aspect of ecological footprint which is mirrored at Findhorn. A possible reason for this additional flying is the 'Jevons Paradox'. Jevons (1865) observed that as a technology became more efficient more resources were consumed. So in a highly insulated house the savings might not be as great as predicted because the owners heat it to a higher temperature, simply because they can afford to do this due to having a more efficient house. Living in a zero-energy or very low-energy house also releases disposable income, which might well be spent on air travel. What is even more curious, given the very good access to public transport at BedZED, is that the lowering of footprint in this area is not greater. This again may be an outcome of affluence, as 83 per cent of owner-occupiers at BedZED had a car, compared to 47 per cent of those in social housing and 43 per cent of shared-ownership households, while no one in privately rented accommodation had a car (BioRegional, 2009: 26). This is a rewriting of an old proverb: 'you can house people near public transport but you can't make them use it'.

Although further research is necessary it may be the different model of housing provision that leads to the footprint differences between the HPP, Findhorn and BedZED. Findhorn and the HHP are initiatives originating with those involved in living in the housing rather than from an outside organisation providing low-energy houses. However, it is still the sense of being a community that BedZED households value most, followed by the design, the sustainability initiatives, and general well-being and internal comfort in the dwellings (BioRegional, 2009: 35). Somehow the community is not yet working together in the way that the other projects do to reinforce the idea of low footprint living. This is an issue explored further in the next section, describing a competition entered by streets in New Zealand to see if they can reduce their ecological footprint over nine months.

Working Together: The HHP and Green Streets

In 2011–2012, Kāpiti Coast District Council (KCDC) ran their second 'Greenest Street' competition with the underlying intention of both improving sense of community and increasing local resilience while reducing environmental impact (KCDC, 2012). The Kāpiti Coast is a narrow strip of land between the Tararua Ranges and the sea, extending from Paekākariki in the South (which has a railway station for local electric trains to Wellington, the capital city of New Zealand) to Ōtaki in the north (which also has a station, served by a long-distance commuter train into Wellington). Crossing this strip from the coast, there is a band of housing settlements, the main State Highway 1, the railway, then a few houses in the foothills of the ranges. This pattern is repeated more or less the whole length of the Kāpiti Coast. The area is served by a few local shops, local services like primary schools, a bus service, and railway stations that are not always spaced as frequently as the residents would like (KCDC, 2008: 36; Raumati Public Transport Action Group, undated). This means there is a high level of car dependence. Community transport is estimated to account for 27 per cent of all community greenhouse gases, excluding agricultural emissions (KCDC, 2008: 25).

The competition aims are for three to five streets to compete over nine months to see which can lower their collective ecological footprint by the greatest amount, with points also being awarded for working together as a local community. The prize (NZ$3,000) is awarded to support further projects in the community, chosen by the winning street. The footprint is assessed by filling in a questionnaire that in 2011 was almost identical to the one used for the HHP assessment, and results were fed into the same calculator. Table 16.4 sets out the 'before' and 'after' footprints for two years of the competition, compared with the HHP, Findhorn and BedZED averages.

Although some minor changes were made to some of the houses in the competing streets, such as installing low-energy light bulbs, some double glazing, insulation and curtains, New Zealand houses are not known for their energy efficiency, or even their comfort in winter (Isaacs, 2006). These were not, therefore, zero- or low-energy developments like those in

Table 16.4 EF comparisons for two years of Greenest Street, the HHP, Findhorn and BedZED

Community	EF – before (gha/person)	EF – after (gha/person)	Average earths needed – before	Average earths needed – after
Street A (2010–2011)	6.48	5.58	3.6	3.1
Street B (2010–2011)	6.30	5.04	3.5	2.8
Street C (2010–2011)	6.48	4.68	3.6	2.6
Street D (2010–2011)	5.22	4.14	2.9	2.3
Street A (2011–2012)	5.94	5.40	3.3	3.0
Street B (2011–2012)	5.94	4.86	3.3	2.7
Street C (2011–2012)	6.48	6.12	3.6	3.4
HHP	—	2.73	—	1.5
Findhorn*	—	2.56	—	1.4
BedZED	—	4.68	—	2.6

* Residents and guests

Table 16.3. Nevertheless, significant reductions were made in the footprints. In the 2010–2011 competition the average reduction was 20 per cent, with Street C achieving a 29 per cent reduction, bringing its footprint in line with that of BedZED. In the following year's competition, the average for all streets was 12 per cent, with Street B achieving an 18 per cent reduction. These values should be compared with the average 15 per cent reduction achieved at BedZED.

What this suggests is that local action and working together for behaviour change can have a significant effect. The types of changes made by the competition residents included food growing at home, joining organic cooperatives or purchasing organic food from conventional sources, collective composting, rainwater collection, grey water recycling, pest eradication, making 'greener' choices when replacing household appliances, collective making of household cleaning materials, sharing equipment and generally trying to reduce car use. The reinforcement that comes from undertaking these actions together seems significant. It is easier to change behaviour as part of a group than as an individual (Anable et al., 2006: 153). This has been recognised by movements, such as Transition Towns, which seek to encourage groups brought together by the identity of belonging to a place to change together (Hopkins, 2008). This approach is also found to be effective in developing countries. In Surabaya in Indonesia, the recent winner of the Greenest Kampung competition (a *kampung* in this case being an urban settlement organised round a shared street) had installed a collective grey water recycling system, and were growing vegetables in pots along the street outside the houses and composting; the women had also started making products from waste packaging for sale. This activity took place in houses densely packed either side of a street

Table 16.5 EF reductions achieved at BedZED by built environment design compared with behaviour change

Service	Typical UK household (gha)	Keen household* (gha)	Percentage reduction in overall EF for keen household*
Building	0.73	0.13	11%
Neighbourhood	1.26	1.22	0.7%
Behaviour**	3.41	2.23	22%
Total	5.40	3.58	34%

* Living at BedZED and behaving so as to reduce environmental impact
** With regard to transport, waste, food and consumer items

about 3 metres wide, normally used by people and slowly travelling motorcycles. Again the prize awarded was money for further projects of a sustainable nature. As an urban environment, because of all the well-cared-for planting, the kampung was also a very attractive place to live. The environment benefitted and the residents benefitted.

It seems that behaviour change can happen in perfectly ordinary and conventional environments providing there is some catalyst to promote change. In fact an earlier analysis of BedZED, summarised in Table 16.5, demonstrates that more can be achieved through behaviour change than through the design of the environment (Vale and Dixon, 2005).

Although the difference between the UK and BedZED footprints attributable to the building is very large (82 per cent) this represents a big reduction in a small part of the overall footprint. The built environment-related changes to the neighbourhood and buildings at BedZED produce an 11 per cent reduction in the overall footprint, but those from the combined behaviour changes are very much larger, at 22 per cent. This is the value of projects like Kāpiti Coast's Greenest Street and the Green Kampungs of Surabaya – they emphasise the ability of everyone to make significant changes now, without waiting for zero-energy developments to become available. Of course, doing both things ultimately leads to the lowest footprint, as Findhorn and the HHP demonstrate, but in both of these low footprints the behaviour component is still the most important.

Collective Activity

The HHP is a success because it set out to be a sustainable or low-impact environment and those living there have held onto this goal. Apart from the fact that it was set up by at least some of the people still living there, it also involves a somewhat different view of ownership. Each household owns their home under a 999-year leasehold and has a private garden. The houses are freely tradable with no veto from the other residents. Under the leasehold the residents also own a share of vital services such as the water systems. The leasehold is important as it is this that binds the residents to certain rules, such as contributing to the on-site business and to food growing. Collectively the residents also lease surrounding land on an agricultural tenancy. This raises another issue, that of collective stewardship of a place to live, rather than the ownership that comes with the current capitalist model. 'Ownership' implies ideas of property rights, exclusive use and being able to do anything within the curtilage of ownership, whereas 'stewardship' contains the idea of responsibility when it comes to use – the difference between your own book and taking care of a book borrowed from the library. Most people automatically buy into the latter idea, whereas not everyone will treat their own books with equal care. When it comes to one-planet living, stewardship of the earth's resources is an important key to values and attitudes that will produce more sustainable behaviour.

The other important thing to emerge from this study is the importance and effectiveness of doing things together and doing things locally with the people you see every day. It is possible to live a western lifestyle and just about achieve one-planet living (based on a fair earth share of around 1.8 gha per person), but this is going to be much easier if your neighbours have the same goal.

References

Anable, J., Lane, B. and Kelay, T. (2006) *An Evidence Base Review of Public Attitudes to Climate Change and Transport Behaviour*, http://tna.europarchive.org/20090605231654/http://www.dft.gov.uk/pgr/sustainable/climatechange/iewofpublicattitudestocl5730.pdf, accessed 25 July 2012

Anderson, P. and Cupitt, J. (undated) *The Robin Hood Line*, http://myweb.tiscali.co.uk/sherwoodtimes/line.htm, accessed 19 July 2012

BBC (2011) 'Wind turbine produces more electricity than village', http://news.bbc.co.uk/local/nottingham/hi/people_and_places/nature/newsid_9355000/9355049.stm, accessed 18 July 2012

BioRegional, (2009) *BedZED Seven Years On*, BioRegional Development Group, Surrey, England

Bradshaw, G. (2011, reprint) *Bradshaw's Railway Map 1907*, Old House Books, Botley, Oxford, England

Centre for Sustainable Economy (undated) 'Ecological footprint quiz', http://myfootprint.org/, accessed 15 July 2012

Ebreo, A., Hershey, J. and Vining, J. (1999) 'Reducing solid waste: linking recycling to environmentally responsible consumerism', *Environment and Behaviour*, vol. 31, no. 1, pp. 107–135

Findhorn Foundation (undated) 'Creating a positive future', http://www.findhorn.org/aboutus/vision/positive/, accessed 21 July 2012

Hanfman, R. (2006) 'The whiskey barrel house', http://www.findhorn.org/2008/01/the-whiskey-barrel-house/, accessed 21 July 2012

HHP, (undated) 'Fiskerton Station (Notts, UK)', http://www.hockerton.demon.co.uk/productsservices/fiskerton.html, accessed 19 July 2012

Hopkins, R. (2008) *The Transition Handbook*, Green Books, Totnes, Devon, England

Isaacs, N. (ed.) (2006) *Energy Use in New Zealand Households*, BRANZ Study Report No. SR 155, BRANZ, Porirua, Wellington, New Zealand

Jevons, W.S. (1865) 'Of the economy of fuel', Chapter 7 in *The Coal Question: An Inquiry Concerning the Progress of the Nation, and the Probable Exhaustion of Our Coal-Mines*, Macmillan and Co., London

KCDC (Kāpiti Coast District Council) (2008) *Kapiti Coast: Choosing Futures – Towards a Sustainable Transport System*, http://www.kapiticoast.govt.nz/Documents/Downloads/Strategies/Sustainable-Transport-Strategy.pdf, accessed 25 July 2012

KCDC (Kāpiti Coast District Council) (2012) 'Being a Green Street', http://www.kapiticoast.govt.nz/Our-District/greenest-street/Being-a-Green-Street/, accessed 26 July 2012

Lappé, F. M. (1971) *Diet for a Small Planet*, Ballantine Books, New York

Phillips, N. and West, A. (2012) 'China to build high-speed rail link to Europe', *The Sydney Morning Herald*, 10 March, http://www.smh.com.au/travel/travel-news/china-to-build-highspeed-rail-link-to-europe-20100309-pvuf.html, accessed 19 July 2012

Raumati Public Transport Action Group (undated) 'Raumati station now', http://www.raumatistation.com/, accessed 26 July 2012

Sustainable Hockerton Ltd. (2012) 'Sustainable Hockerton's wind turbine', http://www.sustainablehockerton.org/vilturbine.html, accessed 31 July 2012

Tinsley, S. and George, H. (2006) *Ecological Footprint of the Findhorn Foundation and Community*, Sustainable Development Research Centre, Forres, Moray, Scotland

Vale, R. (1973) 'Analysis of forms for an autonomous house', Working Paper 10, University of Cambridge, Department of Architecture, Technical Research Division, Cambridge

Vale, R. and Dixon, J. (2005) 'The battle for hearts and minds: sustainable design and sustainable behaviour', Urbanism Down Under Conference, 18–20 August, Wellington, New Zealand

17 Education for Lower Footprints

Sant Chansomsak

Introduction

Education has to be recognized as a critical element in promoting sustainability and a driving force for the change needed for lower footprint living. Learning to live in a more sustainable way is for everyone at every stage of life. Though at least as much learning takes place outside the school system, it is common to focus on education in school as the locus of learning. This leads to the important necessity for schools to be considered places for learning within an ethos of sustainability, particularly during childhood.

Though a school is commonly referred to as a place for learning, it also involves organization and learning activities. To promote lower footprint living, children could learn from both the learning environments and the activities happening within a school. While an educational approach that focuses on learning and acting appropriately to improve or at least sustain the interrelationship between the human and ecological communities is frequently called 'sustainable education', schools that apply 'sustainable' or 'green' design concepts are often referred to as 'sustainable schools'. Because of the reinforcement of unsustainable values and practices in society, the stress and pressure of competition, and the lack of infrastructure and will to commit to implementation, the idea of sustainable education, which has long been evolved from earlier relevant practices, particularly environmental education, slowly but gradually is implemented in educational systems (Sterling, 2001; Federico *et al.*, 2002). On the other hand, although the idea of sustainable schools is relatively new, case studies from around the world affirm that the emerging trend of sustainable school design is a significant consideration for many stakeholders – from government agencies and school administrators to architects and planners – particularly since the new millennium. Many sustainable design strategies, moreover, have become basic criteria for a good school building design.

The reason many schools are increasingly engaged with sustainable principles is that sustainable design provides many advantages, both for the schools and the people who use them. According to data taken from 30 green schools in the US, 'sustainable' schools on average use 33 per cent less energy and 32 per cent less water than conventionally designed schools (Kats, 2006). Similarly, case studies of sustainable schools in the UK show that they consume 30 to 75 per cent less energy than most schools (DfES, 2006). Besides reduction of environmental impact, sustainable schools also support student and teacher health and performance. The Carnegie Mellon building performance programme investigation found that the average health gain from improvement in indoor air quality is 41 per cent, while average productivity improvements are 3.6 per cent from improved temperature control and 3.2 per cent from high-performance lighting systems (Kats, 2006). According to a 2005 survey by the Turner Construction Company, 70 per cent of the executives at organizations involved with green school facilities believe that green schools improve student performance, reduce student absenteeism, and attract and retain teachers, and 87 per cent of the executives also reported that these facilities create a better community image (Turner Green Building, undated).

This chapter aims to show how schools can be designed and used as learning tools for promotion of constructive attitudes and culture. The current practices of sustainable schools are investigated and discussed. However, rather than focusing only on the physical features of sustainable schools, the chapter places the emphasis on changes in attitudes and behaviours – the fundamental requirements for low footprint living.

Sustainable Schools at Present

Based on a review of current sustainable schools, mainly in developed countries, improvement of old schools and creation of new school buildings usually involves several sustainable design strategies, covering a healthy learning environment, energy use, water use and management, materials selection, construction and waste management, site selection and development, accessibility and transport options, participatory systems in design, commissioning and maintenance, and using the school as a learning tool (Gaia Architects and Gaia Research, 2005; CHPS, 2006; Ford, 2007; USGBC, 2007; LPA, 2009). Details of each of these issues, their limitations, deficiencies and barriers, as well as the possible challenges and suggestions for progressing sustainable design have been discussed elsewhere (see Chansomsak and Vale, 2010). Many of these strategies have been put into practice before, but they have less often been considered and developed together.

Among the strategies mentioned, design, commissioning and maintenance and using the school as a learning tool are directly and notably related not only to the physical features of the buildings but also to users' behaviour and educational activities. Though the main purpose of participation in design, commissioning and maintenance is to ensure the stakeholders benefit from the use of the school, another advantage of this process is that it can encourage a sense of belonging and lead to improvement in user performance aimed at achieving the sustainable condition. Monitoring equipment could help to show how buildings and human behaviours interact with the environment, making environmental impacts more visible and easy to comprehend. Particularly for children, involvement in design, operation and maintenance of the buildings is crucial for participation and collaborative skills, and for learning how to interact properly with their surrounding environment. In addition, providing users with opportunities to control their immediate environment, such as operable windows and thermal and lighting controls, also enhances their perceptions of the environmental conditions and assists them in learning how to react and adapt to create suitable environments.

Using the school building as a learning tool for sustainability is the most important issue in bridging the gap between architectural and educational activities. Although the school can be used for teaching about sustainability, its goal should be encouragement of environmentally friendly attitudes and behaviours. Basically, while a school can provide comfortable spaces to facilitate education for sustainability, sustainable strategies such as selection of environmentally friendly materials and systems, renewable energy generators, rainwater collectors and water treatment systems, separate bins and recycling stations, sustainable modes of transportation, and so on can be used as subjects of study and learning tools. Whenever possible schools should promote interactions with the natural environment and reduce impact on it. After the occupants have encountered design solutions that provide high quality health and comfort, they usually feel comfortable and experience mental and physical well-being. Consequently they gain positive experiences from the design, which may lead to recognition of its benefits and higher demand for design for sustainability. Finally, these healthy, efficient, sustainable places will make buildings, dwellings and schools an education in themselves.

In addition to the physical infrastructure of the school, what is taught is also of vital importance. Sharing facilities, resources, appropriate knowledge and skills to contribute to a more sustainable way of living with the local community and communities at large will reduce demand for new facilities and resources and also encourage the children to interact with local people and learn more about local community conditions. Educational activities that aim to improve student values, decisions and actions with regard to sustainability could be based on real local conditions, such as social structures, products and food, as part of the fundamental questions for problem-based learning methods. These activities can be used to encourage the learners in critical thinking and to be aware of the effects of their actions. Through a participatory process of design, commissioning and maintenance, while participants can learn more about the built environment and how to behave appropriately to make good connections with their environment, citizen skills such as inclusive participation, civic engagement, team dynamics, group facilitation and managing democratic networks can also be practised. Subsequently their knowledge, skills, and experiences of learning networks can be developed for appropriate implementation in other situations. Taken together, all this will enhance their ability to create holistic solutions and to negotiate rationally between ecological, social and economic goals for personal and communal actions.

Different from conventional design, sustainable school design encompasses the process of design, construction and continual involvement in the use of a school, with consequent changes. As a process, sustainable school design can be formulated, reformed, transformed and constantly developed to support sustainable educational activities. Sustainable schools should therefore provide the setting for such learning and act as participants in the learning process. Briefly, they can be used as a positive learning tool for sustainable education. Constructive experiences of sustainable schools can educate people about sustainability and sustainable practices, make them aware, and move them to behave responsibly and sustainably. The people who experience sustainable architecture will learn how to interact appropriately with their environments and may extend their perceptions of appropriate design into other areas of their lives.

Behaviour Change and Sustainable Schools

According to current practice, sustainable schools are generally considered a process of creation of a sustainable place and the subsequent use of this to support sustainable learning. However, sustainable conditions do not rely totally on physical, man-made environments; clearly they depend also on what people do to and for their society and environments, including creation and modification of their environments. Thus, making school buildings more sustainable cannot ensure that sustainable conditions will happen, as changing architecture does not on its own change people. In fact, it is a role of schooling to suggest and guide the expected beliefs, perceptions, attitudes and behaviours that a society requires from learners, such as creating a sustainable society. Rather than being limited to focusing only on the change of architecture, changes in attitudes and behaviours, including living with a lower footprint, are also essential.

This is especially the case in the world today, where problems such as climate change, natural degradation, overconsumption, poverty, social inequality and conflict and warfare are accumulating and apparently increasing as global populations continue to grow, while non-renewable natural resources are finite and gradually reduce. The dominant lifestyle of the consumer society is based on the attitude, at odds with living in a finite system, that fulfilment and happiness can be achieved through ever greater material comforts and increased consumption. Accordingly, many authors have commented on these unsustainable behaviours and suggested change in behaviours and patterns of living as a solution (see Rifkin, 1990; McKenzie-Mohr and Smith, 1999; Jackson, 2006; Monbiot, 2006; Steffen,

2006; Goodall, 2007; Kneidel and Kneidel, 2008; Vale and Vale, 2009). Changing of consumption habits and ways of eating, living and working, such as reducing consumption, reusing, repairing, sharing, recycling, purchasing of local food, products and services, selecting environmentally friendly transport, and supporting and participating in sustainable practices are usually recommended. As Kollmuss and Agyeman (2002) have mentioned, one of the greatest barriers to sustainable behaviour is old behaviour patterns. As it is commonly known that old habits die hard, it is much easier to shape appropriate and sustainable behaviours, rather than trying to change habits that are already ingrained. People should have a chance to become familiar with and participate in sustainable practices before they are involved in unsustainable behaviours, and should also continually practice sustainable actions and develop them to become habits. This leads to the necessity of formation and education, which should begin from an early age and regularly repeat the process of strengthening sustainable behaviours. Since a school is a common place for young people to learn and socialize, it should be recognized as one of the first places for cultivating sustainable behaviours in children.

Although several sustainable school projects show the potential for the use of schools to encourage change in attitudes and behaviours, contemporary consideration of sustainable schools usually focuses on the development of the physical features of schools – an external factor that, it is hoped, influences the change in attitudes and behaviours – while a school's involvement with activities that aim to promote internal change of attitudes is frequently overlooked. In fact, many sustainable features incorporated in sustainable schools will not be effective if there is no change in the attitudes and behaviours of the users. Moreover, if sustainable systems and appliances are used improperly, infrequently maintained, allowed to be damaged and left unrepaired, such facilities will not be able to serve their purpose effectively and will probably need new parts or have to be changed. This was certainly the case with many of the innovative sustainable technologies at the BedZED housing project in London (Hodge and Haltrecht, 2009). Even with new and highly sustainable systems and technology, this will lead to higher operating costs and a requirement for untimely budget increases. These factors may put people off sustainable technologies, as being 'too difficult'.

In contrast, attitude and behaviour changes require less budget and fewer resources, and frequently create significant positive effects on sustainability. Several examples in Thailand illustrate the potential for reducing environmental impacts through behaviour, with no or very little change to the school environment (Chansomsak, 2009). For example, the Zero Waste Project at Roong Aroon School in Bangkok is a solid waste management programme that encourages the students to reduce and recycle their waste. Over a two-year period, with the provision of only small recycling booths and separate bins, the project has led to a reduction in the school's waste from 206 to 11 kilograms per day, as measured from February 2004 to April 2006 (Kroo Doot, 2006). Similarly, through promotion of the need to reduce and recycle, the reported reductions in solid waste for three schools in Chiang Mai province in Thailand that participated in the Cleaner Technology for Resource Use Efficiency in Schools (CT-Schools) project varied from 7 to 156 kilograms per day (TEI, 2002). In addition, the average energy use reduction of another four CT-Schools, three in Chiang Mai and one in Bangkok, was 507.75 kilowatt hours per month when compared with the same month before the project (TEI, 2002). These reductions are the result only of behaviour change, such as the promotion of energy conservation and frequent monitoring of energy consumption, without any investment in changes to electrical systems and equipment.

Unlike built environments, which cannot be changed rapidly or frequently, change in behaviour can be adopted straight away. For most schools, there is no need to wait for new and sufficient budgets for investing in changes to the physical environment. Thus this strategy is suitable for every school condition, particularly in most developing countries, where changes to architecture are seemingly difficult due to limited financial support and lack of access to sustainable technology. Because changes in attitudes and behaviours are flexible and easier to modify, they are more easily adjusted to different traditional values and cultures (for example diverse beliefs, religions and social norms) or to new approaches (for example new educational approaches and new teaching methods). Whenever people realize they can do better, behaviour adjustment can occur. In fact, many behaviour changes, such as using fewer resources and less energy, will happen if people cannot access the resources. For example, on Waiheke Island, near Auckland in New Zealand, where there is no central water supply, households rely totally on rainwater collection, and manage to control their water consumption pattern and live within the limit (Vale and Vale, 2000). The film *The Power of Community: How Cuba Survived Peak Oil* (Morgan *et al.*, 2006) provides another example of people's behaviour changing when it has to. The film depicts the situation when Cuba lost access to Soviet oil in the early 1990s and how Cuban people adapted by changing their lifestyles, modes of transport and agricultural system from a highly mechanized and fossil fuel-intensive farming to small, less energy-intensive organic farms and local, urban gardens. Community support and collaboration are the essence of the story. However, from a sustainability point of view, rather than waiting and changing only when it is inevitable, people should shift their attitudes from consumerism to living within the finite limit of resources, controlling their demand and emotional impulses, and changing behaviours towards sustainability as soon as possible.

Another benefit of changing behaviour is that the action itself is also part of the process of development and reinforcement of sustainable actions. Once people achieve a change they may realize that change is not impossible, but only needs some time to get used to (Kneidel and Kneidel, 2008). For example, in changing from private vehicles to public transport, people have to walk to the station, bus stop or pier and wait for public transport to arrive. However, if they are accustomed to it, the short period of exercise and waiting will be manageable and acceptable. Through critical thinking and the processes of self-realization, self-evaluation and feedback, people who engage in sustainable activities may have a chance to see themselves as environmentally or socially concerned. Doppelt (2008: 65) has warned about the false feedback and self-verification that may cause people to make unfair judgements – people usually select information to pay attention to and believe only if it supports their existing view about themselves and the world around them – so a supportive environment is needed. A self-evaluation activity, such as the good behaviour record programme at Ban Sai-youy Community School in Phitsanulok, reminds learners about what they do and how they can improve it. When sustainable actions become approved social norms, learners will have more positive attitudes towards their actions and will be willing to keep practising and improving them to be more productive.

Schools for Lower Footprint Living

In using schools to promote lower footprint living, it is necessary to consider both internal incentives, including strong intentions, right attitudes and personal responsibility, and external incentives, which usually encompass both rewarding positive behaviours and elimination of barriers. The development of the physical environment is an obvious example of external incentives that supports sustainable practices and makes achieving sustainable practices easier. For instance, provision of separate bins, recycling containers and organic waste collection in schools helps to reduce the inconvenience of achieving waste reduction. Rather than only promotion of energy conservation behaviours, the modification of electrical equipment and systems in two CT-schools, in Chiang Mai and Bangkok, was able to reduce energy use of the schools by 2,147 and 12,510 kilowatt hours per month (TEI, 2002). While the first school cut down on the number of lights and running time and changed electrical systems in one of their buildings, the second selected new energy-efficient air conditioners for their renovated classrooms, and reduced the use of air conditioners in every classroom. However, because built environments cannot be changed frequently, existing conditions – such as ineffective use of energy and natural resources in buildings and infrastructure, insecure traffic systems and lack of green or public spaces – often form another barrier to progressing sustainability, and their replacement and renewal is costly in social as well as financial terms (Richardson, 1994). Consequently solutions that improve existing built environments should be carefully considered.

Helping students to accept and realize the benefits of living with a lower footprint and to sustain their behaviour, suitable practices of the organization and other school members, as well as a good collaboration with families and the local community, are required. For example, the TV programme *Jamie's School Dinners* (2005) shows that changing school dinners to be healthier cannot succeed simply by changing a menu. It requires support from multiple stakeholders (see Gilbert *et al.*, 2005). The understanding and commitment of dinner ladies and school principals and the participation of teachers are needed. Support from both schools and government to promote the campaign for change and provide budgets to cover increased costs of food and increased salaries for dinner ladies, who need more time to cook fresh food, are also required. Additionally, it demands that parents understand the necessity of change and encourage change by giving children a chance to eat healthy food at home. The programme also shows the importance of social acceptance. The greater the number of students who were seen to prefer fresh-cooked food, instead of reheated, reconstituted or packaged foods, the greater was the number of initially resistant students that began to change their attitudes.

For such behaviour change to happen, groups like educationalists, designers, policy makers and community members need to come forward and commission many more sustainable buildings and many more sustainable activities. The participation of multi-stakeholders and professions in creating sustainable schools and developing their use as a tool for sustainable education is also an essential part of enhancement of responsible social attitudes and sustainable culture. People who gain positive experiences of such schools could also lead the demand for more sustainable school design in the future. These schools then increasingly become part of constructive sustainable local conditions and one of many external factors influencing responsible sustainable behaviours. Furthermore, to advance the idea of sustainable schools as an educational tool for sustainability also requires more detailed research into how attitudes and behaviours change via the experience of sustainable schools, and how such change affects the demand for sustainable schools. This task should be a long-term project involving architectural, educational and other fields.

Although local situations can be helpful in many ways, they can also be a major obstacle to sustainable practices. Examples of such barriers are economic constraints, lack of facilities and services, discouraging policies, conflicts of ideas and approach, and unsustainable habits and culture, including antisocial behaviour, vandalism and overconsumption. Such unsupportive conditions not only impede contemporary actions, they also limit opportunities for future sustainability practices. When people adhere to misguided beliefs and base their actions on these, the problem is made worse and increasingly hard to solve. To solve this, changing attitudes and behaviours towards sustainability and creating sustainable culture from childhood, as discussed in this chapter, should be promoted and practised.

Acceptance and appreciation of teachers and peers is a common reward used as a motivation for children to begin a sustainable activity or to perform such activity more effectively in schools. To promote healthy eating habits in school, Jamie Oliver, the British chef, gave stickers reading 'I've tried something new' to students involved in the activity. The stickers represented acceptance and appreciation, which are useful rewards to encourage students to change their eating patterns (Gilbert *et al.*, 2005). Rewarding school members who cut down car use (as at Sidwell Friends Middle School, Washington, DC) and giving the money from selling recyclable products back to the students who bring the products to school (as at Phaisali Pittaya School, Phitsanulok, Thailand) are also examples of reward programmes that can help to encourage behaviour changes. In terms of management, allowing renovations to be paid for through savings in energy use can help the school moderate the financial obstacles, if it is too expensive to upgrade or install energy-efficient building elements and systems.

Additionally, providing suitable knowledge as well as building up awareness and emotional inspiration is a major strategy applied in many books and media related to sustainability. Recent films like *Power Shift: Energy + Sustainability* (2003), *The End of Suburbia: Oil Depletion and the Collapse of the American Dream* (2004), *An Inconvenient Truth* (2006), *The 11th Hour* (2007) and *Food, Inc.* (2008) are examples of environment-related media that have helped to arouse concerns. Positive, persuasive or warning messages suggesting social norms are also common. A message is much more productive if it

- is noticeable and suitable for target groups
- presents as close in time and space as possible to the targeted behaviours
- encourages people in positive and approved behaviours rather than informing of harmful effects and destructive actions they should avoid
- emphasizes that many people are already behaving that way (McKenzie-Mohr and Smith, 1999; Goodall, 2007).

Although threatening messages can direct people's attention to crises, they can also lead to unsuitable practices such as antisocial behaviours and using ignorance as a self-defence mechanism. As recommended by McKenzie-Mohr and Smith (1999), a threatening message should be used carefully and specific suggestions should be also provided.

Self-evaluation, assessment and commitment are helpful tools to help people to realize their position and choose appropriate directions for change and improvement (McKenzie-Mohr and Smith, 1999; Doppelt, 2008). Additionally, supportive communication (Doppelt, 2008) and external motivation, such as government policies of polluter-pays, tax cuts for environmental actions, reward programmes and enhancing public recognition of the need to be environmentally concerned (McKenzie-Mohr and Smith, 1999), can be useful in fostering sustainable behaviours.

Although successful achievement of sustainable practices in schools unavoidably relies on local contexts and larger circumstances at the communal level, sustainable schooling and other actions at the local scale are both begun and carried on by efforts made at the individual level. Involvement of the members of a school, family and local community, with the right intentions, suitable objectives and strong commitment, is a crucial component. These responsible actions are usually influenced by personal attitudes and responsibility, and knowledge of the issues and action strategies, skills and opportunities needed to participate and select actions. Personal development thus becomes a main focus and education should be considered a key element in this. As a result, the most important issue in creating and maintaining sustainable schools is not, in the end, the physical features of the buildings. Rather it is related to the actions of people who participate in sustainable schools, undertaken to achieve changes in attitudes to and the culture of sustainability. These changes, in turn, will improve local conditions, increase demand for sustainable schools, and progress the quality of sustainable school design and other related practices.

In conclusion, using sustainable schools to promote lower footprint living should be recognized as a process that coexists with other sustainable practices aiming to create the required change in attitudes and cultures. Its aim is not only to create good architecture, but, more importantly, to create good people with right thinking and actions, who will use, maintain and demand such architecture. Its focus accordingly should be on learners and people who participate in school activities, not merely on the physical building and its surroundings. Since those at school are and will be members of society, their change in attitudes and behaviours will influence the practices of the schools and the larger community to which they belong. Their actions for sustainability thus take effect beyond the school premises, being also a part of society as a whole. Developing sustainable school design according to this approach will cause school design to truly become design *as* sustainability. (For a fuller discussion on the meanings of sustainable architecture, related to architecture *about, for* and *as* sustainability, see Chansomsak and Vale, 2008.) With a process of continual development, sustainable schools and all actions for sustainability will eventually move forwards to become an integrated part of a whole sustainable system.

References

Chansomsak, S. (2009) 'Appropriate Sustainable Design for Schools in Lower Northern Part of Thailand', PhD Thesis, University of Auckland, Auckland

Chansomsak, S. and Vale, B. (2008) 'Sustainable Architecture: Architecture as Sustainability', in G. Foliente, T. Luetzkendorf, P. Newton and P. Paevere (eds), *Proceedings of the 2008 World Sustainable Building Conference, Vol. 2*, Melbourne Convention Centre, Melbourne, Australia, 21–25 September 2008, pp. 2294–2301

Chansomsak, S. and Vale, B. (2010) 'Progressing Practices of Sustainable School Design', *Journal of Green Building*, vol. 5, no. 2, pp. 147–157

CHPS (2006) *Best Practices Manual, Vol. I–VI*, http://www.chps.net/dev/Drupal/node/288, accessed 3 June 2007

DfES (2006) *Design of Sustainable Schools: Case Studies*, The Stationery Office, London
Doppelt, B. (2008) *The Power of Sustainable Thinking: How to Create a Positive Future for the Climate, the Planet, Your Organization and Your life*, Earthscan, London
Federico, C. M., Cloud, J. P., Byrne, J. and Wheeler, K. (2002) 'Kindergarten through Twelfth-Grade Education for Sustainability', in J. C. Dernbach (ed.) *Stumbling toward Sustainability*, Environmental Law Institute, Washington, DC, pp. 607–624
Ford, A. B. (2007) *Designing the Sustainable School*, Image Publishing, Mulgrave, Victoria, Australia
Gaia Architects and Gaia Research (2005) *Lessons from School Buildings in Norway and Germany: Design and Construction of Sustainable Schools, Volume 1 and 2*, Scottish Executive, Edinburgh
Gilbert, G. (director and producer), Thirkell, R. (series editor), Walker, D. (series producer), and Conrad, A. (executive producer) (2005) *Jamie's School Dinners* [TV programme], FremantleMedia, Fresh One Productions, London
Goodall, C. (2007) *How to Live a Low-Carbon Life: The Individual's Guide to Stopping Climate Change*, Earthscan, London
Hodge, J. and Haltrecht, J. (2009) *BedZED Seven Years On*, BioRegional, London
Jackson, T. (ed.) (2006) *The Earthscan Reader in Sustainable Consumption*, Earthscan, London
Kats, G. (2006) *Greening America's Schools: Costs and Benefits*, http://www.cap-e.com/ewebeditpro/items/O59F12807.pdf, accessed 16 July 2007
Kneidel, S. and Kneidel, S. (2008) *Going Green: A Wise Consumer's Guide to a Shrinking Planet*, Fulcrum Publishing, Golden, CO
Kollmuss, A. and Agyeman, J. (2002) 'Mind the Gap: Why Do People Act Environmentally and What are the Barriers to Pro-Environmental Behavior?' *Environmental Education Research*, vol. 8, no. 3, pp. 239–260
Kroo Doot (pen name) (2006) *2 years of 'Zero Waste' Project* (in Thai), http://www.roong-aroon.ac.th/aw_Kaya2.pdf, accessed 9 April 2007
LPA (2009) *Green School Primer: Lessons in Sustainability*, Images Publishing Group, Mulgrave, Victoria, Australia
McKenzie-Mohr, D. and Smith, W. (1999) *Fostering Sustainable Behavior: An Introduction to Community-Based Social Marketing*, New Society Publishers, Gabriola Island, BC, Canada
Monbiot, G. (2006) *Heat: How to Stop the Planet Burning*, Allen Lane, London
Morgan, F. (director and producer), Blessing, T. (producer), Quinn, M. (writer and producer) and Murphy, E. (writer and producer) (2006) *The Power of Community: How Cuba Survived Peak Oil* [film], The Community Solution and AlchemyHouse Productions Inc.
Richardson, N. (1994) 'Making Our Communities Sustainable: The Central Issue is Will', in ORTEE (ed.), *Sustainable Communities Resource Package*, Ontario Round Table on Environment and Economy, Toronto, Ontario, Canada, pp. 15–43
Rifkin, J. (ed.) (1990) *The Green Lifestyle Handbook*, Henry Holt and Company, New York
Steffen, A. (ed.) (2006) *World Changing: A User's Guide for the 21st Century*, Harry N. Abrams, New York
Sterling, S. (2001) *Sustainable Education: Re-Visioning Learning and Change*, Green Books, Totnes, Devon, England

TEI (2002) *Cleaner Technology: Guidelines for Cleaner Technology for Resource Use Efficiency in Schools* (in Thai), Thailand Environment Institute, Bangkok

Turner Green Building (undated) *2005 Survey of Green Building Plus Green Building in K-12 and Higher Education*, http://www.turnerconstruction.com/greensurvey05.pdf, accessed 17 June 2007

USGBC (2007) *LEED for Schools for New Construction and Major Renovations (Approved 2007 Version)*, http://www.usgbc.org/, accessed 10 July 2007

Vale, B. and Vale, R. (2000) *The New Autonomous House: Design and Planning for Sustainability*, Thames and Hudson, London

Vale, R. and Vale, B. (2009) *Time to Eat the Dog? The Real Guide to Sustainable Living*, Thames and Hudson, London

18 Footprints and Income

Ella Lawton

Employment is a source of income, social relationships, identity and individual self-esteem (Winkelmann and Winkelmann, 1998) and the amount of money earned – income – allows an individual to exercise different types of lifestyles with varying levels of resource consumption. Once people earn more than it costs to fulfil their fundamental needs, they are fulfilling wants (Max-Neef, 1991). Individuals may even strive to fulfil wants before needs if the mix of social pressures is right. Wants are closely linked to the novelty of people's possessions, experiences and consumption, which are an outcome of their lifestyle choices and the urban form in which they live. The choices individuals make regarding the type of lifestyle they lead in association with their income will impact on the likely size of their ecological footprint.

Income affects an individual's wealth and reflects their chosen lifestyle. 'Lifestyle' is a characteristic bundle of behaviours that make sense to both others and oneself in a given time and place, including social relations, consumption, entertainment and dress (Assadourian, 2010; Spaargaren and Vliet, 2000). Increasing or decreasing an individual's personal income can have both positive and negative effects on their lifestyle. For example, earning an income is not only a means of making money but for many an important part of their social wellbeing. The workplace provides a space to share ideas, feel empowered and create community (Winkelmann and Winkelmann, 1998). Losing one's job, and hence income, could have negative effects on a person's social relations (Lawlor *et al.*, 2009).

The aim of this chapter is to explore how individuals can earn an income that allows them to fulfil their fundamental human needs while requiring no more than their fair earth share of resources. This chapter will explore the links between an individual's income, how they spend their time, the wellbeing of the whole community and the effect on the world's resources.

The Link between Income and Ecological Footprint

Numerous case studies show that, generally, the more money a country or individual has the more natural resources they consume and the higher their ecological footprint. In Canada, 'the ecological footprint of the richest 10% is nearly two-and-a-half times that of the poorest 10%' (Canadian Centre for Policy Alternatives, 2008). One fifth of the global population, living in the highest-income countries, accounts for 86 per cent of private consumption expenditure, while the poorest fifth accounts for a little over 1 per cent (Tilford, 2000).

New Zealand research shows the distribution of resource consumption across a range of different lifestyles. New Zealand is known as the 'melting pot' of the Pacific, with a large variety of cultures, and within these cultures there are a range of socio-economic backgrounds. Market research by Caldwell and Brown (2007) resulted in the book *8 Tribes: The Hidden Classes of New Zealand*. The book attempts to put an end to the myth of the 'typical New Zealander' by providing a snapshot of the varying cultural and socio-economic backgrounds that in turn cause distinct ways of thinking and living. The book's 'eight tribes', each named after a New Zealand location, have a range of values, cultures and wealth profiles which cover the vast majority of New Zealanders (Caldwell and Brown, 2007). In the list below, the first figure is the annual household income of each tribe, in New Zealand dollars.

- *$130,000* The North Shore tribe are typified by their love of winning, of being richer, higher up the ladder at work, smarter, more fashionable, more influential and more beautiful than the people around them.
- *$105,000* The Remuera tribe are also wealthy; old wealth. They rest on old values including their obligation to contribute to the world.
- *$125,000* The Grey Lynn tribe are likely to earn almost as much as the North Shore tribe but spend their money in a different way. Their affluence allows them to live in comfort and have access to things they love – art, travel, a vibrant social life, well-made things.
- *$90,000* The Balclutha tribe is the heartland tribe – down-to-earth, practical, conservative people from the provinces. The Balclutha tribe are efficient with their resources, making do with what they have, and have a strong sense of community. They are reasonably 'well-to-do', particularly as their income is subsidised by the goods they receive straight from the land.
- *$70,000* Cuba Street tribe members are opportunists and masters of the new. They are the young culture-makers, business innovators especially in the tech industries but may be seen as a little 'weird' by the mainstream. The Cuba Street tribe could be earning a considerable amount of money.
- *$60,000* The Papatoetoe tribe is the home of the 'kiwi' working man and woman whose approach to life is characterised by down-to-earth common sense, a focus on life's essentials and a mistrust of intellectualism. This tribe is thrifty; they are in the lower income bracket but have practical skills to ensure a little money can go a long way.

- **$75,000** The most important thing in life to the Raglan tribe is freedom – freedom from authority, financial constraints and possessions. The Raglan tribe will not waste money on things that do not matter, but those things that do matter will be of the best quality and well looked after. They may have high earning power and this provides the luxury to work less.
- **$50,000** The Otara tribe is made up of ethnic minorities, often immigrants, whose lives centre on family and a community of people from the old countries. The average income of an Otara tribe household is likely to be relatively low; they tend to work long hours fulfilling roles in low-paying, labour-intensive jobs.

The North Shore, Balclutha and Grey Lynn tribes are the most populous, while the Otara and Raglan tribes are increasing in size. The background research to *8 Tribes* also found that the tribes can generally be grouped into three income categories: the North Shore, Remuera and Grey Lynn tribes have high incomes, Cuba Street, Balclutha, Raglan and Papatoetoe are somewhere around the average, while Otara earn below average income.

The background market research for the classification of the eight tribes was used as the input for a New Zealand-specific ecological footprint calculator produced by the research team leading the New Zealand Footprint Project (Lawton *et al.*, 2012). The average footprint for each tribe was then converted from New Zealand hectares into global hectares. Figure 18.1 shows the number of planets each tribe would need if everyone on the planet lived the same way they do, and, for comparison, the fair share that would need only one planet.

As the lifestyles (tribes) are shown from left to right in Figure 18.1, the average income of each of these groups generally increases, from Otara

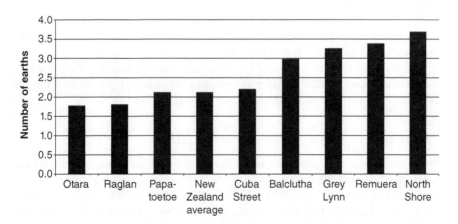

Figure 18.1 Distribution of resource use amongst New Zealand's 'eight tribes'. One earth is an indication of a fair earth share, a lifestyle which, if adopted by the world's population, would fit within one planet

with an average annual household income of $NZ50,000, to North Shore households with an income of $NZ130,000. The only groups whose ecological footprint and income may not align are the Raglan Tribe and the Grey Lynn tribe.

The Grey Lynn tribe think of themselves as conscientious consumers and therefore thoughtfully try to reduce their footprint. However, their eagerness to look and act green is still outweighed by their above-average income and their drive to achieve comfort and access the things they love via air travel. Therefore their resource requirement is still well above those with less green tendencies but much lower incomes.

The Raglan tribe tend to have the strongest preference to work less and live simply, close to the natural environment. They are not so conscientious about reducing their footprint but their lifestyle requires a lower footprint. Their values encourage food self-sufficiency and minimalist living. Their choice of lifestyle revolves around their natural surroundings, which reduces their travel footprint. On the other hand they have expendable income, so for all the good they do there are some luxuries, such as taking overseas trips every few years, which increase their ecological footprint considerably.

Buying goods, no matter how 'green', is likely to increase an individual's footprint. For example, imagine that a new generation of car engines is introduced which halved fuel consumption. Driving would then be cheaper and that would save money, but it is money which would almost certainly be spent on something else or used to drive further (Wilkinson and Pickett, 2009: 219). A comparison of two different-sized households shows that 'the quickest way to reduce embodied energy is not to try to use materials that have low energy, but quite simply to live in smaller houses' (Vale and Vale, 2009: 135). The same goes for appliances: 'new devices are not necessarily more energy intensive than old ones, but they are either much bigger, in the case of televisions, or they are completely new – not so much replacing an old appliance as adding to the total of appliances' (Vale and Vale, 2009: 217).

The idea of the eight tribes is being used in the New Zealand Footprint Project (Lawton *et al.*, 2010) as a way to convey messages to specific audiences relating to how they might reduce their ecological footprint.

The Impact of Wealth on Resource Inequality

Consumerism plays a central role in the increase of inequality, the need to earn money and the resulting increase in ecological footprint, since emotional insecurity contributes to a hunger for material status symbols. The need for recognition and acceptance fuels the drive to acquire possessions – possessions that will 'make you somebody'. Ultimately, this is a far more important motivating force than a fascination for the things themselves (Norberg-Hodge, 1992).

The drive to 'win', rise up the pecking order and outdo each other has created a situation in which societies are destroying the environmental, social and economic systems that they rely on in an attempt to fulfil their desired lifestyles (Barrett *et al.*, 2006). Those individuals who can afford to have gone beyond satisfying their fundamental needs. There seems to be a belief among many that the amount of money they have affords them the right to satisfy their needs and wants however they can, with no concern for environmental or social consequences.

Within each country, people's health and happiness are related to their incomes. Rich people tend, on average, to be healthier and happier than poor people in the same society. But comparing richer countries, it makes no difference whether on average people in one society are twice as rich as people in another. What matters in rich countries may not be your actual income level and living standard, but how you compare with other people in the same society. Average standards do not seem to matter; what does matter is simply whether you are doing better or worse than other people – where you come in the social pecking order (Wilkinson and Pickett, 2009: 13). The gap continues to increase. Many people live in a society of extremes, where the incomes of the very rich and the very poor continue to pull away from each other. In the UK this is influenced by many factors, but wage inequality is at the heart of them. It is a corrosive, destabilising issue that is linked to a range of social problems (Lawlor *et al.*, 2009).

For the increasing urban population worldwide (United Nations Population Fund, 2007), inequality of income correlates not only with growing social issues but also with inequality of access to natural resources. Inequality is not only making people yearn to have more money but also denying them access to the resources they need to provide themselves with food and shelter.

It Depends on the Satisfier

> Material goods play a symbolic role in our lives. The 'language of goods' allows us to communicate with each other – most obviously about social status, but also about identity, social affiliation, and even – through giving and receiving gifts for example – about our feelings for each other.
>
> Jackson, 2009: 136

Income provides a path to fulfilling human needs with a range of satisfiers; however, it is an individual's responsibility to choose satisfiers that do not negatively affect social and environmental systems. Universal human needs are finite, few and classifiable. They are the same in all cultures and in all historical periods (Maslow, 1943; Max-Neef, 1991; Tay and Diener, 2011). What changes, both over time and across cultures, is the way or the means by which needs are satisfied.

Manfred Max-Neef (1991), a Chilean economist, proposes that there are nine needs, comprising subsistence, protection, affection, understanding, participation, idleness, creation, identity and freedom. Max-Neef and others (Jorge, 2010; Spaargaren and Vliet, 2000; Walter, 2012; Zorondo-Rodríguez et al., 2012) propose that needs are expressed through satisfiers and it is the satisfiers which vary according to historical period and culture. Hence satisfiers are what render needs historical and cultural, and economic goods are their material manifestation.

Satisfiers behave in two ways: they are modified according to the expected change that comes with history and vary according to culture and circumstance. Each economic, social and political system adopts different methods for the satisfaction of the same fundamental human needs. In every system, needs are satisfied (or not satisfied) through the generation (or non-generation) of different types of satisfiers. It could be said that one of the aspects that define a culture is its choice of satisfiers. Cultural change is, among other things, the consequence of dropping traditional satisfiers for the purpose of adopting new or different ones (Jorge, 2010; Max-Neef, 1991). Capitalism has exploited the use of objects and artefacts as a way to 'satisfy' the needs of the masses. In industrial capitalism, the production of economic goods, along with the system of allocating them and the psychology of marketing them, has conditioned the type of satisfiers that consumers want (Max-Neef, 1991: 25). Economic goods (artefacts, technologies) are modified according to episodic rhythms (vogues, fashions) and diversify according to cultures and, within those cultures, according to social strata (Max-Neef, 1991: 28; Spaargaren and Vliet, 2000).

There is growing awareness that certain objects or artefacts marketed as satisfiers have larger or smaller ecological footprints than others. Where and how they are produced, consumed and disposed of has more or less of an impact on local and far-off environments and communities. For some there is a shift to choosing their satisfiers more carefully, either selecting those with 'green' credentials (Doerr, 2007; United States Environmental Protection Agency, 2004) or through questioning whether physical satisfiers fulfil their fundamental needs (Becker, 2012; Bruno, 2010).

Increasingly the question of how to be a 'good' consumer is being asked by individuals and in the mainstream media (Aisner et al., 2008). Being a 'good' consumer includes taking responsibility for those things we buy and paying a 'fair' price for them. On the other hand, the extent to which being 'good' lowers the ecological footprint of goods and services is questionable and needs third-party verification. There are numerous verification labelling programmes, the best known for social justice is Fairtrade (Fairtrade International, 2012) and for environmental responsibility is the ISO 14000 standards (ISO, 2012). However, verification costs money and in some cases being a 'good' producer might also cost more. As a result, these additional costs are often (though not always) passed on to the purchaser, meaning that being 'good' costs more. In a counter-argument, if individuals are really only buying those things they need, then money is saved overall and a fair price can be paid for 'good' consumer products.

Moving to an Income-Based Society

Satisfiers of fundamental human needs will change as cultures transform, particularly as cultures move from subsistence to consumer lifestyles. Ladakh is a high mountain region and the most northern state of the Republic of India. Before changes brought by tourism and modernisation, the Ladakhis were self-sufficient, psychologically, spiritually and materially (Norberg-Hodge, 1992). The introduction of western satisfiers has changed their resource wants from those things they grew and created themselves to what others are seen to have (Norberg-Hodge et al., 2011).

Over the past few decades the state of Ladakh has gone through many changes; the biggest are the introduction of western advertising and the uptake in consumerism. A young Ladakh girl wrote that, 'Before 1974, Ladakh was not known to the world. People were uncivilised. There was a smile on every face. They didn't need money. Whatever they had was enough for them' (Norberg-Hodge, 1992: 3). Pre-consumerism Ladakhis also had time to sit and socialise. Their communal work, which allowed everyone to provide themselves with enough to fulfil their basic needs, required time, but substantially less than the time individuals now require to earn money to provide themselves with the same things (Norberg-Hodge et al., 2011).

The increasing emphasis on a western style of education pulled people away from land-based agriculture into the city, where they become dependent on a money-based economy. Traditionally in Ladakh there was no such thing as unemployment. But in the modern sector there is now intense competition for a very limited number of paying jobs. As a result, unemployment is a serious problem (Norberg-Hodge, 1991).

In government statistics the 10 per cent of Ladakhis who work in the modern sector are listed according to their occupations; the other 90 per cent – housewives and traditional farmers – are lumped together as 'non-workers'. Their work is not included as part of the Gross National Product. They do not earn money for their work, so they are no longer seen as productive. Farmers and women are coming to be viewed as inferior, and they themselves are developing feelings of insecurity and inadequacy. The outcome is a vicious circle in which individual insecurity contributes to a weakening of family and community ties, which in turn further shakes individual self-esteem (Norberg-Hodge, 1992).

Time to Make a Living

> Where's all the life we supposedly made at work? For many of us, isn't the truth closer to 'making a dying'? Aren't we killing ourselves – our

health, our relationships, our sense of joy and wonder – for our jobs? We are sacrificing our lives for money – but it's happening so slowly we barely notice. Eventually we may have all the comforts and even luxuries we could ever want, but inertia itself keeps us locked into the nine-to-five pattern.

Moss and Toltz, 2011

In 1992 Robin and Dominguez wrote *Your Money or Your Life*. The aim of the book was to show people their relationship with the money they earn. The relationship encompasses more than just their earning, spending, debts and savings; it also includes the time these functions take, as Illich also pointed out (Illich, 1974). The need for 'stuff' and therefore the need to earn an income also requires a substantial amount of time. Robin and Dominguez highlight that life is finite; when people go to work they are trading their finite time here on earth for money. While money has no intrinsic reality, life does. It is tangible, finite and precious. Money is something that the majority of people consider valuable enough to spend easily a quarter of their allotted time here on earth getting, spending, worrying about, fantasising about or in some other way reacting to (Robin and Dominguez, 1992).

Robin and Dominquez (1992) offer an example showing the true amounts of money and time that are directly related to having a job. Time and money are needed for the cost of escape entertainment, for job-related illness as well as for vacations to recover from work. The example shown in Table 18.1 below shows the impact of these work-related expenses on an ordinary 40-hour week, and on an income of $US440 per week.

Table 18.1 Hours actually devoted to work versus earnings

	Time (hours/week)	Pay ($/week)	Pay ($/hour)
Basic job (before adjustments)	40	440	11
Adjustments			
Commuting	7.5	−50	—
Clothing	1.5	−15	—
Meals	5	−20	—
Relaxing at home	5	−20	—
Escape entertainment	5	−20	—
Vacation	5	−20	—
Job-related illness	1	−15	—
Time and money spent on maintaining a job	70	280	4

Source: Moss and Toltz, 2011: 8

Table 18.1 shows that having a job is not just a 40-hour-per-week time commitment – it could be as much as a 70-hour time commitment. Although the pay cheque says you earn $US440 per week, once work-related expenses are taken into consideration this falls to $US280 per week, or 64 per cent of the original amount. When the total amount earned is divided by the true total hours required, the hourly pay drops well below the minimum wage, to only $US4 per hour.

The Statistics New Zealand time use survey 2009–2010 (Statistics New Zealand, 2011) shows that on average 12 per cent of an individual's total time is spent 'earning an income'. If only those hours spent awake are considered, this represents almost 20 per cent or one fifth of people's waking lives spent in an effort to afford to live for the other four fifths.

What is also interesting is what New Zealanders do with the rest of their time. On average New Zealanders sleep 8 hours and 48 minutes a day (or 528 minutes). This has been removed from the calculations because it could be argued that this is a natural requirement and something few people can alter. Figure 18.2 below provides a picture of what, on average, New Zealanders do with the remaining 912 minutes of their day. It was found that almost 60 per cent of the day was consumed by just five activities: in order of time spent, paid employment, watching TV, eating and drinking, socialising and conversing, and preparing food and drink.

These figures are an average for the whole country, so include people of non-working age, those with illness or those without paid employment. If the same chart were to be created for males between the ages of 25 to 44 years old, the use of time is quite different. Including time spent sleeping, this group on average spends 21 per cent of their time working. Excluding sleep, 25- to 44-year-old males spend on average 31 per cent of their time working and another 34 per cent of their lives engrossed in mass media

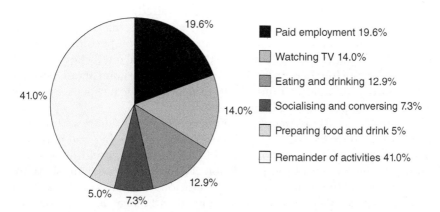

Figure 18.2 How a New Zealander spends time (excluding sleeping), 2009–2010

Source: Data from Statistics New Zealand (2011)

and 'free-time' activities, namely watching TV. This particular social group spend two thirds of their lives either earning an income or watching TV (an effective tool for telling the masses the types and quantities of satisfiers required to fulfil their needs and 'keep up with the Joneses'). In turn this encourages them to work more, increase their income, buy more stuff and watch more TV.

Feel-Good Work

The type of work an individual ends up doing will have two impacts, one on themselves, in terms of how much they enjoy their job, and one on society, in terms of the net worth of the job and the value the products and services created to the local and global community.

The Buddhist point of view takes the function of work to be at least threefold: to give a man a chance to utilise and develop his faculties; to enable him to overcome his ego-centredness by joining with other people in a common task; and to bring forth the goods and services needed for a becoming existence. The consequences that flow from this view are endless. To organise work in such a manner that it becomes meaningless, boring, stultifying or nerve-racking for the worker would be little short of criminal; it would indicate a greater concern with goods than with people, an evil lack of compassion and a soul-destroying degree of attachment to the most primitive side of this worldly existence. Equally, to strive for leisure as an alternative to work would be considered a complete misunderstanding of one of the basic truths of human existence, namely that work and leisure are complementary parts of the same living process and cannot be separated without destroying the joy of work and the bliss of leisure (Schumacher, 1968).

When individuals fortunate enough to have the opportunity to choose are deciding what field to work in, the general advice is to aim towards something that they are passionate about, something that they enjoy and can feel good about. For a few people their dream job becomes a reality; however, many others get whatever job they can. Those not lucky enough to find that 'dream job' may do work that they do not feel good about, perhaps for income security, while others focus primarily on the pay packet, aiming as high as they possibly can whatever the cost.

A New Zealand Department of Labour report, *Work Values and Quality of Employment* (Johri, 2005), explored the question of whether bad jobs were better than no jobs. The report found that low quality employment is not randomly distributed in the population. People with low or no skills are more likely to be in temporary or precarious low income work, with exposure to health and safety risks, limited influence on their job and lacking career development opportunities. Being in low quality employment has impacts not only on the workers, but also more widely on their family, for example with respect to children's activities and work-life balance issues.

Evidence from German panel data shows that being unemployed has a significant negative effect on life satisfaction (Winkelmann and Winkelmann, 1998); however, low quality employment was the main reason unemployment persisted among labour market re-entrants. So although being unemployed makes people unhappy, they would rather that than have a bad job. The emphasis should therefore be on policy aimed at getting unemployed people into a good job rather than any job.

In 2009 the not-for-profit New Economics Foundation (NEF) published the report entitled *A Bit Rich* (Lawlor *et al.*, 2009) which asked two questions. The first was 'What impact do different careers have on the rest of society?' and the second 'Does the income received correspond to this?' Using the principles and valuation techniques of Social Return on Investment analysis, the NEF quantified the social, environmental and economic values that ten different careers produce – or in some cases undermine. The report found that 'while collecting salaries of between £500,000 and £10 million, leading city bankers destroy £7 of social value for every pound in value they generate'. For tax accountants the figure is considerably higher, destroying £47 for every £1 paid in wages. At the other end of the spectrum, waste recycling workers were the most valuable of the professions included in the study: 'for every £1 of value spent on wages, £12 of value will be generated' (Lawlor *et al.*, 2009: 3–4). The report concluded that 'the least well-paid jobs are often those that are among the most socially valuable – jobs that keep our communities and families together. But the market does not reward this kind of work well and such jobs are consequently undervalued or overlooked' (Lawlor *et al.*, 2009: 2).

New Zealand's Baseline

It is highly likely that for most people the amount of money they earn affects the amount of goods and services they consume, and therefore their ecological footprint. In addition, the amount individuals work and the type of work they do often affect the amount of time they have to supply themselves with the goods and services they need. The amount of 'free' time people have affects their lifestyles and, depending on their backgrounds and values, influences how their income is spent, which in turn affects their ecological footprint. Table 18.2 was produced using national average data and shows how income (Statistics New Zealand, 2007), time (Statistics New Zealand, 2011) and ecological footprint (Lawton, 2012) compare across nine footprint categories. The aim is to show whether time, money and ecological footprint are apportioned similarly, and how individuals are deciding to spend their time in relation to footprint and money.

There are some similarities between the proportion of time and money individuals spend within certain ecological footprint categories and the proportion of resources they require. The figures in the travel section of Table 18.2 show that on average New Zealanders spend 14 per cent of their

Table 18.2 Comparison of the proportion of money spent, time required and ecological footprint of activities undertaken by the 'average' New Zealander

EF category	Components*	Expenditure after tax (%)	Time (%)	Ecological footprint** (%)
Food and drink	Food and drinks, including alcohol and time spent growing food in the garden	18.1%	15.4%	36.6%
Travel	Transport by car, bike, bus, train, ferry and pleasure craft	14.1%	7.3%	12.0%
Consumer goods	Household contents and services, tobacco, clothing and footwear, personal care, household cleaning, personal effects, insurance, credit services, interest payments, contribution to savings, money given to others, expenditure incurred whilst overseas	36.3%	39.7%	31.8%
Holidays	Accommodation, package holidays, domestic holidays, national and international flights	1.6%	n/a	4.5%
Energy	Household energy	3.8%	n/a	3.6%
Housing	Rentals, home ownership, budgeting, household maintenance	16.8%	3.9%	3.6%
Infrastructure	Local authority rates and payments	2.2%	n/a	3.4%
Government	Local and national	Previously subtracted***	n/a	0.5%
Services	Health, water, refuse disposal and recycling, communication, education, fines and religious services	7.1%	4.9%	4.2%
Total	—	100.0%	71.3%	100.0%
Additional non-EF/ money	Unpaid employment, socialising and conversation, residual categories, other employment	n/a	9.1%	n/a
Subtotal	—	n/a	80.4%	n/a
Paid employment	—	n/a	19.6%	n/a
Total	—	100.0%	100.0%	100.0%

* The components are set out in the New Zealand Ecological Footprint template (see Lawton, 2012)

** The ecological footprint has been calculated using 'global hectares'

*** The amount of money that the government takes are those taxes subtracted before these calculations are made and depends on total annual income. In New Zealand an individual's income up to $NZ38,000 is taxed at 19.5%, over $NZ38,000 and up to $NZ60,000 is taxed at 33% and over $NZ60,000 at 39% (Inland Revenue, 2012)

Sources: Data from Statistics New Zealand (2007, 2011); Lawton (2012)

income and 12 per cent of their total ecological footprint on transport. But they only spend 7 per cent of their total time travelling. To lower the ecological footprint of travel requires taking transport options such as walking or biking, or public transport. These are likely to take slightly more time but are also likely to decrease the proportion of spending that goes on transport.

The table also shows that 15 per cent of a New Zealander's time is spent in ways associated with food: eating, drinking and spending time in the garden (which is assumed to be spent tending to vegetables rather than non-edible plants). A similar portion of a New Zealander's income is spent on food, at 18 per cent. However, in contrast, the land associated with the growing, cooking, processing, packaging, transporting and eating food totals 37 per cent of a New Zealander's ecological footprint. Again, to reduce the food footprint at least some time will be required but this could also save money. The most effective ways to lower the food footprint are to reduce or eliminate the need for commercial land to grow food – for example through better use of backyards – use less packaging and processing, and reduce or eliminate the footprint of travel required when purchasing or retrieving food, for example by walking or cycling to the supermarket.

These two examples show that, to a certain extent, New Zealanders are choosing to spend money on food and transport rather than expending more time and reducing their ecological footprint. Conversely, lowering their ecological footprint through using slower means of transport or spending time gardening is likely to save money. In accordance with Figure 18.2, it would be possible to find additional time to do these things by reorganising how time is spent, that is, by spending less time in paid employment or watching TV.

Too Much Money

So what do people do if they have a job they love, and with that job comes an income that is well beyond what is required to fulfil their fundamental human needs? Being a conscientious consumer is a good start but, as seen for the Grey Lynn tribe in Figure 18.1, this is not enough to move individuals all the way to achieving a fair share lifestyle. Another beneficial move is to have a smaller house (Vale and Vale, 2009: 135). This decreases the amount of room for 'stuff' and offers lower costs for materials and operation. But then what? Below is a list of other possible options if people find that they have too much money and want to lower their ecological footprint.

- Work less: spend more time undertaking activities that reduce EF.
- Employ others to do the things they have little time for, such as planting and maintaining a vegetable garden and making clothing, furniture and other household consumables by hand to save on transport, packaging and industrial production.

- Give money to help others to fulfil their fundamental human needs.
- Choose needs satisfiers with a low ecological footprint, even if they cost more.
- Throw away the television. Spend the time increasing their understanding of how to reduce their ecological footprint and share this knowledge with others.
- Make every minute count; time is a finite resource too.

The Choice of a Nation

In 2007 New Zealand had 45.2 million global hectares of bioproductive land, of which 37.1 million global hectares were required for the production of exports, totalling 82 per cent of New Zealand's available biocapacity (Global Footprint Network, 2011). In return, New Zealand received $48,275 million from the export of goods and services (Statistics New Zealand, 2009), $1,303 in export earnings for every hectare of bioproductive land used. Assuming New Zealand continues to export the same kinds of goods and services, New Zealand will reach full biocapacity when its total export earnings exceed $58,815 million. In 2011 it earned $58,157 million; between 2006 and 2011 earnings increased by 4.1 per cent per year (Statistics New Zealand, 2012). Assuming this trend continued, in 2012 New Zealand earned an estimated $60,528 million. It is at this point that New Zealand will be using up more biocapacity than can be naturally regenerated. This means there will be reduced resource availability, ecological systems will degrade and land will need additional inputs to sustain a given level of productivity. Note that these calculations only include exports and do not include the land required to produce goods and services to support the New Zealand population.

As a nation New Zealand has to make some decisions about whether to continue to degrade the ecological systems of a 'clean green' country (Ministry for the Environment, 2001). If the decision is to try to reduce the impact of the export of goods and services there are only two options. One is to stop the increase in exports and therefore means reducing export income; the other is to earn a living producing much lower footprint goods and services.

Conclusion

The amount of money that individuals earn allows them to follow different types of lifestyles with varying levels of resource consumption. The *8 Tribes* case study shows that, on average, the income bracket that someone is in correlates with the size of their ecological footprint. The goods and services individuals and communities consume are satisfiers for fundamental human needs; the decisions individuals need to make are about the types of satisfiers they will use to fulfil their needs. Some reduction in footprint can be made by moving to 'greener' products and services but the general rule of thumb is that, the more products and services people buy, the bigger their footprint. Reducing some of the largest components of the New Zealand footprint, such as food and transport, is likely to require time. Footprint and expenditure of time and money are intricately connected, and time and money can often be used interchangeably to lower a person's ecological footprint.

References

Aisner, J., Barry, M., Batten, R. and Bhattacharya, C. B. (2008) 'The good consumer: buying ethical is not as straightforward as it seems', *The Economist*, 17 January

Assadourian, E. (2010) 'Transforming cultures from consumerism to sustainability', *State of the World*, The Worldwatch Institute, Washington, DC

Barrett, J., Birch, R., Baiocchi, G., Minx, J. and Wiedmann, T. (2006) 'Environmental impacts of UK consumption: exploring links to wealth, inequality and lifestyle', paper presented at the IABSE Henderson Colloquium, *Factor 10 Engineering for Sustainable Cities*, Cambridge, England, http://www.istructe.org/IABSE/Files/Henderson06/Paper_01.pdf, accessed 1 August 2012

Becker, J. (2012) *Becoming Minimalist*, http://www.becomingminimalist.com/2012/05/08/becoming-minimalist-start-here/, accessed 29 May 2012

Bruno, D. (2010) *The 100 Thing Challenge*, William Morrow Paperbacks, New York

Caldwell, J. and Brown, C. (2007) *8 Tribes: The Hidden Classes of New Zealand*, Wicked Little Books, Wellington, New Zealand

Canadian Centre for Policy Alternatives (2008) 'Richest 10% create bigger ecological footprint', *Projects & Initiatives: Climate Justice Project, Growing Gap*, 24 June, http://www.policyalternatives.ca/newsroom/news-releases/richest-10-create-bigger-ecological-footprint, accessed 10 August, 2011

Doerr, J. (2007) 'John Doerr sees salvation and profit in greentech' [video], *TED Talks*, http://www.ted.com/talks/lang/en/john_doerr_sees_salvation_and_profit_in_greentech.html, accessed 16 July 2012

Fairtrade International (2012) 'About us', *Fairtrade International*, http://www.fairtrade.net/about_us.html, accessed 12 July 2012

Global Footprint Network (2011) *National Accounts: New Zealand 2007*, Global Footprint Network, Oakland, CA

Illich, I. (1974) *Energy and Equity*, Harper and Row, New York

Inland Revenue (2012) 'Find out about: tax rates and codes – income tax rates for individuals', http://www.ird.govt.nz/how-to/taxrates-codes/itaxsalaryandwage-incometaxrates.html, accessed 16 July 2012

ISO (2012) *ISO 14000: Environmental Management, Management System Standards*, http://www.iso.org/iso/iso_14000_essentials/iso14000, accessed 12 July, 2012

Jackson, T. (2009) *Prosperity without Growth: The Transition to a Sustainable Economy*, Sustainable Development Commission, London

Johri, R. (2005) *Work Values and the Quality of Employment: A Literature Review*, Department of Labour, Wellington, New Zealand

Jorge, M. (2010) 'Patients' needs and satisfiers: applying human scale development theory on end-of-life care', *Curr. Opin. Support Palliat. Care*, vol. 4, no. 3, pp. 163–169

Lawlor, E., Kersley, H. and Steed, S. (2009) *A Bit Rich: Calculating the Real Value to Society of Different Professions*, New Economics Foundation, London

Lawton, E. (2012) 'The New Zealand ecological footprint template', *The New Zealand Footprint Project*, Otago Polytechnic Centre for Sustainable Practice, Wanaka, New Zealand

Lawton, E., Vale, R., Vale, B. and Lawton, M. (2010) 'Footprinting urban form and Kiwi lifestyles', paper presented at the Transitions to Sustainability Conference, University of Auckland, New Zealand, 30 November–3 December 2010

Lawton, E., Vale, R., Vale, B. and Lawton, M. (2012) 'The New Zealand Footprint Project', *International Journal of Environment and Sustainable Development*, in press

Maslow, A. H. (1943) 'A theory of human motivation', *Psychological Review*, vol. 50, no. 4, pp. 370–396

Max-Neef, M. A. (1991) *Human Scale Development: Conception, Application and Further Reflections*, The Apex Press, New York

Ministry for the Environment (2001) *Valuing New Zealand's Clean Green Image*, Ministry for the Environment, Wellington

Moss, C. and Toltz, L. (2011) 'Summary', http://ymoyl.wordpress.com/summary-of-your-money-or-your-life/, accessed 25 January 2012

Norberg-Hodge, H. (1991) *Ancient Futures: Learning from Ladakh*, Sierra Club Books, San Francisco

Norberg-Hodge, H. (1992) *The Pressure to Modernise, The Future of Progress*, Green Books, Totnes, Devon, England

Norberg-Hodge, H., Gorelick, S. and Page, J. (writers and directors) (2011) *Economics of Happiness* [film], available from International Society for Ecology and Culture, Berkeley, CA

Robin, V. and Dominguez, J. (1992) *Your Money or Your Life*, Viking Penguin, New York

Schumacher, E. F. (1968) 'Buddhist economics', *Resurgence*, vol. 1, no. 11, January–February

Spaargaren, G. and Vliet, B. V. (2000) 'Lifestyles, consumption and the environment: the ecological modernization of domestic consumption', *Environmental Politics*, vol. 9, no. 1, pp. 50–76

Statistics New Zealand (2007) 'Household Economic Survey: Year ended 30 June 2007', http://www.stats.govt.nz/browse_for_stats/people_and_communities/Households/HouseholdEconomicSurvey_HOTPYeJun07.aspx, accessed 16 July 2012

Statistics New Zealand (2009) 'National Accounts: Year ended March 2009', http://www.stats.govt.nz/browse_for_stats/economic_indicators/NationalAccounts/info-releases.aspx, accessed 16 July 2012

Statistics New Zealand (2011) 'Time use', http://www.stats.govt.nz/browse_for_stats/people_and_communities/time_use.aspx, accessed 25 January 2012

Statistics New Zealand (2012) 'National Accounts: Year ended March 2011', http://www.stats.govt.nz/browse_for_stats/economic_indicators/NationalAccounts/NationalAccounts_HOTPYeMar11.aspx, accessed 25 January 2012

Tay, L. and Diener, E. (2011) 'Needs and subjective well-being around the world', *Journal of Personality and Social Psychology*, vol. 101, no. 2, pp. 354–365

Tilford, D. (2000) 'Sustainable consumption: why consumption matters', http://www.sierraclub.org/sustainable_consumption/tilford.asp, accessed 2 July 2012

United Nations Population Fund (2007) *State of World Population 2007: Unleashing the Potential of Urban Growth*, United Nations Population Fund, New York

United States Environmental Protection Agency (2004) *Let's Go Green Shopping*, United States Environmental Protection Agency, Washington, DC

Vale, R. and Vale, B. (2009) *Time to Eat the Dog? The Real Guide to Sustainable Living*, Thames and Hudson, London

Walter, A. (2012) 'Human behaviour and adequacy of satisfiers in the light of sustainable development', MPRA Paper No. 38800, posted 14 May 2012, Munich Personal Research Papers in Economics Archive, http://mpra.ub.uni-muenchen.de/38800/, accessed 18 June 2012

Wilkinson, R. and Pickett, K. (2009) *The Spirit Level: Why More Equal Societies Almost Always Do Better*, Bloomsbury Press, New York

Winkelmann, L. and Winkelmann, R. (1998) 'Why are the unemployed so unhappy? Evidence from panel data', *Economica*, vol. 65, pp. 1–15

Zorondo-Rodríguez, F., Gómez-Baggethun, E., Demps, K., Ariza-Montobbio, P., García, C. and Reyes-García, V. (2012) 'What defines quality of life? The gap between public policies and locally defined indicators among residents of Kodagu, Karnataka (India)', *Social Indicators Research*, http://link.springer.com/article/10.1007/s11205-012-9993-z, accessed 27 November 2012

19 Sustainable Urban Form

Fabricio Chicca

It is difficult to see why the environment is still an issue. Companies around the world have already realized the importance and the possible financial gains of adopting green marketing or environmentally friendly approaches (Rees, 1994: 123–124; Oyewole, 2001: 239; Mcmanus and Haughton, 2006: 118; Fuerst and McAllister, 2011: 45). They are also aware of the probable drawbacks associated with non-green policies (Linstroth, 2009: 81; Cronin *et al.*, 2011: 158). Public opinion has progressively become more conscious of the environment, and on occasion even agrees that priority should be given to environmental policies when they conflict with economic growth, although it is not totally clear to what extent this priority has to be given (Oyewole, 2001: 239; Ginsberg and Bloom, 2004: 79; Brown, 2007: 5). Some consumers are still not prepared to sacrifice their other desires in order to be green (Ginsberg and Bloom, 2004: 79). Although there are some controversial aspects of public engagement in green consumption, public pressure towards green policies continues to increase (Noiseux and Hostetler, 2010: 571; Cronin *et al.*, 2011: 159). The integration of environmental policies into marketing theory and practice is understood as part of the societal marketing concept (Peattie and Charter, 2003: 727). Corporate Social Responsibility (CSR), where environmental strategies are normally inserted, is a key part of companies' policies and has strategic importance for many of them. Environmental policies are also known as the 'triple-bottom line' concept, or '3P' (planet, people and profit). Firms have gradually been introducing the '3P' actions into their management guidelines (Brown *et al.*, 2006: 20). Almost every corporation mentions on its website some degree of the triple-bottom line policies, normally described as the 'sustainability report' (Ginsberg and Bloom, 2004: 80; Linstroth, 2009: 81). The ways in which companies communicate these strategies to investors and consumers are also highly important, along with the media coverage that they produce (Bhattacharya and Xueming, 2006: 1).

Unfortunately, the word 'sustainability' has been so corporately and widely used that it is at risk of meaning nothing (Lehmann, 2010: 65). Another problem concerns the meaning of the term 'sustainable development': 'sustainable' and 'sustainable development' are used almost interchangeably (Hotten, 2004, cited in Lehmann, 2010: 65). A more carefully defined difference between these terms may be expressed as sustainability being an end, while sustainable development is a process to achieve that end (Dovers, 2003, cited in Lehmann, 2010: 65). Indeed, there is no single definition for sustainability (Oyewole, 2001: 239). Cities and urban areas are, in fact, unsustainable. They have become dependent on distant supplies of resources, highly damaging to the environment. Moreover, the sense of community has been eroded instead of reinforced (Berg et al., 1989: 104). A generic definition of 'sustainable development', and the most widely used, is that of the World Commission on Environment and Development, also known as the Brundtland Commission: 'development that meets the needs of the present without compromising the ability of future generations to meet their own needs' (Haughton and Hunter, 1994: 16). In an urban context, in order to become sustainable, urban regions have to leave behind old values and aims, and, moreover, need a radical change in the relationship between inhabitants and environment (Rees, 1995: 344). The process of transforming an urban region into a truly sustainable area goes far beyond the obvious issues of pollution, public transportation, conservation of green areas, recycling, renewable energies, public services and waste management. Urban areas will need to articulate and put together urban sustainable developments with a balanced distribution of agri-ecological, industrial and urban activities (Leff, 1990: 55–56). Such developments must aim to be resourceful in their use of energy and user friendly regarding their function as places to live (Elkin et al., 1991: 12). A sustainable urban region involves business and people attempting to improve their natural, built and cultural environment at both the local and regional level. It works to support the aim of a global sustainable development (Haughton and Hunter, 1994: 27). The whole urban productive chain has to be fully committed to sustainable production, but it cannot be concerned only with the environment. It has to be aware of the needs and the welfare of the people in that environment (Oyewole, 2001: 240).

The economic model today relies on the idea that open and free markets will ensure sustainability (Rees, 1995: 347). Adam Smith (1723–1790) had an important role in establishing the modern economic model and the relationship between wealth and the environment. He is normally considered the first economist (Conway, 2009: L155). He stated that self-interest in a market system produces a solution to an economic problem (Common and Stagl, 2005: 309). In a free-market system there is nothing wrong with people acting in their self-interest; acting this way will bring benefits to all of society, enriching everyone including the poor (Conway, 2009: L132; Heinberg, 2011: L726). The 'invisible hand' of the market serves to benefit the whole of society. In theory the market would distribute the generosity of nature and the outcome of human labour as fairly as possible. However, because of industrialization and colonialism the market has worked to favour those who make money as their primary occupation (bankers, traders, industrialists and investors; Heinberg, 2011: L757). The 'invisible hand' is in fact shorthand for the law of supply and demand (Common and Stagl, 2005: 3, 310; Conway, 2009: L129–152). According to the idea of supply and demand, whenever a natural resource becomes scarce, the price rises, naturally controlling and reducing its use and stimulating the search for substitutes (Rees, 1995: 347).

After Adam Smith, another philosopher introduced a new interpretation of the economic phenomenon. Karl Marx (1818–1883), a German philosopher, proposed a new name for the entire system which began in the middle ages: 'capitalism'. According to his interpretation, the economic system was designed to expand in an infinite way, meaning it is inherently unsustainable (Heinberg, 2011: L784). The system, in his opinion, had no tendency to reach balance in the market; in fact the way it operates underlies the trade cycle of boom and bust (Wolff, 2011).

> The need of a constantly expanding market for its products chases the bourgeoisie over the whole surface of the globe. It must nestle everywhere, settle everywhere, establish connections everywhere ...
> Marx and Engels, 1969

> Capitalism is, in its very essence, a process of colonization of countries and people, with the exploitation process as its backbone.
> Hersh and Brun, 2000: 107

A more moderate view arose in the late nineteenth century. It was a response to the incompatible tendencies of capitalism and Marx's theory (Marxism). This new economic view was introduced by sociologist Lester Ward (1841–1913), psychologist William James (1842–1910), philosopher John Dewey (1859–1952) and the physician Oliver Holmes (1809–1894) and was entitled 'social liberalism'. This new system would prevent uncontrolled concentration of wealth and the state was to be responsible for tackling social problems such as unemployment and workers' abuse. The social liberals also supported the theory of progressive increase in tax and restrictions against monopolies (Heinberg, 2011: L803). These were the core of John Maynard Keynes' (1883–1946) ideas. He was probably the most influential economist of the time. According to Keynes, taxation and government investment should be the answers to controlling the economy. Keynes was successful with his theory during the depression of the 1930s, and Keynes' ideas were commonly credited with influencing the New Deal programmes of Franklin Roosevelt (Conway, 2009: L709–725; Heinberg, 2011: L812).

Between 1930 and 1970, the world saw the contest between social liberals (Keynesians), Marxists, and neoclassical economists, who were partially marginalized in the polarized Marxist-Keynesian contest. The neoclassical view was essentially different from that of the social liberals. They believed that the intervention of the state obstructed the efficiency of the market. After the collapse of the Soviet Union at the end of the 1980s Marxism lost its credibility, and this created space for the rapid ascension of neoclassical economics under the neoliberals. The United States and British Governments in the 1980s, under Ronald Reagan and Margaret Thatcher respectively, relied heavily on advice from neoliberal thinkers, most of them influenced by Milton Friedman (1912–2006) and Friedrich von Hayek (1899–1992; Heinberg, 2011: L812).

Friedman's ideas, also called 'monetarism', declared that the State should only control money flowing around the economy. This, he held, ultimately controls inflation, which should be the priority over unemployment. Inflation is a major risk to an economy, probably because it directly affects consumption, making it hard for the credit system to expand. In contrast, Keynes had imagined that unemployment was the major risk to an economy (Conway, 2009: L795).

Essentially, monetarism re-embraces the ideas of the classical economists about total market self-organization. However, different from Adam Smith's point of view, the monetarists and Keynesians do not foresee a future stable stage of an economy because of resource limits. They believe in perpetual growth, and the divergence between the two views was about how this was to be achieved: through state intervention or through leaving the market to find the best way (Heinberg, 2011: L821).

Apparently the idea of perpetual growth without any limit or restriction has taken over modern society. This idea of perpetual increase, already consolidated in most national and local economies, can also be found in urban design and the expansion of cities. Urbanization is a process which started in the eighteenth century but which has no limit. The assumption is that cities can indefinitely increase in area while there is population growth. Ultimately this means that while there are more births than deaths cities are able to expand (Davis, 1965: 18–21).

It is hard to determine which factor is the most important trigger in the processes of urbanization and urban expansion. Industrialization began attracting people from the countryside from the eighteenth century onwards (Clark, 2003: L401). Since the industrial revolution the world's economy has been growing rapidly; indeed population, food production, industrial production, and resource consumption are growing more and more rapidly (Meadows *et al.*, 2004: L368). The idea of perpetual urban growth is convenient for the real estate market. The intentional relationship between economic growth and urban growth has been helping real estate companies to gain an important role as urban planners and developers. The current production model holds that it is essential that investments are made not only in production processes, but also in the built environment. In fact, the built environment is an intrinsic part of the economic model today. It is essential for consumption (Christophers, 2011: 1348–1349). The economic model directly affects decision makers, therefore shaping the expansion of urban fabric all over the world (Frank, 2004: 146–147). The real estate market and its expansionist idea have conveniently been using the concept that growth is not limited by the environment (Rees, 1995: 347). In addition, the unrealistic view adopted by economists that the economy has to grow and perpetually expand (Heinberg, 2011: L190–275) gives real estate developers the opportunity to consolidate in modern society the idea that cities are growth machines (Portney, 2003: L1410). Urban growth, therefore, is perfectly aligned with the ideas of *Homo economicus* and 'growthmania' (Rees, 1995: 343; Conway, 2009: L3524).

The Unsustainable Real Estate Approach to Urban Growth: The Myth of Profit

Several concepts have been extremely important in maintaining the real estate market as an area of notable importance and position in the market economy. It is difficult to identify the most important concept which the real estate market has used to prevail, but, overall, the most important concepts for real estate are perhaps general neoliberal ideas of deregulation, the free market and perpetual expansion. In theory, expansion of the urban fabric and the urbanization process would provide a reasonable and sizeable market for real estate developers all over the world. Although the phenomenon of globalization has been seen in the real estate market, some local specificities still exist.

The direct reduction in interference in spatial development promoted by states at all levels, national, regional and local, according to the neoclassical principles of deregulation, flexibility and privatization has dramatically decreased the importance of the local level in the process of urban planning (Coy, 2006: 122). The lack of local or national government interference and profound economic changes are the main reasons why real estate companies and big private investors have been able sharply to increase their control over urban evolution and development (Coy, 2006: 122). Because of urbanization and urban growth, the expansion of the real estate market is (temporarily) assured. This expansion provides a sizeable market for all real estate players. However, for the real estate market this simple expansion and its absolutely remarkable numbers are not enough. Simple urban demand does not entirely guarantee prosperity. Each single project requires a predefined financial result. Each company has its own predefined financial criteria for assessing the financial performance of projects. Essentially the companies aim for a certain amount of profit in a certain period of time, although some more complex and sophisticated indices, such as EBITDA (earnings before interest, taxes, depreciation, and amortization), ROE (return on equity) and ROI (return on investment), can be found in real estate developers' market reports. Therefore, a market in which the predefined level of profit is not achievable will not appeal to developers. Developers and investors may determine the expansion of the urban fabric according to the potential for obtaining profit from their developments. Big players in the real estate market are normally large international corporations listed on stock exchanges all over the world. Globalization has allowed, almost freely, substantial capital flows between markets. The characteristics of real estate investors and stock markets' conditions have made the requirements of financial performance very inflexible. Investors in the real estate market largely demand financial performance from their investments all over the world.

In order to achieve investors' demands, real estate developers have based an important part of their strategy on site acquisition. Prospecting for and acquisition of sites are positioned at the very beginning of the real estate chain of production, with the requirement being the acquisition of sites on which projects can be developed. Excluding legal matters and zoning restrictions, financial unfeasibility is probably the main reason that developers reject sites. The criteria are straightforward: not enough profit, no acquisition. However, some other financial criteria are also important. Large developers tend to discard sites for reasons in which the net revenue is not relevant. First, some small sites are rejected because the marketing budget associated with them would be insufficient. Developers normally calculate the marketing budget according to the expected project revenue; a small site will have a small revenue, hence a small marketing budget. A small marketing budget can lead the project to financial failure because it cannot keep up, for instance, against better-structured and larger advertisement campaigns from competitors. In addition, small projects are unable to cover the basic general and administration costs (G&A) of large corporations. Therefore, small sites are relegated to small developers, which struggle to compete against bigger and better-structured competitors. The site acquisition method adopted by most of the developers is not related to the aspirations of cities. In fact, as important players in cities' expansions, real estate developers often have aims which are very different from cities' aspirations. Developers, as private institutions, want profit. The real estate industry's site selection may produce uninhabited sites, and consequently underdeveloped regions, with low density. This raises all the problems of empty sites in the middle of the cities, such as increased urban infrastructure expenditure, public transportation unfeasibility and energy waste with private transportation.

The process of land acquisition is so vital for developers that some companies have more than five years of project launches guaranteed by land stock. This process of acquisition and concentration of site stock is normally called 'land banking' by real estate developers. It may play an important role in cities' expansion. A good project launch strategy can lead to an overvaluation of land stock. Put simply, developers may start to sell distant projects, forcing urban development towards these distant projects, consequently passing over better located sites. This is commonly known as real estate speculation. This is also an example of financial strength acting through real estate developers to shape urban expansion. In a progressive deregulated market, residents of the cities experience whatever urban situation results from the real estate developers' expansion (Portney, 2003: L1412). Profit and the economic arrangements made to achieve it are therefore undoubtedly factors in determining the organization of cities' expansion.

Recent decades have seen an increase in mobility and in the influence of globalization. This has reinforced changes at local and regional levels towards a more globalized lifestyle. Businesses have become more decentralized, low density areas have been created, through gated communities for instance, making urban sprawl one of the most remarkable urban characteristics of the twentieth century (Coy, 2006: 122; Haughton and Hunter, 1994: 81).

The idea of any sort of constraint or limitation on expansion will directly affect and prejudice the real estate market. The politically deregulated conditions combined with land banking policies have led cities into patterns of fragmentation. Public authorities have therefore been left with very little to offer in order to balance socioeconomic, environmental and spatial disparities (Rees, 1994: 122; Coy, 2006: 123).

New urban developments and their investors and developers are part of, and are taking financial advantage of, the current economic model (Portney, 2003: L1380). The real estate market has opportunistically accepted and applied the theory that resource depletion is not a fundamental problem (Rees, 1994: 123). Developers and authorities have been relying on the assumed simple mechanics of the free market to ensure sustainability (Rees, 1995: 347). Hypothetically, technology should be able to replace resource scarcity; according to the Nobel laureate Robert Solow (1974), 'If it is very easy to substitute other factors for natural resources, then there is in principle no "problem". The world can, in effect, get along without natural resources ...' (Rees, 1994: 123; Rees, 1995: 347; Common and Stagl, 2005: 258).

Despite its expressed concerns with the environment, eco-friendliness and green building, urban development continues to follow the paradigm of neoliberal economic regulation, treating cities as growth machines (Rees, 1994: 123). In this scenario, the true pursuit of sustainability is impossible (Portney, 2003: L1365), but urban development seems to be exploring the popular increase in demand for green products. Real estate developers, as much as other market sectors, have been moving towards green policies. Perhaps because of the increase in importance of cities, with their resource demands, and the continuous process of urbanization since the nineteenth century, the real estate market has become more aware of its environmental impacts (Stern *et al.*, 2007, cited in Fuerst and McAllister, 2011: 45). Developers have also realized the benefits of and potential for financial gain in adopting green marketing, which normally increases prices or sales (Noiseux and Hostetler, 2010: 571; Fuerst and McAllister, 2011: 65–66).

The real estate market's environmental strategies may be purely a market approach, or simply the result of mandatory ecological regulations imposed by authorities (Fuerst and McAllister, 2011: 65). At the current time the definitions of sustainability employed by the majority of developers simply reflect an effort to operationalize sustainability in a way that does not disrupt the market (Portney, 2003: L115). This green strategy is essentially aimed at capturing the growing share of the market made up of environmentally concerned consumers (Oyewole, 2001: 239). Whenever a green product is developed by the real estate industry, in order to compensate for any additional construction cost, investors require higher income or reduced risk, or a combination of the two. Ultimately this means that investors or real estate developers will be fully compensated for the possible extra costs of providing 'green' features (Noiseux and Hostetler, 2010: 571; Fuerst and McAllister, 2011: 50).

The deregulated economic climate has led some companies to develop their own approaches to sustainability, with a basis in market rules of supply and demand (Frame and Newtom, 2007: 574). The process of decoding the market consumer, also known as the target consumer, is a helpful instrument to marketers in understanding the sorts of green attributes that should be incorporated in the market mix (Ginsberg and Bloom, 2004: 80). When the target consumers are defined, companies analyse whether they form a large enough group to be profitable, and decide accordingly on the inclusion of appropriate green characteristics in the product (Polonsky, 2005: 130). Subsequently, companies study their competitors and how they are approaching the market, and analyse whether there is space for more green products. The growth of the green market has led to the advent of voluntary market-based environmental certifications. A range of labels from different countries cover most of the real estate market products, including Leadership in Energy and Environment Design (LEED; United States), Green Star (Australia), Green Globes (United States) and Building Research Establishment Environmental Assessment Method (BREEAM, United Kingdom). The certification market has sharply increased and is becoming more recognized all over the world (Glenn, 2011: 167; Fuerst and McAllister, 2011: 45). The certification market is, indeed, another mechanism for increasing profit. Labels yield positive effects on real estate property values (Zheng *et al.*, 2011: 4). It is notable that there have been reports of developers making fraudulent claims about obtained certification at the beginning of construction (Fuerst and McAllister, 2011: 48). These environmental certifications have become so popular that some cities have been using them as a *de facto* environmental building standard, ignoring the market basis of these labels (Stansberry, 2011: 98). Local authorities, sometimes misinformed, have often provided incentives for certificated buildings. Essentially it means another financial incentive for developers

(Fuerst and McAllister, 2011: 49). At the local level, an authority is often more concerned about economic stability than environmental stability (Girardet, 2007: 13). Therefore a deeper analysis of the current certification of 'sustainability' seems unlikely. In practice products which claim to be green or sustainable should be subjected to an increased level of inspection. Self-certification and market-based labels are no longer enough to guarantee a sustainable product (Linstroth, 2009: 78).

The proliferation of 'green' labels points to the need for the definition of sustainability to be more specific. Currently, the definition varies from company to company and institution to institution, and in addition the lack of mathematical parameters has provided an enormous range of possibilities for the real estate industry to exploit the marketability of this idea. Developers have conveniently appropriated the concept of energy efficiency, for instance, to self-label their products as 'sustainable' or 'green'. Real estate developers have also realized the value inherent in parts of cities with large green areas, and have attempted to target these areas for the development of projects for the high end of the market. Green areas can easily be associated with ecological or eco-friendly products. This process of privatization of green areas is legal, and has been accompanied by the negligence of the public authorities. From observation, in places like Brazil, regions close to cities which were predominantly green, instead of being preserved as a green belt, have been subdivided into large house plots and sold.

Conclusion

Perhaps as important as the interference in urban design caused by economic models is the collective change in the public mindset. In fact, at present, more land is allocated to roads, highways, garages and parking lots than to housing in most normal patterns of development (Warren, 1998: 16). The real estate market has had an important part to play as an urban planner in recent decades. In addition, as an economic sector it occupies a central position. Its urban product, made up of new developments, demands high levels of resources. However, orthodox economic analysis has become so abstract that some important aspects of environmental impact studies have been severely compromised (Rees, 1994: 122). The economic model used today has to change in order to include holistic sustainability analyses, instead of isolated financial well-being analyses.

The other major change that should occur is the inclusion of factors currently treated as externalities in the parameters used to measure prosperity. An externality occurs when products or consumption by one part of society directly affects another part. In addition, the concept can be used when a product's financial value is not affected, for instance, by its environmental impact. The cost of recovering the environment is today usually paid by a third party, especially social and environmental costs. Under the current system an environmental catastrophe caused by urban expansion may be seen as positive economically. Because of externalities, developers are not responsible, and the economy will be boosted by the costs expended to repair the damage (Heinberg, 2011: L850). Internalizing externalities would mean prices that reflect full costs, including the costs of environmental and social impact (Meadows *et al.*, 2004: L2794). The suggestion that external costs should be incorporated in prices is known as 'Pigouvian theory' (Common and Stagl, 2005: 417). It would lead to understanding among consumers of the real costs of their behaviour. Indeed, overall wealth should have its sustainability assessed (Stiglitz *et al.*, 2010). This cannot be neglected; companies cannot morally use the term 'sustainable' if they are worsening the situation of any given group in society (Oyewole, 2001: 249). Similarly it is not acceptable that, in order for one region to maintain a high quality of life, another has to be kept below the poverty line (Mcmanus and Haughton, 2006: 116), as illustrated so clearly by the finite land-area-based ecological footprint. This concept attests to the depth that is needed in the urban sustainability concept, which necessarily demands conservation of resources and environmentally friendly strategies at all stages (Johri and Sahasakmontri, 1998: 265). It should be concerned not only with the environment, but also with the people in the environment (Oyewole, 2001: 240).

So far, the mainstream corporate approach to urban sustainability has totally failed to undertake analyses that go to the core of the problem. Real sustainability will require a pattern shift in the way that business is done. Real estate companies have been operating opportunistically and failing to genuinely reduce their impact (Rees, 1995: 344; Polonsky, 2005: 125). The attention that has been paid to social and economic aspects of sustainable development may indicate its real value and concerns. The concepts of ecological footprint and regional carrying capacity are frequently ignored by real estate developers and their products. The ecological footprint clearly shows how the ecological impacts of new urban developments go far beyond the physical limits of those developments. Real estate developers do not tackle the problem of overconsumption of land required to supply the lifestyle based on fossil fuel consumption supported by their designs. In fact, the process of economic and municipal deregulation has been facilitating commerce and trade which allows populations to appropriate carrying capacity from elsewhere through product consumption.

The introduction of the ecological footprint into the real estate market would help to disseminate the idea of limits. It would graphically illustrate to consumers the amount of land needed to support a lifestyle in a real estate 'sustainable' product. The concept of sustainability should contain the idea of promoting equity, instead of promising greater production (Mcmanus and Haughton, 2006: 115). Regional and global sustainability is theoretically achievable if all regions live within their own carrying capacities (Rees, 1994: 126). However, it seems highly unlikely that the real estate industry will voluntarily take any steps to move towards any form of real sustainability assessment. Developers may begin to address the concept of technical sustainability, which reflects building materials and construction methods. However, the idea of sustainable behaviour, meaning the behaviour of residents, is probably regarded as a consideration outside the developers' field of action. This may lead to the conclusion that a truly sustainable project has to combine both concepts. Ultimately it means that market-oriented sustainable projects are indeed ineffective at achieving sustainability. Green real estate products are not able to initiate or develop sustainable behaviour among their residents (Noiseux and Hostetler, 2010). The case studies in this book also suggest this: more top-down or industry-developed projects seem to have higher footprints than projects developed by their users (see Chapter 16, for example).

Developers and investors are becoming more aware of sustainability, but if their green products do not drive better sales performance or even market share, the strategy will be demoted (Ginsberg and Bloom, 2004: 81). Sustainability is only another attribute for consumers and developers, incorporated into products to be considered alongside others (Polonsky, 2005: 126). For consumers, when they are required to trade off product attributes with helping the environment, the environment almost never wins out, even where a 'green' house has been their stated aim (Ginsberg and Bloom, 2004: 79). A similar situation may occur when the trade-off proposed is price. For developers the parameters are simpler still. If the green product does not perform according to financially predefined criteria, it is very likely that developers will choose not to engage in such a strategy again (Cronin *et al.*, 2011: 174).

The overriding problem is that the economic model which demands profit and which has helped to shape modern cities is incompatible with real sustainability. Throughout the development process, from land acquisition and land banking to construction and marketing, developers have only financial objectives. The idea of profit as a main goal, overriding the idea of sustainability, has led modern society into an unsustainable situation. In ecological terms, cities are nodes of pure consumption living parasitically on their resource bases (Rees, 1994: 128; Odum, 1989: 49). Cities have been following the neoclassical concept of expansion without reference to environment and resources, but it is highly unlikely that this can be continued long-term.

References

Berg, P., Zuckerman, S. and Magilavy, B. (1989) *Sustainable Green Cities: A Green City Program for San Francisco Bay Area Cities and Towns*, Planet Drum Books, San Francisco

Bhattacharya, C. and Xueming, L. (2006) 'Corporate social responsibility, customer satisfaction, and market value', *Journal of Marketing*, vol. 70, pp. 1–18

Brown, D., Dillard, J. and Marshall, R. S. (2006) *Triple Bottom Line: A Business Metaphor for a Social Product*, Universitat Autònoma de Barcelona, Departament d'Economia de l'Empres, Barcelona

Brown, P. R. (2007) 'The importance of water infrastructure and the environment', in V. Novotny and P. R. Brown (eds) *Cities of the Future: Towards Integrated Sustainable Water and Landscape Management*, IWA Publishing, London

Christophers, B. (2011) 'Revisiting the urbanization of capital', *Annals of the Association of American Geographers*, July 27, pp. 1347–1364

Clark, D. (2003) *Urban Growth/Global City*, ebook, Routledge, New York

Common, M. and Stagl, S. (2005) *Ecological Economics: An Introduction*, Cambridge University Press, Cambridge, England

Conway, E. (2009) *50 Economics Ideas*, ebook, Quercus, London

Coy, M. (2006) 'Gated communities and urban fragmentation in Latin America: the Brazilian experience', *GeoJournal*, vol. 66, pp. 121–132

Cronin, J. J., Smith, J. S., Gleim, M. R., Ramirez, E. and Martinez, J. D. (2011) 'Green marketing strategies: an examination of stakeholders and the opportunities they present', *Journal of Academy of Marketing Science*, vol. 39, no. 1, pp. 158–174

Davis, K. (1965) 'The urbanization of the human population', in R. LeGates and F. Stout (eds) *The City Reader*, 4th edition, Routledge, Abingdon, England

Elkin, T., McLaren, D. and Hillman, M. (1991) *Reviving the City: Towards Sustainable Urban Development*, Friends of the Earth, London

Frame, B. and Newtom, B. (2007) 'Promoting sustainability through social marketing: examples from New Zealand', *International Journal of Consumer Studies*, pp. 571–581

Frank, L. D. (2004) 'Economic determinants of urban form: resulting trade-offs between active and sedentary forms of travel', *American Journal of Preventive Medicine*, vol. 27, no. 3, supplement, pp. 146–153

Fuerst, F. and McAllister, P. (2011) 'Green noise or green value? Measuring the effects of environmental certification on office values', *Real Estate Economics*, vol. 39, pp. 45–69

Ginsberg, J. M. and Bloom, P. (2004) 'Choosing the right green-marketing strategy', *MIT Sloan Management Review*, pp. 79–84

Girardet, H. (2007) *Creating Sustainable Cities*, Green Books, Bristol, England

Glenn, M. (2011) 'Salt Lake City turns old building into green buildings', in R. L. Kemp and C. J. Stephani (eds) *Cities Going Green*, McFarland & Company, Jefferson, NC

Haughton, G. and Hunter, C. (1994) *Sustainable Cities*, Regional Studies Association, London

Heinberg, R. (2011) *The End of Growth: Adapting to Our New Economic Reality*, ebook, New Society Publishers, Gabriola Island, BC, Canada

Hersh, J. and Brun, E. (2000) 'Globalisation and the Communist Manifesto', *Economic and Political Weekly*, vol. 35, pp. 105–108

Johri, L. and Sahasakmontri, K. (1998) 'Green marketing of cosmetics and toiletries in Thailand', *Journal of Consumer Marketing*, pp. 265–281

Leff, E. (1990) 'The global context of the greening of cities', in D. Gordon (ed.) *Green Cities: Ecologically Sound Approaches to Urban Space*, Black Rose Books, Montreal

Lehmann, S. (2010) *The Principles of Green Urbanism: Transforming the City for Sustainability*, Earthscan, London

Linstroth, T. (2009) 'Green from inside out', in M. Melaver and P. Mueller (eds) *The Green Building Bottom Line*, McGraw-Hill, Chicago, pp. 77–108

Marx, K. and Engels, F. (1969) *Manifesto of the Communist Party, vol. 1*, Progress Publishers, Moscow

Mcmanus, P. and Haughton, G. (2006) 'Planning with ecological footprints: a sympathetic critique of theory and practice', *Environment and Urbanization*, vol. 18, pp. 113–126

Meadows, D., Randers, J. and Meadows, D. (2004) *Limits to Growth: The 30-Year Update*, ebook, Chelsea Green Publishing Company, White River, VT

Noiseux, K. and Hostetler, M. (2010) 'Do homebuyers want green features in their communities?' *Environment and Behavior*, vol. 42, pp. 551–580

Odum, E. P. (1989) *Ecology and Our Endangered Life-Support Systems*, Sinauer Associates, Sunderland, MA

Oyewole, P. (2001) 'Social costs of environmental justice associated with the practice of green marketing', *Journal of Business Ethics*, pp. 239–251

Peattie, K. and Charter, M. (2003) 'Green marketing', in M. Baker (ed.) *The Marketing Book*, Butterworth-Heinemann, Oxford, England, pp. 726–756

Polonsky, M. (2005) 'Green marketing', in R. Staib (ed.) *Environmental Management and Decision Making for Business*, Palgrave Macmillan, New York, pp. 124–135

Portney, K. E. (2003) *Taking Sustainable Cities Seriously*, ebook, MIT Press, London

Rees, W. (1994) 'Ecological footprint and appropriated carrying capacity: what urban economics leaves out', *Environment and Urbanization*, pp. 121–130

Rees, W. (1995) 'Achieving sustainability: reform or transformation?' *Journal of Planning Literature*, pp. 343–361

Solow, R. (1974) 'The economics of resources or the resources of economics', *American Economic Review*, vol. 64, pp. 1–14

Stansberry, M. (2011) 'Eugene and other cities create energy efficient buildings', in R. Kemp and C. J. Stephani (eds) *Cities Going Green*, McFarland & Company, Jefferson, NC, pp. 98–100

Stiglitz, J., Sen, A. and Fitoussi, J.-P. (2010) *Mis-Measuring Our Lives: Why GDP Doesn't Add Up*, The New Press, New York

Warren, R. (1998) *The Urban Oasis*, McGraw-Hill, New York

Wolff, J. (2011) 'Karl Marx', in E. N. Zalta (ed.) *The Stanford Encyclopedia of Philosophy*, Summer 2011 Edition, http://plato.stanford.edu/archives/sum2011/entries/marx/, accessed 24 November 2011

Zheng, S., Wu, J., Kahn, M. E. and Deng, Y. (2011) *The Nascent Market for 'Green' Real Estate in Beijing*, Institute of Real Estate Studies, National University of Singpore, Singapore

Part V
Conclusions

20 'I Wouldn't Start from Here...'

Brenda and Robert Vale

There is a very old joke about a country man asked by a townie in a car for directions to the city, who replies, 'If I were you, I wouldn't start from here.' What he is saying is that the best way to reach the particular destination is to start from a different place. There is a sad echo here of the path to fair share living. In the 1970s much was written about the need for society, particularly modern society with its focus on economic growth, to change direction and reorganise itself to live within the limits of the carrying capacity of the earth (Meadows *et al.*, 1972; Goldsmith *et al.*, 1972). Such a reorganisation involved replacing fossil fuel use with renewable energy from the sun and wind (Daniels, 1964; Golding, 1976); it involved rethinking food production so it was local and organic (Leach, 1975); it involved rethinking settlement patterns so that trade was local rather than international (Schumacher, 1973); and it involved rethinking the built environment so it was walkable and low- or zero-energy in life-cycle operation (Vale and Vale, 1975). Perhaps most importantly, it involved rethinking human values to emphasise community and doing things together with mutual local dependence (Boyle and Harper, 1976). This was a recipe for fair earth share living, and one that has been echoed by all the authors in this current book. Had we started from there and done all these things then, the world would be a very different place now. Achieving fair earth share living today is going to be a much harder task.

The biggest problem we have now and did not have then is not climate change, though this is an important issue, but population. Since the 1970s world human population has approximately doubled, from 3.7 billion to 7 billion. In the same period the world's tiger population has declined from over 35,000 to just over 3,000 (WWF, c. 2010), meaning the tigers would need an increase of 1,160 per cent to restore the population to its original level. This encapsulates the problem: as the human population rises the land available to supply the resources needed by all the species on earth diminishes. It is just that people are better than tigers at getting their hands on the remaining resources. The authors of the chapters in this book have used a value of around 1.8 global average hectares per person as the fair share of land available. However, it is obvious that if the human population continues to grow then this value is going to fall. In the 1950s, when the world population was below 3 billion, we could all have lived the lifestyle of 1950s Europe; at present we could all live as people do in Cuba, and perhaps almost all could live like those in Findhorn and the HHP in the UK. If the world population climbs to 9 billion then the available footprint may well mean life is going to be much more like that in Kampung Naga. Not only does the global population have to curb its use of resources if collapse is to be avoided, it also has to curb itself, urgently.

At present the economic growth so strongly promoted by politicians is moving humanity steadily into the past, in terms of the kind of life that the whole population could have. This raises another big issue – that of economics. A rising population means a rising market, which in conventional economic terms is a good thing. To make a profit you have to sell stuff, and this means either inventing things people haven't already got, or finding new people who don't yet have the stuff, and preferably both. However, there is a limit to how long anyone can go on doing this within a finite system. This book argues that this point has already been reached and the need to define a new economic system that does not rely on growth is very urgent. There are just not the resources for everyone in the world to have a big house, a fast car and foreign holidays, and the sooner we realise this, the better. The ecological footprint makes very clear that those who do have these things have them at the expense of the far larger number of people who do not; an issue of resources and equity. If we do dramatically reduce consumption of resources and the number of people then the human population may be able to ride out resource depletion and the additional problems it brings.

It seems that we cannot look to the politicians to provide solutions. Politicians throughout the world, so many of whom are lawyers or economists, support the economic system that requires continuous material growth. Our lawmakers do not understand that, while human laws, both legal and economic, are made up and can be changed, the second law of thermodynamics cannot be repealed. Unfortunately, our politicians have no grasp of physics. There is, however, a small hope. We can all do much to reduce our footprint now through our behaviour choices, and the examples in this book show that this seems much easier to do when we do it together. Changing behaviour and choosing the no- or low-footprint option is often the cheap thing to do. This leaves money over to invest in some of the more expensive low-footprint technologies such as renewable energy generating systems. We can also make far better use of the resources we already have. In doing these things together we may also recover much that we have lost in the modern world, in terms of community, local dependence and self-help, and just doing things with other people. The only way to live in a fair share world is to give up the idea of being consumers and instead become our own producers. The point here is that it is no good waiting for someone else to create a fair share world for you to live in; the only way this is going to happen is as a result of collective endeavours everywhere; collective endeavours which include you.

References

Boyle, G. and Harper, P. (eds) (1976) *Radical Technology*, Pantheon Books, New York

Daniels, F. (1964) *Direct Use of the Sun's Energy*, Yale University Press, New Haven, CT

Golding, E.W. (1976, first published 1955) *The Generation of Electricity by Wind Power*, E. & F. N. Spon, London

Goldsmith, E., Allen, R., Allaby, M., Davoll, J. and Lawrence, S. (1972) 'A blueprint for survival', *The Ecologist*, vol. 2, no. 1

Leach, G. (1975) *Energy and Food Production*, International Institute for Environment and Development, London

Meadows, D. H., Meadows, D. L., Randers J. and Behrens W. W. (1972) *The Limits to Growth*, Earth Island, London

Schumacher, E. F. (1973) *Small is Beautiful: A Study of Economics as if People Mattered*, Blond and Briggs, London

Vale, B. and Vale, R. (1975) *The Autonomous House*, Thames and Hudson, London

WWF (c. 2010) 'Tiger population', http://wwf.panda.org/what_we_do/endangered_species/tigers/about_tigers/tiger_population/, accessed 31 July 2012

Index

agriculture xiv, 11, 13, 14–15, 40, 43, 45, 49, 70, 159, 165, 194, 202, 217, 219, 293
air-conditioner(s) 189, 207, 229–231, 281
air pollution 135, 141, 190
airport 60, 117, 125–130, 168
appliance(s) 24, 28–29, 70, 74, 85–87, 90, 93–94, 155, 165–166, 170, 172, 175, 189, 190, 207, 254, 271, 279, 290
animal land (see grazing land)
arable land 14–15
Auckland 50, 59–60, 64, 92, 117–120, 123, 125–131, 241–246, 248, 256

bach 166, 173
bamboo 216–217, 221
BedZED 267–272, 279
bicycle (see cycling)
bio-capacity 4, 12, 15, 17, 29, 43–44
bio-diesel 61
biodiversity 5, 9, 98–99, 241
Broadacre City 64
Buddhism 185, 191, 192, 196, 202, 296
bus (see public transport)

carbon 44, 46, 76, 122, 149, 167, 195, 265, 267, 269
  credits 148
  dioxide 6, 10, 17, 221
  emissions 13, 16, 22–24, 28, 40, 44–48, 93, 105, 148, 149, 190, 231
  footprint 17, 149, 195
  intensity 23, 44, 122
  neutrality 148, 155
  offsets (see carbon neutrality)
  sequestration 24, 46, 49
  sink 13, 17, 24
  tax 28
Cardiff 58–60, 69, 85
cargo passenger ships 111, 152
CBD 59, 143, 159, 243
climate change 9, 15, 79, 141, 143, 190, 195, 278, 319
clothes (clothing) 26, 29, 75, 77–79, 87–88, 170–172, 186–190, 209, 216, 226–227, 233–234, 294, 298–299
computer games (see electronic goods)
concrete 22, 92–93, 117–119, 121–123, 125, 212
Confucianism 191, 193–194
Consumer Price Index 160
consumerism 73–76, 96, 280, 290, 293
cradle-to-cradle 80
cropland 13, 15, 18–19, 21–22, 24, 40, 42, 49–50
Cuba 17, 38, 280, 320
cultural indicators 99–102, 107–108

cycling 22–23, 28,59–61, 67–68, 70, 131, 168–170, 225, 230, 266, 299

densification 14, 27, 58
developing countries 4, 73, 76, 271, 280
Dewi Sri 219–220
domestic electricity (see household electricity)

Earthsong 242
eBay 76
eco-village 19–21, 26, 194, 268
ecological indicators 98–99, 107
ecosystem 3–16, 29, 75–76, 80, 99, 106, 134, 195
ecotourism 96–97, 99–100, 104, 108, 111
electricity 17, 23, 44–46, 84, 86–88, 90–93, 120, 122, 124, 127–128, 150–152, 160, 172, 176, 205, 207–208, 211–212, 217–218, 246–247, 256, 263
  distribution 263
  grid 262
  mix 124, 165
  renewable 60, 78, 148, 153, 165, 167, 169, 242
electronic goods 26, 28–29, 67, 75, 78–79, 172, 176
energy land 17–19, 44–46, 49, 53, 89, 160, 163, 165–170, 172, 175–176, 211
energy intensity 5, 22, 124
  value 170, 172–173, 176
equitable ownership/share (see equity)
equity xiii, 4, 15–16, 27, 74, 187, 194, 308, 313, 320

fashion 78
fertilizers 19, 44
Findhorn 19–21, 70, 264, 267–272, 320

fishing 14, 18, 20, 43–46, 48–49, 52, 135, 137–139, 163–164, 217, 219–220, 262, 266–267
flying 59–64, 66–67, 69–70, 111, 152, 264, 266, 269
Food and Agricultural Organization of the United Nations (FAO) 37, 41, 46, 48, 160, 206
foodshed 48–50
forest land 10, 14, 18, 24, 44–46, 48, 160, 165, 175–176, 215
forestry 40–41, 43, 48–50
free market 140, 148, 305, 308, 310
freight 40, 47–48, 52, 130–131
fuel consumption 44, 61–63, 65–66, 68–69, 230–231, 234, 290, 313

Genuine Progress Indicator (GPI) 55
global carrying capacity 3–4
Global Footprint Network (GFN) 44
globalisation 73, 137, 144
golf 107, 190
governance 134–140, 143–145
grazing land 14, 41, 49–50, 53, 163, 176
green consumption 304
Greenest Kampung 271
Greenest Street 195, 270, 272
greenhouse gas emissions 105, 117, 126, 141–143, 149
grey water 195, 205, 208–209, 211–212, 271
Gross Domestic Product (GDP) 75, 102, 105–106
growing land (see cropland)
Gwynedd 58–59

health-care services 173
high density city 69
hinterland 3, 12, 15

Hockerton Housing Project (HHP) 242, 262–274
holidays 65, 67, 109–110, 159, 168, 173, 175, 231, 290, 294, 298, 320
household 201, 205–206, 208, 213, 219, 226–227, 243, 249, 264–266, 268–270, 272
   electricity 46, 74, 88, 92–93, 246
   energy 87, 165–166, 207, 228, 230, 256, 298
   food production 210, 247, 256
   goods 73, 75, 77–78, 87, 170–172, 189, 271, 298
   size 87, 204, 226–229, 232, 248, 256, 290
   travel 65–66, 230, 232
   waste 85
   water use 87, 209, 262, 280
Household Energy End-use Project (HEEP) 85–86, 166
Hong Kong 13, 152
hybrid vehicle 14, 61, 63, 66
hydro (see renewable electricity)

income 12–13, 15–17, 19–20, 27, 29, 59, 69, 74, 101–102, 111, 177–178, 187, 190, 194, 196, 225–226, 230, 263, 268–269, 287–303, 311
indigenous people 98–99, 101, 103
inequality (see equity)
insulation 87, 89, 91, 93, 150–151, 270
internet shopping 75
iPod (see electronic goods)

Jevons paradox 14, 269

land banking 309–310, 314
landfill 24, 148, 154, 164, 170, 172, 176, 219=222
LEED 141–142, 311
livestock unit 41

medium density 240, 242–245, 248–250, 252–257
methane 42, 44, 49, 221–222
Middle Way (see Buddhism)
monetarism 307
mortgages 76

NABERS 142–143
National Happiness Index 75
natural capital 4, 14, 104, 134–141, 143–145
negentropy 5–8, 12
New Economics Foundation (NEF) 207
New Theory Agriculture 201–203, 212
newspaper (see paper)
nutrients 10–11, 52, 220

organic farming 70
organic food 22, 27, 262, 266, 271
organics 25, 153–154
Otago Central Rail Trail (OCRT) 108
overseas trips (see holidays)

palm leaves 216
passive systems 23, 28, 70, 151, 242, 254
paper 24–27, 29, 67, 153–154, 171–172, 176
participatory design 277
peasants 187
photovoltaics 91, 205, 207–208, 211, 242, 246–248, 254, 262
Pigouvian theory 313
plastics 17, 22, 25, 76, 78, 153–154
   rain tank 92, 212
processing energy 45
property rights 140–141, 144–145, 273
public transport xiv, 23, 28, 58–60, 62, 65–70, 127, 131, 152–153, 160, 168–170, 173, 175,

178–179, 190, 230, 232, 264, 268–270, 298–299, 305, 309
PV (see photovoltaics)

quality of life 27, 69, 75, 79, 81, 156, 159, 177, 179, 185, 190–192, 194–196, 313

railway (see train)
rain 217, 248
   collecting 87, 205, 208–210, 212, 240, 242, 246, 247–248, 250, 252–254, 256, 271, 277, 280
   tank 92, 242, 256
recycling 8, 11, 25–26, 220, 264, 268, 271, 277, 279, 281, 297–298, 305
refrigeration 45–46, 85–87, 90, 254
renewable 45
   electricity 14, 17, 23, 46, 64, 90–91, 148, 151–153, 160, 165, 169, 172, 176
   energy 28, 44–45, 49, 62, 66–67, 79, 89, 90, 110, 160, 167, 169, 195, 262, 264, 268, 277, 319–320
residential energy 85–86, 90
resource stocks 4, 76
rice 193, 202, 207, 217, 219–221
rubber 24

satisfiers 291–293, 296, 300
second-hand 75–76, 79, 195, 263
self-determination 137, 139
self-employment 233
shipping 11, 47–48, 111, 152–153
shop house 225
social liberalism 306
social responsibility 138, 144–145, 304
solar water heating (see water heating)
steel 22, 185
   rails 120–123

Sufficiency Economy 201–203, 211–212
supermarkets 46, 48, 163
sustainable yield (see yield)

Taoism 191
textiles 24, 100
tigers 319
timber 89, 93, 117, 121–123, 125, 165. 175, 217
train xiv, 58, 60, 62–68, 127–128, 131, 152–153, 168–169, 173, 264, 266, 268–270, 298
tram (see public transport)
Transition Towns 81, 271
trolley bus (see public transport)
Transport Oriented Development (TOD) 59
Triple-bottom line 304
Tube house 225
Tulou 193
TV (see electronic goods)

unemployment 178, 186, 293, 297, 306–307
urban cycle 68–69
urban intensification (see intensification)
urbanisation 187, 193–194, 196, 201

vacations (see holidays)
Vancouver 12–15, 17–28, 30, 127
vegan 27, 52, 266
vegetarian 15, 20, 22, 27, 50, 52, 269
Victoria University of Wellington (VUW) 147–156
Village Homes 242

waste 4–13, 19, 24–25, 76, 79–80, 120, 153–154, 171
   construction 164–165, 176
   food 39, 50, 52–53, 154

footprint 85, 106
heat 86
water 91, 109, 217
water heating 27, 86–90, 93
   solar 70, 88, 91, 242, 246–247, 254
water saving fixtures 205, 209, 211–212
wind energy 14, 64, 70, 90–91, 110, 155, 160, 165, 176, 262–263, 267

Yanhe Village Project 194–195
yield 13, 26, 40–44, 47–49, 98–99, 33–135, 248

zero energy house 165, 266
zero waste project 279